PROTEIN SCIENCE AND ENGINEERING

CYSTATINS: PROTEASE INHIBITORS, BIOMARKERS AND IMMUNOMODULATORS

Protein Science and Engineering

Additional books in this series can be found on Nova's website
under the Series tab.

Additional E-books in this series can be found on Nova's website
under the E-books tab.

PROTEIN SCIENCE AND ENGINEERING

CYSTATINS: PROTEASE INHIBITORS, BIOMARKERS AND IMMUNOMODULATORS

JOHN B. COHEN
AND
LINDA P. RYSECK
EDITORS

Nova Science Publishers, Inc.
New York

For permission to use material from this book please contact us:
Telephone 631-231-7269; Fax 631-231-8175
Web Site: http://www.novapublishers.com

Additional color graphics may be available in the e-book version of this book.

Library of Congress Cataloging-in-Publication Data

Cystatins : protease inhibitors, biomarkers, and immunomodulators / editors,
John B. Cohen and Linda P. Ryseck.
p. ; cm.
Includes bibliographical references and index.
ISBN 978-1-61209-343-7 (hardcover)
1. Cystatins. 2. Cysteine proteinases--Inhibitors. I. Cohen, John B. II. Ryseck, Linda P.
[DNLM: 1. Cystatins. QU 55]
QP609.C94C94 2010
572'.76--dc22

2010051528

Published by Nova Science Publishers, Inc. † New York

CONTENTS

PREFACE

This new book presents current research in the study of cystatins, including the evolution, members, structure, and mechanisms of action of the cystatin superfamily; the inhibitory effects of cystatins directly on cysteine proteases; cystatin C in acute coronary syndromes and acute kidney injury and the atherosclerosis and lipid-lowering therapy of statins

Chapter 1- The cystatins constitute a large group of evolutionary related proteins with diverse biological activities and wide occurrence in tissues, targeting primarily papain-like cysteine proteases and parasitic proteases like cruzipain, but also mammalian asparaginyl endopeptidase. The cystatin superfamily comprises a large group of the cystatin domain containing proteins divided into three main groups depending on the presence of single or multiple 'cystatin domains', and the presence or absence of a signal sequence. Family I- the stefins are cytosolic, 11 kDa single chain proteins with no disulphide bonds or carbohydrates, Family II- the cystatins are mainly extracellular, 13-15 kDa single chain proteins with two disulphide bonds but no carbohydrates and Family III- the kininogens are multidomain, blood plasma glycoproteins.

The subfamilies possess few distinctive features with regards to amino acid composition. The alignment of sequences of cystatins reveals common features, significant to the structure and activity of the proteins like the amino terminus glycine residue and two β-hairpin loops, one in the middle and one in the C-terminal segment of the protein. The first loop contains a QXVXG sequence conserved in almost all inhibitory members of cystatins, whereas the second loop contains a P-W motif, which is also highly conserved. All the cystatins are specific, non-covalent, competitive, reversible, tight binding inhibitors which inhibit the target enzymes in micromolar to picomolar range forming tight equimolar complexes with proteases.

Myriad functions attributed to cystatins are associated with cell proliferation, differentiation, ageing, antigen presentation, immunomodulation, etc. They are also being hailed as anti-angiogenic agents. Cystatins are at the front-line of defense against pathogens that secrete proteases as virulence factors.

Maintenance of appropriate equilibrium between free cysteine proteases and their complexes with inhibitors is imperative for proper function of all living systems/tissues. The importance of cystatins is underlined by the pathological conditions that arise upon loss or mutations of cystatin genes/function like sepsis, cancer, rheumatoid arthritis, purulent bronchiectasis, multiple sclerosis, muscular dystrophy, pancreatitis, diabetes, etc. Profound changes in the lysosomal system seem to be an early event in "at-risk" neurons of AD brains. Cystatins exhibit high propensity for fibrillation.

The present article will discuss the cystatin superfamily in general (evolution, members, structure, mechanism of action) and the already ascribed and putative biologic functions of its members.

Chapter 2- The Trypanosomatidae family, assorted on the Kinetoplastida order, consists of distinct genera of eukaryotic monoflagellated protozoa. Among the trypanosomatids with a digenetic life cycle, some species stand out: *Trypanosoma cruzi*, *Trypanosoma brucei* and several *Leishmania* species. These parasites are the causative agents of Chagas' disease, African sleeping sickness and leishmaniasis, respectively. Likewise, some species belonging to the *Phytomonas* genus can induce serious diseases in plants, which indicates the economical importance of these trypanosomatids, a problem especially affecting developing countries. Trypanosomatid cysteine proteases have been implicated in several processes including proliferation, differentiation, nutrition, host cell infection, and evasion of the host immune responses. For instance, *Leishmania* spp. possess three major cysteine proteases of the papain family (designated clan CA, family C1), namely, the cathepsin L-like CPA and CPB and the cathepsin B-like CPC, which are directly linked to the parasite survival inside macrophage cells and modulation of host immune response. In addition, cruzipain is a major cysteine protease, expressed in all developmental forms of *T. cruzi*, being highly immunogenic in patients with chronic Chagas' disease. Since cysteine proteases are present in trypanosomatids and their catalytic properties can vary considerably from those of the host enzymes they are considered important targets for new chemotherapeutical intervention. Typical cysteine protease inhibitors like E-64, leupeptin and K777 are able to inhibit both the cell-associated and released cysteine proteases produced by trypanosomatid cells in different extensions. Moreover, the superfamily of cystatins (stefins, cystatins and kininogens), which are endogenous proteins and tight-binding reversible competitive inhibitors of clan CA, family C1 papain-like cysteine inhibitors, interfere with different physiological aspects of trypanosomatid cells. In this sense, the present chapter will summarize the knowledge about the inhibitory effects of cystatins directly on cysteine proteases produced by pathogenic trypanosomatids or indirectly by potentiating the host immune cells.

Chapter 3- Calpains are neutral calcium-dependent cysteine proteases that have been extensively studied in mammalians and that exist in two major isoforms, m-calpain and μ-calpain, which require millimolar and micromolar concentrations of calcium ions, respectively, for their activation. Calpains are involved in several physiological events in eukaryotic cells. However, significant activation of calpains can be detected under several pathological conditions including cancer, neurological disorders, spinal cord injury, atherosclerosis, diabetes and cataract. In order to control these human disorders, the scientific community and pharmaceutical industries have developed bioactive compounds with capability to inhibit the calpain activity. Calpain-like molecules are also produced by pathogenic microorganisms, especially the protozoan parasites belonging to the Trypanosomatidae family. The commercial calpain inhibitors have been tested in order to block some crucial events in the trypanosomatid cells as well as their interaction with their hosts. These studies were encouraged by the publication of the complete genome sequences of three human pathogenic trypanosomatids, *Trypanosoma brucei, Trypanosoma cruzi* and *Leishmania major*, which allowed several *in silico* analyses that in turn directed the identification of numerous genes with interesting chemotherapeutic characteristics. In this sense, a large family of calpain-related proteins was described: 12 genes were identified in *T. brucei*, 15 in *T. cruzi* and 17 in *L. major*. It is interesting to note that, with few exceptions, most organisms outside the animal kingdom have only a single calpain gene, while in the trypanosomatids there is a surprising expansion of genes, which may reflect parasite plasticity to face distinct environments, such as the mammalian host and the insect vector. Calpain-related molecules produced by trypanosomatids are involved in virulence and relevant physiological processes, such as cytoskeleton rearrangement, proliferation, cellular differentiation and interaction with host structures. Interestingly, homologous of calpain have been detected in non-pathogenic trypanosomatids, suggesting a possible conservation of these important molecules during the evolution of the Trypanosomatidae family. Current therapy against both *Trypanosoma* and *Leishmania* is suboptimal due to toxicity of the available therapeutic agents and the emergence of drug resistance. In this sense, parasite cysteine proteases are regarded as a promising target in the therapeutic treatment of trypanosomiasis and leishmaniasis. This review will survey the available information on trypanosomatid calpain-related proteins and prospects for exploitation of this class of cysteine proteases as a novel drug target.

Chapter 4- With the development of highly active antiretroviral therapy (HAART), chronic kidney disease has become a relevant cause of morbidity in individuals infected by HIV. In this context, cystatin C is emerging as an interesting biomarker both for the evaluation of glomerular filtration rate (GFR) and the detection of drug-induced kidney injury.

In this chapter, the authors will first focus on serum cystatin C as a GFR marker in HIV infected patients. Authors will compare the respective advantages and limitations of serum creatinine and cystatin C in the context of HIV infection and will review the very first clinical studies on the use of cystatin C in this specific setting. Secondly, the authors will discuss the potential interest of urine cystatin C as a biomarker to detect renal tubular injuries associated with nucleotide reverse transcriptase inhibitors therapies. The authors will conclude by examining the questions that need to be answered in order to clarify the real added value of cystatin C for the management of HIV infected patients.

The United Nations Programs on HIV/AIDS (UNAIDS) estimated that 33 Millions of people were living with an HIV-Infection in 2007 [1]. The development of Highly Active Antiretroviral Therapies (HAART) has resulted in an improved survival among HIV seropositive individuals. With advancing age and HAART-related metabolic effects (hypertension [2], diabetes mellitus [3], dyslipidemia [4]), Chronic Kidney Disease (CKD) has become one of the major comorbidities in HIV-seropositive individuals [5]. Prevalence of CKD (defined by an eGFR below 60 mL/min/1.73m^2) in HIV infected is variable from 3 to 24 %. [6, 7] Risk Factors for CKD in HIV-infected patients are female sex, black race [8], Acquired Immuno-Deficiency Syndrome (AIDS), lower CD4 nadir, older age, HCV infection, hypertension, diabetes mellitus [6,9], and injection drug use [10]. Exposure to HAART is associated with an increased risk for CKD. Most frequently incriminated antiretroviral drugs are tenofovir, didanosine, atazanavir, lopinavir and indinavir. [11,12]

CKD is associated with an increased risk of both mortality [13,14] and cardiovascular events (CVE) [15] in HIV-seropositive individuals. The odds ratio for CVE is 1.2 for every 10 mL/min per 1.73m^2 decrease in eGFR. Proteinuria is a risk factor of AIDS (Hazard ratio 1.31) and death, an increase in serum creatinine is associated with an increase risk of AIDS defining illness. [16] The etiologies of CKD in HIV-infected individuals are multiple and presented in table 1. [17]

Chapter 5- Background: Cystatin C(CC) could contribute adding value to traditional cardiovascular risk factors in the prediction of adverse events in Acute Coronary Syndromes (ACS). Aim of present chapter is to assess the evidence on the prognostic value of CC in ACS patients, by reviewing current literature.

Methods and Results: by Pub Med, Embase, Ovid, 29 papers were identified, and 9 longitudinal observational studies in which serum CC was investigated as prognostic marker in ACS, were selected. The reference populations allowed to classify studies in: Group A) 4 studies with 50-60% of ACS patients and a sample size of 450-1030 patients; Group B) 3 studies on non ST Elevation Acute Coronary Syndromes (NSTEACS) patients and a sample size of 380-1120 patients; Group C) 1 study with 160 ST Elevation Myocardial infarction (STEMI) and NSTEACS patients; Group D) 1 study with 71 STEMI. Outcome: The end-point was "time to": 1) cardiovascular death, non fatal myocardial infarction (MI), or stroke, with a median follow up of 3 years; 2) major adverse cardiovascular events (MACE), with a variable median follow up of 3 years, 1 year, and 6 months; 3) death or recurrence of MI, with a median follow up of 3 years, 6 months. Prognostic role: According to results in: -Group A CC levels >1.3 mg/L were a significant risk factor for fatal and non fatal cardiovascular events (HR estimates ranging from 1.72 to 2.27); -Group B, for CC levels >1.25 mg/L, the risk of death was about 12 times greater than that of patients with lower CC levels; in a second study, for CC levels >1.01 mg/L a significant prognostic value on death was found (HR=4.07,(CI:2.16-7.66)). According to the recurrence of AMI evaluated in two studies, only one assessed a significant prognostic role (HR=1.95(CI:1.05-3.63)). For another study , according to a composite endpoint of fatal and non fatal cardiac events, with a CC >0.93 mg/L, a HR of 1.57(CI:1.04-2.49) was shown. For Groups C and D similar results on the prognostic value of CC on MACE were obtained: HR=9.43(CI:4.0-21.8) for CC levels >1.05 mg/L, HR =2.17(CI:1.07-6.98) for CC levels >0.96 mg/L.

Discussion: Despite the low number and the poor level of evidences, there is a general agreement on the prognostic value of CC in ACS, encouraging further studies on homogeneous cases series of STEMI and NSTEMI. In this patients the further exploitation of the marker as therapeutic target could improve their management for secondary prevention purposes.

Chapter 6- The kidney is particularly vulnerable to various drugs. Early screening of drug-induced acute kidney injury (AKI) is therefore critical for its clinical management, leading to a better outcome of clinical treatment. For the pharmaceutical industry, drug-induced AKI is a major concern in the early stage of preclinical safety evaluations. One major limitation in early detection of AKI has been the low detection power of traditional biomarkers, such as creatinine and blood urea nitrogen. Recent advances in basic and clinical research have provided several valuable biomarkers for early detection of AKI for the preclinical safety evaluation of drugs, clinical trials, and early therapeutic intervention. Serum cystatin c (CysC) has been identified as an attractive alternative biomarker for estimation of the glomerular filtration rate. Additionally, several biomarkers, such as kidney injury molecule-1 (KIM-1), neutrophil gelatinase-associated lipocalin, interleukin-18 and liver type fatty acid binding protein, have been discovered and their usefulness has been evaluated in cross-sectional studies. Recently, the Predictive Safety Testing Consortium's Nephrotoxicity Working Group, a collaboration between biotech and pharmaceutical industries, the US Food and Drug Administration (FDA), the European Medicines Agency (EMEA) and academia, published a report concerning the qualification of seven urinary nephrotoxic biomarkers, including total protein, albumin, KIM-1, clusterin (CLU), β_2-microglobulin, CysC, and trefoil factor 3, for particular uses in regulatory decision-making. Furthermore, the International Life Sciences Institute Health and Environmental Sciences Institute reported an extensive data package on the four urinary nephrotoxic biomarkers glutathione S-transferase α (GSTα), GSTμ, renal papillary antigen-1 and CLU to the FDA and the EMEA. This chapter describes the usefulness of these biomarkers with respect to clinical and preclinical usage and the possible mechanisms that underlie their alterations in serum and/or urine. Finally, the authors discuss the future view of AKI biomarkers.

Chapter7- Acute kidney injury (AKI) is defined as an abrupt and sustained decrease in kidney function. There are a lot of definitions in the literature, which explain the large variations in the reported incidence. It is well recognized for its impact on the outcome of patients, as it increases morbidity and mortality. Diagnosis of AKI is always difficult, particularly in the early stage of the disease. Numerous definitions and parameters have been used but the gold standard in clinical practice remains the creatinine clearance. Recently, in order to develop early biomarkers, cystatin C (CysC) was proposed. CysC is a protease inhibitor produced in a constant manner by nucleated cells. This molecule is passively filtrated by the glomerule and quite completely catabolized in the proximal tubules. It has the advantage of not being influenced by age, sex, race or muscular mass. Its excretion increases after reversible and mild dysfunction and may not necessarily be associated with persistent or irreversible damage. However, it is primilarly a sensitive marker of reduction in glomerular filtration rate but it cannot differentiate between different types of AKI. Some studies demonstrated the superiority of CysC over plasma creatinine while other studies did not, depending on the population. On the other hand, urinary CysC seems superior to conventional and new plasma markers, but its main disadvantage is its instability in the urine samples. In conclusion, the authors can use CysC like an additional argument of renal failure but it does not seem to be superior to plasma creatinine in the early diagnosis of AKI in all situations.

Chapter 8- The search for new serum markers of aging, atherosclerosis, and predictors of cardiovascular emergencies is important in contemporary society. Recently, new non-lipid markers of atherosclerosis and predictors of cardiovascular events were introduced. These markers are related to inflammation and macrophage stimulation, such as cystatin C, matrix metalloproteases (MMPs), and chitotriosidase. Increased serum cystatin C concentration, an alternative measure of renal function, is now suggested as a strong predictor of cardiovascular events. The authors compared new non-lipid atherosclerosis indexes with common inflammatory (hs-CRP) and lipid markers in elderly persons and patients with atherosclerosis and ischemic heart disease (IHD) who have undergone coronary bypass surgery.

Cystatins are known to be very potent endogenous inhibitors of cysteine proteases of the papain superfamily. They form equimolar, tight, and reversible complexes with human cysteine proteases (cathepsins B, H, K, L and S), and express different cellular functions as proteins (cell proliferation, degradation of extracellular matrix, etc.). In humans, cystatin C, localized predominantly in the extracellular space, was used for early detection of impaired kidney function, as well as a marker in several inflammatory and tumor diseases. The question of whether cystatin C is an atherogenic or protective protein, as well as its possible role as a marker or predictor in IHD is still not clear.

Using an ELISA method, the authors have shown that in healthy persons aged 25-45, cystatin C concentration is the highest in cerebrospinal fluid and much lower in urine and especially in bile (cerebrospinal fluid>saliva>serum>urine>bile). A similar distribution was shown for procathepsin B, which is an enzymatically inactive precursor of the mature cysteine protease cathepsin B: cerebrospinal fluid>saliva>serum>urine. However, the relative concentration of cystatin C in serum was higher (~100-fold) than in urine, as compared to the procathepsin B level (~ 5-fold).

The serum cystatin C concentration was shown to increase significantly in elderly persons of 45-65 years old with a high risk of IHD (with normal serum and urine creatinine level), and especially in patients of the same age with atherosclerosis and IHD before coronary bypass surgery. The elevated serum level of cystatin C, MMPs, and chitotriosidase activity were observed in an elderly group with atherosclerosis and IHD, and lipid-lowering therapy by statins significantly decreased the hs-CRP concentration and the activity of MMPs, , but not the cystatin C level. After coronary bypass surgery, there was a rapid and marked increase in hs-CRP and a mild increase in cystatin C concentration. Moreover, 30 days, and one year after surgery, cystatin C levels were still elevated.

Protease inhibitors are generally regarded as atheroprotective, because proteases participate in matrix degradation, a process regarded mostly as atherogenic. The cystatin C most likely has its origin in different cell types present in aortic lesions, such as SMCs, endothelial cells, and macrophages, known to produce cystatin C. The role of cystatin C compared to hs-CRP and the activity of MMPs, as possible predictors of cardiovascular events is discussed.

Chapter 9- Glomerular filtration rate (GFR) is a key parameter to evaluate the function and thereby the quality of the transplanted kidney. Direct measures of the renal elimination of different exogenous GFR (e.g. inulin clearance) are the "gold standard" for assessing GFR. These techniques are however rarely implemented in routine clinical practice. As an alternative, a number of easy-to-use mathematical equations, incorporating different anthropometrical variables in addition to biological parameters, have been developed to predict ('estimated GFR'), rather than to directly measure GFR ('true GFR').

International guidelines recommended relying on serum creatinine for GFR estimation (1). It has become, however, increasingly evident that serum creatinine, alone or even incorporated into estimating equations, is not an ideal marker of the renal graft function. As a result, interest has arisen regarding alternative endogenous marker. Among them, cystatine C tends to be regarded as a better marker of GFR than serum creatinine in a variety of different patient populations, including renal transplant patients.

In this chapter, the authors will first review the main clinical studies that have questioned the relevance of serum-creatinine based GFR estimates in renal transplantation. They will then present the existing evidence favoring serum cystatine C over serum creatinine as a GFR marker in this context. Finally, the authors will discuss the different challenges that still have to be addressed in order to definitely legitimate a widespread use of cystatine C as a routine index of renal graft function.

Chapter 10- Robustness of an analytical chromatographic method for separation of cystatin c has been verified. Changes in many parameters were carried out, such as, wavelength, column oven, mobile phase composition, chromatographic column.

Imperative changes have altered the efficiency of the chromatographic separation; such changes include pH alteration of the mobile phase as well as alkyl sulfonate molarity changing.

All robustness conditions showed no major effect on the chromatographic separation of the analyte except with the changes related to TFA and alkyl sulfonate ion pair reagents. Peak area RSD, asymmetry and No. of theoretical plates were < 0.7%, < 1.2 and > 10,000, respectively. Results obtained using mobile phase after 6 months of storage have proven its stability and possibility of use. Gradient elution mode was utilized to elute cystatin c with a UV detection of 224 nm. Ace and Waters C8 (150 x 4.6 mm i.d., 5 μm) as chromatographic columns were used.

Chapter 11-Cystatins are protease inhibitors that are specifically active against lysosomal cysteine proteases. Cystatins regulate proteases by the formation of reversible high affinity complexes. Members of this superfamily are classified into three subfamilies based on their amino acid homology 1) Type 1 cystatins (cystatin A and B) have a single cystatin domain, intracellular, and lack secretory signal, 2) type 2 cystatins are mainly secretory and comprised of cystatin C, D, E/M, F, S, SN, and SA, 3) type 3 cystatins are those with multi cystatin domains and represent kininogens, the plasma proteins. Cystatins are essential to maintain cell homeostasis. Impairment of cystatins and their substrate proteases leads to pathological conditions including cancer. While some of the cystatins are over-expressed in some cancers they are also down-regulated. Cystatin A is over-expressed in lung, breast, head and neck and prostate cancers and serves as an important prognostic biomarker. Cystatin B expression is elevated in lung, breast, prostate, and hepatocellular carcinoma while being down regulated in oesophageal cancer. In contrast to Type 1 cystatins, Type 2 cystatins are mostly down regulated in breast, lung, cervical, prostate, and brain cancers. Type 2 cystatins are inactivated by various mechanisms including deletion, promoter hypermethylaion, deacetylation, and mutations in the substrate binding regions. Type 3 cystatins, kininogens, have been shown to play a suppressive role in colon cancer. The functions of cystatins in different cancers are not limited to protease inhibition alone but also include cell cycle regulation, apoptosis, and cancer cell adhesion. In this chapter, details on the mechanisms of cystatin gene inactivation and their role in different cancers are summarized

In: Cystatins: Protease Inhibitors ...
Editors: John B. Cohen and Linda P. Ryseck
ISBN: 978-1-61209-343-7

Chapter 1

CYSTATINS: THE MULTIFACETED PROTEASE INHIBITORS

Medha Priyadarshini and Bilqees Bano*
Dept. of Biochemistry, F/O Life Sciences,
Aligarh Muslim University,
Aligarh, U.P., India

ABSTRACT

The cystatins constitute a large group of evolutionary related proteins with diverse biological activities and wide occurrence in tissues, targeting primarily papain-like cysteine proteases and parasitic proteases like cruzipain, but also mammalian asparaginyl endopeptidase. The cystatin superfamily comprises a large group of the cystatin domain containing proteins divided into three main groups depending on the presence of single or multiple 'cystatin domains', and the presence or absence of a signal sequence. Family I- the stefins are cytosolic, 11 kDa single chain proteins with no disulphide bonds or carbohydrates, Family II- the cystatins are mainly extracellular, 13-15 kDa single chain proteins with two disulphide bonds but no carbohydrates and Family III- the kininogens are multidomain, blood plasma glycoproteins.

The subfamilies possess few distinctive features with regards to amino acid composition. The alignment of sequences of cystatins reveals common features, significant to the structure and activity of the proteins like the amino terminus glycine residue and two β-hairpin loops, one in the middle and one in the C-terminal segment of the protein. The first loop contains a QXVXG sequence conserved in almost all inhibitory members of cystatins, whereas the second loop contains a P-W motif, which is also highly conserved. All the cystatins are specific, non-covalent, competitive, reversible, tight binding inhibitors which inhibit the target enzymes in micromolar to picomolar range forming tight equimolar complexes with proteases.

Myriad functions attributed to cystatins are associated with cell proliferation, differentiation, ageing, antigen presentation, immunomodulation, etc. They are also being

* Corresponding author.

hailed as anti-angiogenic agents. Cystatins are at the front-line of defense against pathogens that secrete proteases as virulence factors.

Maintenance of appropriate equilibrium between free cysteine proteases and their complexes with inhibitors is imperative for proper function of all living systems/tissues. The importance of cystatins is underlined by the pathological conditions that arise upon loss or mutations of cystatin genes/function like sepsis, cancer, rheumatoid arthritis, purulent bronchiectasis, multiple sclerosis, muscular dystrophy, pancreatitis, diabetes, etc. Profound changes in the lysosomal system seem to be an early event in "at-risk" neurons of AD brains. Cystatins exhibit high propensity for fibrillation.

The present article will discuss the cystatin superfamily in general (evolution, members, structure, mechanism of action) and the already ascribed and putative biologic functions of its members.

Introduction

1.1. Proteases: General

Proteolytic enzymes, also known as proteases/peptidases, are enzymes that catalyze the breakdown of proteins by hydrolysis of peptide bonds. Proteases are essential for the survival of all kinds of organisms, and are encoded by approximately 2% of all genes [Rawlings et al., 2004]. Proteases are customarily classified as exopeptidases when they hydrolyze only the N- or C- terminal bonds in proteins and endopeptidases when they hydrolyze internal peptide bonds. However, proteases and proteinases are synonymously used in the literature. On the basis of their action mechanism they are either serine, cysteine or threonine proteases (amino-terminal nucleophile hydrolases), or aspartic, metallo and glutamic proteases (with glutamic proteases being the only subtype not found in mammals so far). Besides their indigenous roles of protein degradation relevant to food digestion and intracellular protein turnover, a number of other functions are ascribed to them lately, like in the control of large number of key physiological processes such as cell-cycle progression, cell proliferation and cell death, DNA replication, tissue remodelling, haemostasis (coagulation), wound healing and immune response [Turk, 2006].

Cysteine proteases (CPs) are the proteins with molecular mass about 21-30 kDa, showing the highest hydrolytic activity at pH 4-6.5. CPs are present in all living organisms. They are synthesized in a precursor form in order to prevent unwanted proteolysis and later subjected to cotranslational and posttranslational modifications to convert them into catalytically active mature enzymes [Turk et al., 2000]. Till now, 21 families of CPs have been discovered, almost half of them in viruses and rest of them in bacteria, protozoa, fungi, plants and mammals [Barrett et al., 2001]. The first clearly recognized and extensively investigated cysteine protease is papain isolated from the latex of plant *Carica papaya*. Mammalian CPs are divided into 4 main groups; namely

1. Lysosomal cathepsins
2. Caspases
3. Calpains
4. Legumain

Cathepsins comprise an important section of the papain family of CPs, sharing similar amino acid sequences and folds. There are eleven human cathepsins known at the sequence level [Turk et al., 2001]. Out of which seven viz. cathepsins B, H, L, C, O, F and X are ubiquitous such that they have a broad tissue distribution, but they may be involved in more specialized processes. Cathepsins K, V and S are more tissue specific with cathepsin K expressed in osteoclasts only, cathepsin V in thymus and testis, and cathepsin S in spleen and lung [Turk et al., 2000]. Cathepsin K is also expressed by breast carcinoma cell, mature macrophages, and multinucleate giant cells adjacent to amyloid deposits in brain [Punturieri et al., 2000; Zaidi et al., 2001; Rocken et al., 2001]. Cathepsins are all relatively small monomeric proteins with molecular mass (Mr) in the range of 24-35 kDa, with exception of cathepsin C, which is an oligomeric enzyme with Mr around 200 kDa [Turk et al., 2002]. All mature cathepsins are glycosylated at usually one or more glycosylation sites except cathepsin S. Human cathepsins play very important role in intracellular protein turnover in lysosomes and in processing and activation of other proteins including proteases, in antigen processing and presentation and in bone remodelling. However, their specific and individual functions are often associated with their restricted tissue localization [Brix et al., 2008]. Lysosomal CPs have been found to be critical for rheumatoid arthritis, osteoarthritis and osteoporosis [Vasiljeva et al., 2007], neurological disorders [Nakanishi, 2003], pancreatitis [van Acker et al., 2002], cancer [Gocheva and Joyce, 2007], cardiovascular diseases [Lutgens et al., 2007]. Cathepsins also participate in apoptosis, although there role is still not clear [Turk and Stoka, 2007]. In some pathological conditions like ischemia, hypervitaminosis and on exposure to UV radiations lysosomal enzymes are released in extracellular space and produce extensive damage to the extracellular matrix.

Calpains and Caspases are cytoplasmic thiol proteinases. Calpains depend on Ca^{2+} for activity and participate in many intracellular processes like turnover of cytoskeletal proteins, cell differentiation and regulation of signal peptides. They are ubiquitously distributed and have been implicated in acute neurological disorders, Alzheimer's disease, muscular dystrophy and gastric cancer [Huang and Wang, 2001]. Caspases are cysteine dependent aspartate specific proteinases involved in cytokine maturation, apoptosis signalling and in apoptosis mediation [Goyal, 2001].

Legumains are cysteine-dependent asparagine endopeptidases. They are involved in MHC class II–restricted antigen presentation [Manoury et al., 1998] and local negative regulation of osteoclasts formation and activity [Choi et al., 1999].

Regulation of Lysosomal Thiol Proteinase Activity

Despite indespensable, the immense hydrolytic potential of cathepsins can be detrimental to living systems if not kept under strict reins. Failures in biological mechanisms controlling protease activities result in many diseases such as neurodegeneration, cardiovascular diseases, osteoporosis, arthritis and cancer. Cells have evolved several distinct mechanisms for the regulation of excessive CP activity via proper gene transcription, maintenance of the rate of protease synthesis and degradation and most importantly the interaction of CPs with the proteins that inhibit them, viz. cysteine proteinase inhibitors or thiol proteinase inhibitors (CPIs or TPIs) or more commonly cystatins.

The cystatin superfamily comprises a large group of the cystatin domain containing proteins present in wide variety of organisms including humans. Cystatin inhibitory activity is vital for the regulation of normal physiological processes by limiting the potentially inappropriate activity of their target proteases, cathepsins, mammalian legumain and some calpains [Alvarez-Fernandez et al., 1999; Crawford, 1987].

1.2. Discovery of the Cystatin Superfamily

Hayashi et al., [1960] for the first time reported the presence of a factor capable of inhibiting the clotting activity of a thiol protease in mammalian system. The first isolated and partially characterized protein inhibitor of CPs was from chicken egg white and was shown to inhibit papain, ficin [Fossum and Whitaker, 1968; Sen and Whitaker, 1973] and cathepsins B and C [Keilova and Tomasek, 1975]. Later for the same protein term cystatin was proposed because of its unique property of arresting the activity of CPs [Barrett, 1981]. The first intracellular protein inhibitor of papain, cathepsin B and H was isolated and partially characterized from pig leucocytes and spleen [Kopitar et al., 1978]. The amino acid sequences of chicken cystatin [Turk et al., 1983; Schwabe et al., 1984] and human stefin (stefin A) from the cytosol of polymorphonuclear granulocytes [Machleidt et al., 1983] were determined confirming structural differences between these two homologous proteins. Contemporarily, isolation of CPs from sera of patients suffering from autoimmune diseases by Turk et al. [1983] and unearthing of its sequence homology with chicken cystatin, christened it as human cystatin [Brzin et al., 1984], soon renamed as human cystatin C (HCC) [Barrett et al., 1984]. Sequences of bovine and human kininogens were also determined almost at the same time [Nawa et al., 1983; Muller-Esterl et al., 1985]. The birth of the concept of cystatin "superfamily" was heavily based on the observation that multiple cystatin-like sequences were present in kininogens and that stefins were related to both cystatins and repeats of kininogens [Ohkubo et al., 1984]. This data and First International Symposium on Cysteine Proteinases and their Inhibitors (Portoroz, Yugoslavia (now Slovenia) September 1985, organized by V. Turk) were crucial for nomenclature and classification of the cystatin superfamily [Barrett et al., 1986].

1.3. Classification of the Cystatin Superfamily

The first classification of the cystatin superfamily into three families was based on at least 50% sequence identity, inhibition of their target enzymes and presence or absence of disulphide bonds [Barrett et al., 1986]. Three distinct families of the protein inhibitors comprise: *family 1* or *the stefin family*, *family 2* or *the cystatin family* and *family 3* or *the kininogen family*. The first two families are single domain inhibitors whereas the kininogens are composed of three domains, two being inhibitory [Figure 1]. Athough the three families differ considerably their tertiary structures are conserved and exhibit the cystatin fold that is formed by a five stranded anti-parallel β-sheet wrapped around a five turn α-helix [Figure 2] [Bode et al., 1988]. Later, the term 'type' was introduced and the mammalian cystatins were divided into types 1, 2 and 3 [Rawlings and Barrett, 1990]. However, an increasing number of cystatins from various sources introduced new subdivision of the cystatins into four families

[Rawlings and Barrett, 1990], the fourth family consisting of non-inhibitory homologues of two cystatin-like domains, such as human α_2 SH-glycoprotein (feutin) and histidine-rich glycoprotein [Brown and Dziegielewska, 1997]. The cystatin superfamily also comprises of phytocystatins. According to recently proposed classification of peptidase inhibitors into families and clans [Rawlings et al., 2004] cystatins are assigned to family I25 which consists of three subfamilies, I25A (stefins), I25B (cystatins), I25C (are mostly not protease inhibitors). The classification of cystatin superfamily will again be discussed in light of new studies under the head of evolution.

1.4. General Properties of Cystatin Superfamily

Type 1 Cystatins (Stefins)

The type 1 cystatins belong to the subfamily I25A [Rawlings et al., 2004]. Stefins are acidic single chain proteins lacking disulphide bonds and carbohydrates, composed of ~100 amino acid residues with Mr of 11 kDa [Turk et al., 2008]. They are primarily intracellular cytoplasmic proteins of many cell types. However, they have been found in extracellular fluids as well [Abrahamson et al., 1986]. In mammals, two members of the stefin family, stefin A (cystatin A or α) and stefin B (cystatin B or β) have been identified [Barrett et al., 1986]. In addition stefin C has been characterized from bovine thymus as the first tryptophan containing stefin with a prolonged N-terminus [Turk et al., 1993] and stefin D from pigs [Lenarcic et al., 1996]. Atleast three different stefin A variants are encoded within the mouse genome [Tsui et al., 1993].

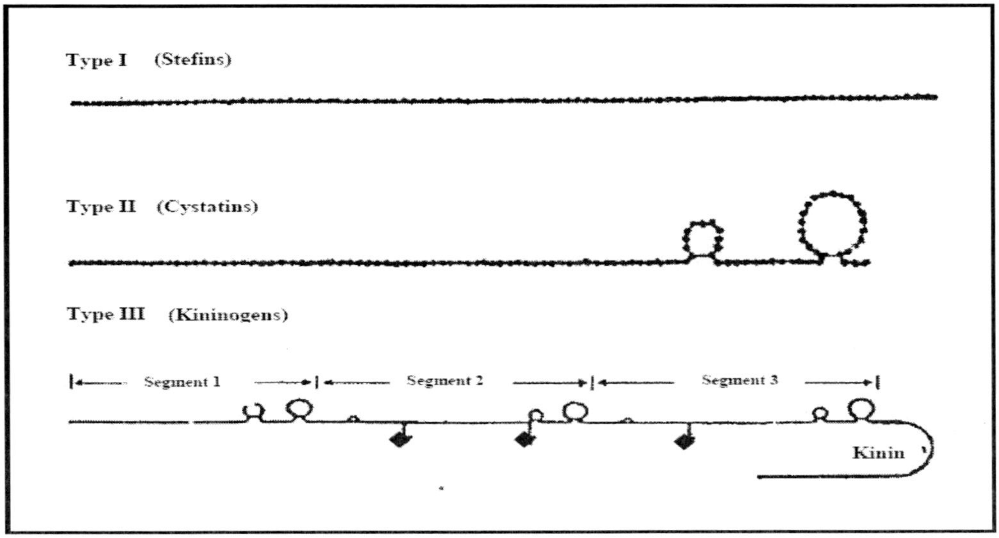

Figure 1. A diagrammatic representation of the chain structures of proteins in the cystatin superfamily. The stefins are single chain proteins without disulphide linkages. The cystatins are also single chain but possess two disulphide bonds. The structure indicated for the kininogens is that of L- and T-kininogen; H-kininogens have a longer carboxyl terminal extension. There is an additional disulphide link from segment 1 to the kinin segment. The symbol marks potential sites for the attachment of the carbohydrate side chains.

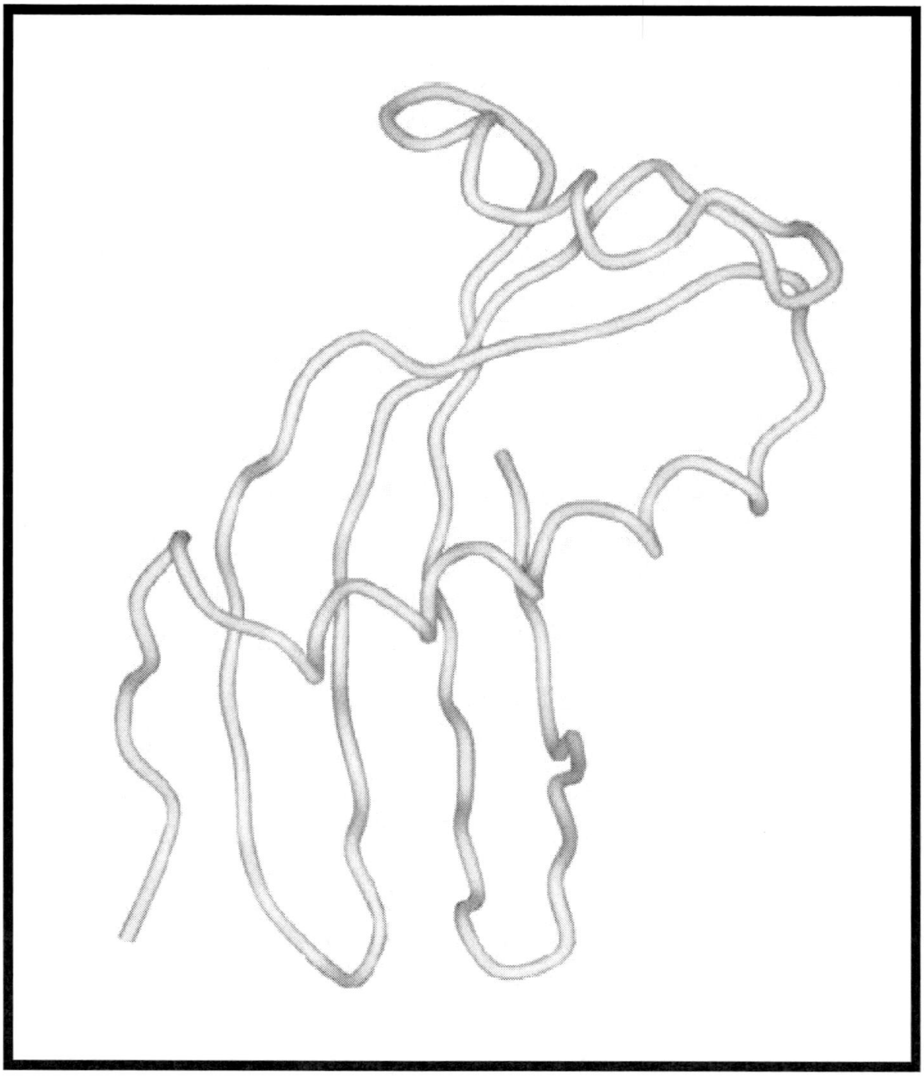

Figure 2. Fold of cystatin C. Cystatin C chain trace is shown in orientation which positions the N-terminal "elephant trunk" and the first and the second hairpin loops to the bottom from left to right [Adapted from Bode et al. 1988, EMBO J, 2593-2599].

Cystatin A

It is an inhibitor of cathepsin B in human skin discovered by Fraki [1976]. Later on Jarvinen [1978] studied it as 'acid cysteine proteinase inhibitor' (ACPI) because of its acidic pI at 4.7-5.0. Brzin et al. [1983] purified an inhibitor from blood leucocytes, and named it as "stefin'. The amino acid sequence was determined by Machleidt et al. [1983]. Green et al. [1984] characterized same type of CPI from human liver and later renamed it as cystatin A. Cystatin A occurs in multiple isoelectric forms with predominantly acidic pI values in the range 4.5-5.0 [Hopsu-Havu et al., 1983a]. Rinnie et al. [1978] detected cystatin A in extracts of squamous epithelia from oesophagus. It was also found in dendritic reticulum cells of the lymph nodes [Rinnie et al., 1983], seminal plasma [Minakata and Asano, 1985], saliva, bovine skin [Turk et al., 1995] and in a number of epidermoid carcinomas [Rinnie et al., 1984]. *Cystatin α* is assumed to be a species variant of cystatin A found in rats. This protein was characterized by Jarvinen [1976] as a specific inhibitor of CP from rat skin having Mr of 13 kDa. Cystatin α is generally found on the epidermal layer [Jarvinen et al., 1978] and various other squamous epithelia [Rinnie et al., 1978]. Because of its high level expression in skin, human stefin A is presumed to control CPs in the skin. Some cathepsins play a crucial role in the antigen presentation process indicating that the interactions between stefin A and cathepsins contribute to the species dependent diversity of the endosomal compartments which participate in the immune response [Mihelic et al., 2006].

Cystatin B

Cystatin B was detected as an inhibitor of cathepsin B and H in human tissues by Lenney et al. [1979]. It has been purified from human spleen and liver [Jarvinen and Rinnie, 1982; Green et al., 1984]. Cystatin B is relatively basic protein with pI values of 6.25 and 6.35 for its two forms [Green et al., 1984]. It forms dimer [Green et al., 1984] which shows no inhibitory activity. With ubiquitous distribution it appears to be general inhibitor in the cytoplasm. *Cystatin β* a species variant of cystatin B was isolated from rat liver [Finkenstaedt, 1957; Lenney et al., 1979] with pI values ranging from 5.04 to 5.6 [Kominami et al., 1981]. It has even distribution in tissues and is more abundant than cystatin α in all tissues except skin. In this it resembles cystatin B of human variant.

Stefin C

Stefin C is unique among the inhibitors from stefin family which was found in multiple forms resulting from the cleavage of Asn 5-Leu 6 bond of the inhibitor. Its molecular weight (MW) is calculated to be 11,546 (101 amino acid residues). It was found to be acidic with pI values from 4.5 to 5.6 [Turk et al., 1993].

Type 2 Cystatins (Cystatins)

The human type 2 cystatins are grouped in subfamily I25B of the cystatin family [Rawlings et al., 2004]. Cystatins are also acidic (with exception of human cystatin C) comprising of ~115 amino acids (Mr ~13-15 kDa), containing two disulphide bonds (formed by four conserved cysteine residues) [Grubb et al., 1984]. These are usually non-glycosylated with exceptions of cystatin E/M [Ni et al., 1997; Sotirpoulou et al., 1997], cystatin F [Ni et al., 1998] and cystatin S [Eshard et al., 1990] which are glycoproteins. They are synthesized with 20-26 residues long signal peptides and are mainly extracellular, secreted proteins, occurring at relatively high concentrations in body fluids [Abrahamson et al., 1986; Turk et al., 2008]. Similar to stefins, the cystatins contain the conserved QXVXG region in the central part of the molecule and the P-W pair in the C-terminal part of cystatins [Turk and Bode, 1991]. Chicken cystatin and HCC represent founding members of this family [Turk and Bode, 1991; Abrahamson et al., 1986]. Human type cystatins include cystatin C, D, S, SA and N with about 50% or less sequence identity [Turk and Bode, 1991; Abrahamson et al., 1986; Balbin et al., 1994].

Cystatin C

Originally cystatin C was termed as γ-trace or post-γ-globulin isolated from human cerebrospinal fluid because of its basic nature and γ electrophoretic mobility [Barrett et al., 1984; Brzin et al., 1984]. It was also found in the urine in renal failure patients [Butler and Flynn, 1961] and ascetic and pleural fluids [Hochwald and Thornbecke, 1962]. Cystatin C was also detected in saliva, normal serum [Cejka and Fleischmann, 1973] and seminal plasma [Colle et al., 1976]. Preferentially abundant in cerebrospinal fluid, seminal plasma, milk, synovial fluid, urine, and blood plasma [Abrahamson et al., 1986] cystatin C has also been detected intracellularly in brain cortical nerves normal and neoplastic neuroendocrine cells in the adrenal medulla [Lofberg et al., 1982], thyroid and pituitary [Lofberg et al., 1983].

Cystatin D

Cystatin D a member of human cystatin multigene family and was cloned from a genomic library using cystatin C cDNA probe [Freije et al., 1991]. The inhibitor consists of 122 amino acids residues (Mr 13,885). The deduced amino acid composition includes a putative signal peptide and has 51- 55% homology with either cystatin C or secretory gland cystatins S, SA and SN. It is a relatively neutral protein with pI in the range of 6.8- 7.0 [Freije et al., 1991]. It is expressed in parotid glands, saliva and tears [Balbin et al., 1994]. This tissue restricted expression is in marked contrast with a wider distribution of all other family 2 cystatins.

Cystatin S

Human saliva contains several low MW acidic proteins which include CPIs [Isemura et al., 1984b].The first salivary inhibitor purified and sequenced was SAP-I (salivary acid protein) by Isemura et al. [1984a], which was renamed as 'cystatin S'. It contains no phosphate, in contrast to other salivary proteins. This inhibitor has also been isolated from human submaxillary, submandibular and sublingual glands and found to be present in the serous cells of the parotid and submaxillary glands [Isemura et al., 1984b]. The protein has also been found in tears, serum, urine, bile, pancreas and bronchus [Isemura et al., 1986].

Variants of Cystatin S

Several molecular variants of cystatin S have been studied by Isemura et al. [1986] which differ in their N-terminal sequence and pI values. Differences in pI values resulted from phosphorylation of residues Ser3 and Ser1 in salivary cystatin [Isemura et al., 1991]. *Cystatin SN:* Originally known as cystatin SV or SA-1 [Abrahamson et al., 1986]. The protein consists of 121 amino acid residues (MW 14,316). The pI values are in the range of 6.6-6.8. *Cystatin SA* consists of 122 amino acid residues (MW 14,351) having acidic pI value of 4-6 [Isemura et al., 1991]. Cystatin SA isolated from saliva had N-terminal residue Glu [Isemura et al., 1986].

Cystatin E

Human cystatin E from amniotic fluid and fetal skin epithelial cell was identified and recombinant cystatin E isolated [Ni et al., 1997]. Human cystatin M is expressed by normal mammary cells and a variety of human tissues [Sotirpoulou et al., 1997]. Both proteins are identical and were renamed as cystatin E/M (MEROPS, the peptidase database). Recently, the expression of cystatin M/E was found to be restricted to the epidermis [Cheng et al., 2006] and is most probably identical to cystatin E/M.

Cystatin F

Cystatin F (leukocystatin) (MW 14,543) is primarily found in peripheral blood cells, T cells, spleen, dendritic cells and selectively, in hematopoietic cells [Ni et al., 1998; Halfon et al., 1998]. Cystatin F has an additional disulphide bridge, thus stabilizing the N-terminal part of the molecule [Ni et al., 1998]. It is the only cystatin synthesized and secreted as an inactive disulphide-linked dimeric precursor which becomes active following reduction to monomeric from [Schuttelkopf et al., 2006].

Type 3 Cystatins (Kininogens)

Kininogens, the precursors of kinin, are large multifunctional glycoproteins in mammalian plasma and other secretions. Three different types of kininogens have been identified: high molecular weight kininogen (HK), low molecular weight kininogen (LK) and T-kininogen an acute phase protein found only in rats [Cadena and Colman, 1991; Muller-Esterl, 1987]. Human HK and LK are single-chain proteins each composed of an N-terminal heavy chain, the kinin segment and a C-terminal light chain. The light and heavy chains are interconnected by disulphide bridges. The heavy chains and kinin segments of both kininogens have identical amino acid sequences while the light chains are different [Cadena and Colman, 1991; Salvesen et al., 1986a]. The heavy chain is composed of three cystatin domains [Figure 1], D1-D3 [Salvesen et al., 1986b] with only D2 and D3 possessing papain inhibitory and D2 possessing calpain inhibitory activities. Both inhibitory domains of LK and HK are grouped in subfamily I25B of the cystatin superfamily [Rawlings et al., 2004].

Other Type 2 Cystatins

There are a number of other cystatins or cystatin related proteins, which are structurally related to cystatins with no inhibitory activity against papain like enzymes [Turk and Turk, 2008]. CRES (Cystatin Related Epididymal Spermatogenic) protein [Sutton et al., 1999], testatin (expression restricted to mouse pre-Sartoli cells) [Tohonen et al., 1998], cystatin SC and cystatin TE-1 (expressed in testis and epididymis, respectively) [Li et al., 2002], and several other genes were found expressed specifically in the male reproductive tract [Hamil et al., 2002; Xiang et al., 2005; Shoemaker et al., 2000], indicating the existence of a new subgroup in the type two cystatins [Cornwall et al., 2003; Sutton-Walsh et al., 2006]. These CREStatins show homology to cystatins, with the exception of the two hairpin loops responsible for the cysteine protease inhibition. Their role could be regulation of proteolysis in the reproductive tract as well as protection against invading pathogens, as shown by cystatin 11 [Hamil et al., 2002]. The CRES protein tend to form oligomers [Horsten et al., 2007], similar to cystatin C [Janowski et al., 2001; Wahlbom et al., 2007] and stefin B [Zerovnik et al., 2002; Jenko-Kokalj et al., 2007]. Another type 2 cystatin, cystatin 10, expressed in cartilage, localized in prehypertrophic and hypertrophic chondrocytes is known to be an inducer of chondrocyte maturation followed by apoptosis [Koshizuka et al., 2003]. A novel cystatin type 2 protein namely CLM expressed widely in normal tissue playing role in hematopoietic differentiation or inflammation, different from CRES was characterized by Sun and coworkers [2003].

NEW MEMBERS OF THE CYSTATIN SUPERFAMILY

The feutins and *histidine-rich glycoproteins (HRG)* comprise fourth family of cystatins. The feutin family consists of two tandem cystatin domains. Bovine feutin was first characterized by Pedersen in 1944, and its relation to cystatin superfamily described in 1988 [Elzanowski et al., 1988]. Human feutin (α_2-HS glycoprotein) was confirmed in 1987 [Dziegielewska et al., 1987; 1990; Dziegielewska and Brown, 1995]. Since then, protein and/or cDNA sequences have been reported for human, cow, pig, rat, mouse, Habu snake, feutins [Brown and Dziegielewska, 1997]. Almost all the feutin sequences contain 12 cysteine residues, showing homology to the cystatins and cystatin domains in kininogens [Dziegielewska and Brown, 1995]. HRG has been characterized in the plasma of man, mouse, rabbit, cow and pig [Leung, 1993], sharing good sequence homology with human and bovine H-kininogen [Koide et al., 1986]. A large number of proteins have been discovered recently, which possess cystatin domains and may even exhibit CPI activity e.g. latexin [Aggarwal et al., 1996]. However, feutin, HRG and latexin all seem to lack CPI activity.

Thyropins constitute a new family of papain-like CP inhibitors [Lenarcic and Bevec, 1998], classified as family I31 [Rawlings et al., 2004]. The p41 invariant chain (Ii)-fragment of the MHC class II-Ii complex 104, 105 and equistatin from the sea anemone [Lenarcic et al., 1997] are best characterized members of this family. Thyropins show inhibitory activity against CPs and also towards aspartic and metalloproteases [Mihelic and Turk, 2007; Lenarcic and Turk, 1999]. *Tick cystatins: Syalostatin L* [Kotsyfakis et al., 2006] and *syalostatin L2* [Kotsyfakis et al., 2007] have been characterized from salivary glands of the tick *Ixodes scapularis*. Both show 75% sequence identity and inhibit cathepsin L with a Ki of 4.7 nM and cathepsin V with Ki of 57 nM. *Staphostatins* are specific inhibitors of staphylococcal CPs. Three members of this family have been described – staphostatins A and B from *Staphylococcus aureus* and staphostatin A from *Staphylococcus epidermidis* [Filipek et al., 2003]. *Clitocybin* is a new type of CPI from a mushroom appearing to be related to fungal lectins and hence a new family of CPIs is suggested for them called mycocypins [Brzin et al., 2000]. *Chagasin* is a cysteine proteinase inhibitor from *Trypanozoma cruzi* inhibiting both cruzipain and papain, but has no homology with cystatins [Monteiro et al., 2001].

Phytocystatins: In plants, inhibitors of CPs are known as phytocystatins. They contain the QXVXG region of type 2 cystatins, but also resemble stefins in the absence of disulphide bonds [Arai et al., 2002], providing a transitional link between type 1 and type 2 cystatins. The structure of the plant inhibitor oryzacystatin, determined by NMR spectroscopy, shows the same cystatin fold as the animal cystatins [Nagata et al., 2000]. There are numerous phytocystatins expressed and characterized on the protein level from corn [Abe et al., 1992], rice [Chen et al., 1992], soyabean [Lalitha et al., 2005], sugarcane [Oliva et al., 2004] and others. C-terminal extended phytocystatins were found as bifunctional inhibitors of papain and legumain [Martinez et al., 2007]. In addition, a "multicystatin" containing two cystatin like domains were isolated from cowpea leaves [Diop et al., 2004]; tomato leaves [Wu and Haard, 2000]. Also there are certain plant proteins like monellin which lack the CPI activity but have a cystatin like three dimensional structure [Grzonka et al., 2001]. Phytocystatins and other inhibitors are important for plant defence response to insect predation, may act to resist infection by some nematodes [Koiwa et al., 1997], play a crucial role in response to various conditions [Diop et al., 2004; Brzin and Kidric, 1995] and show great potential tools for genetically engineered resistance of crop plants against pests [Aguiar et al., 2006]. (For extensive review of general properties of cystatin superfamily members see Turk and Turk, 2008; Turk et al., 2008,).

Variant Cystatins

Divergent cystatins showing significant homology to stefins, cystatins and kininogens have been expressed/purified and characterized from venom of African puff adder *(Bitis arietans)* [Evans and Barrett, 1987]; from perilymph of flesh fly larvae [Suzuki and Natori, 1985]; from *Drosophila melanogaster* [Delbridge and Kelly, 1990]. Some of the mammalian and non mammalian sources from where CPIs have been isolated are summarized in Table 1.

1.5. Evolution

The first two proposed evolutionary dendrograms for CPIs were made based on a small number of members of the cystatin superfamily [Muller-Esterl et al., 1985; Salvesen et al., 1986a]. The new proposed evolutionary dendrograms followed the evolution of the proteins of the cystatin superfamily along four lineages, with special attention that duplication of cystatin like segments has played important contribution to the understanding of the evolution of cystatins. According to the scheme of Muller-Esterl et al. [1985] constructed on the basis of sequence homology, the diversity of CPI has evolved from two ancestral building blocks 'A' and 'B'. The stefin progenitor represents the whole superfamily comprising a single 'A' unit. Cystatin acquired a second element B, possibly by gene fusion, thus forming 'AB' unit. Gene triplication of the archetype inhibitor generated the kininogen heavy chain which contains 3 cystatin like copies $(AB)_3$. The proposed evolutionary pathway also contained a 'missing link', a two cystatin domain protein that evolved from the cystatins by duplication, with two candidates for such a protein: feutin and HRG. This scheme however seem unlikely most importantly because neither domain in feutins/HRG is inhibitory but two domains of kininogens have inhibitory activity. If feutin/HRG were the 'missing link', then the kininogens which have evolved from the two domain protein would have to re-evolve their protease inhibitory activity and sequences [Brown and Dziegielewska, 1997]. Brown and Dziegielewska [1997] proposed the following scheme for cystatin superfamily evolution, with features similar to Muller-Esterl et al., [1985] scheme but with a new missing link, a two cystatin domain protein in which both the domains were functional cysteine proteinase inhibitors.

From it, the kininogens, feutins, and HRG could have evolved separately or perhaps in parallel and retained or lost their protease-inhibitory activity and active site sequences. This scheme [Figure 3] draws support from the observation of conserved sequences immediately around the cysteine at the C-terminus of the feutins, H-kininogens and HRG, again suggesting a common origin for these three proteins. Based on this Lee et al., [2009] have grouped feutins, HRG and kininogens in a single family, type 3 cystatins.

Table 1: CPIs from some mammalian and non mammalian sources

Source	Tissue	Reference
Beef	Spleen	Brzin et al., 1982
Bovine	Brain	Aghajanyan et al., 1996
	Hoof	Tsushima et al., 1996
	Colostrums	Hirado et al., 1985
Dog	Colostrum	Poulik et al., 1981
	Parotid gland and Kidney	Sekine and Poulik, 1982
Guinea pig	Skin	Jarvinen, 1976
Horse show crab	Hemocytes	Aggarwal et al., 1996
Human	Liver	Green et al., 1984
	Spleen	Jarvinen and Rinnie, 1982
	Placenta	Rashid et al., 2006
Rabbit	Liver	Pountremoli et al., 1983
	Skin	Udaka and Hayashi, 1965
		Hayashi, 1975; Tokaji, 1971
Rat	Brain	Kopitar et al., 1983
Snake (Chinese Mamushi)	Serum	Akoi et al., 2009
Fasciola gigantica	Various tissues	Tarasuk et al., 2009
Ixodes scapularis	Salivary gland	Kotsyfakis et al., 2006 and 2007
Staphylococcus aureus and *epidermidis*	-	Filipek et al., 2003
Eriocheir sinensis	*In silico* gene analysis	Li et al., 2010
Angiostrongylus cantonensis	-	Liu et al., 2010
Ornithodoros moubata	Saliva	Salát et al., 2010
Trypanosoma cruzi	-	Monteiro et al., 2001
Fasciola hepatica	-	Khaznadji et al., 2005
Goat	Kidney	Zehra et al., 2005
	Brain	Sumbul and Bano, 2006
	Lung	Khan and Bano, 2009
	Pancreas	Priyadarshini and Bano, 2010
Spirometra erinacei	-	Chung and Yang, 2008
Yellow croaker	Spleen	Li et al., 2009

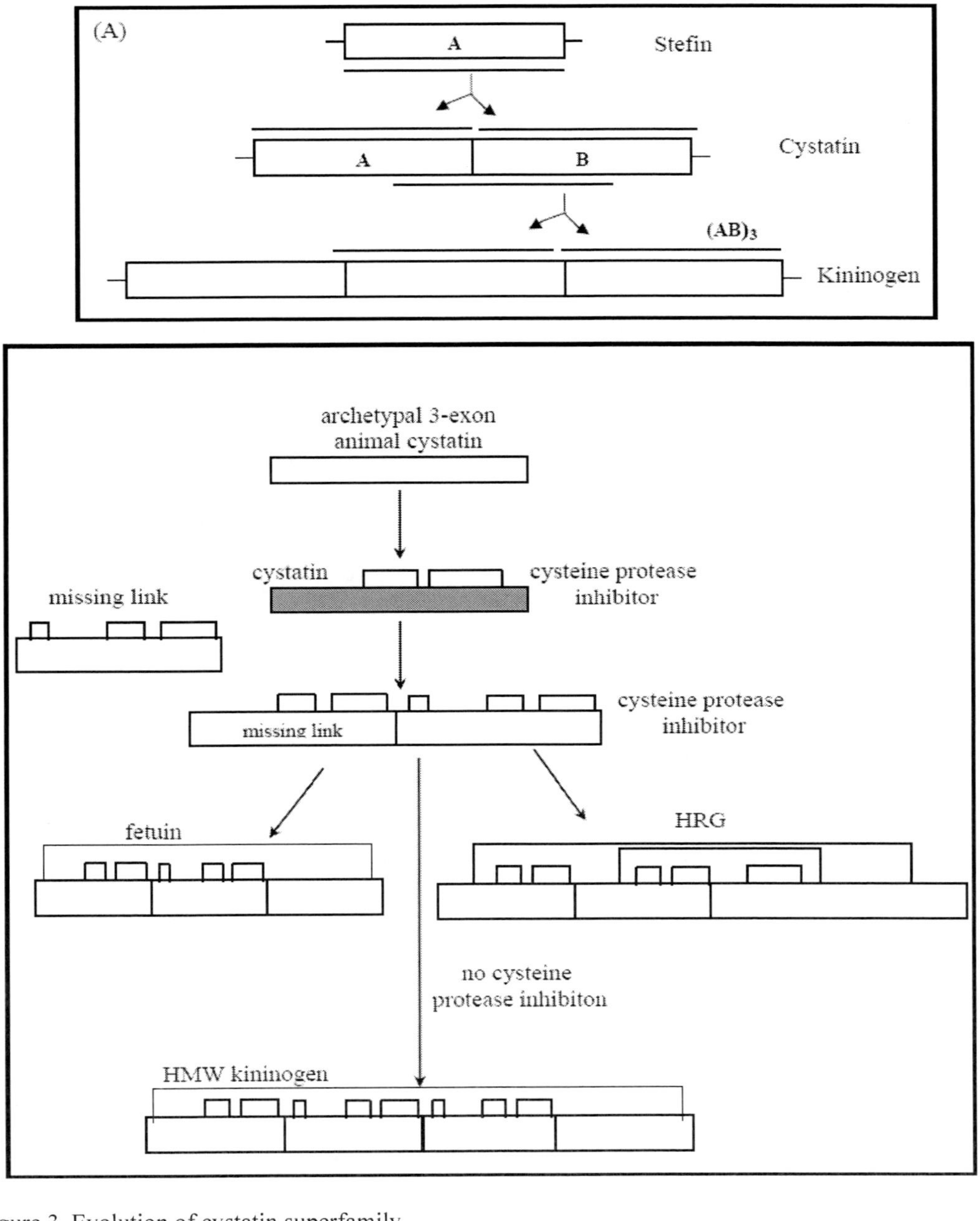

Figure 3. Evolution of cystatin superfamily.
A. Scheme from Muller-Esterl et al. [1985] B. Scheme proposed by Brown and Dziegielewska [1997].

However, recently, Kordis and Turk [2009] identified *in silico* the full complement of the cystatin superfamily in more than 2100 prokaryotic and eukaryotic genomes. According to this study, an analysis of the phyletic distribution of the cystatin superfamily shows the presence of only two ancestral lineages, stefins and cystatins, in eukaryotes and prokaryotes. In contrast to stefins which remained as a single gene or as small multigene families, cystatins underwent more complex and dynamic evolution by numerous gene and domain duplications. The classification and evolution of cystatin superfamily inferred in the pre-genomic era, needs to be revised [Kordis and Turk, 2009]. As discussed, the previously proposed evolutionary scheme assumed a near simultaneous origin and diversification of stefins, cystatins and kininogens and feutins occurring approximately a billion years ago [Rawlings and Barrett, 1990]. However, only two ancestral lineages, stefins and cystatins, are present throughout the eukaryotes [Kordis and Turk, 2009], indicating that the above assumption is wrong. The kininogens and feutins are much younger and restricted to the vertebrates. Thus, the multidomain cystatins are not monophyletic, but originated independently by domain duplication, several times in diverse eukaryotic lineages. Earlier the position of plant cystatins was assumed to be unique and they were regarded as structural intermediates between stefins and cystatins [Rawlings and Barrett, 1990]. However, as shown by Kordis and Turk [2009] this assumption can no longer be maintained, because they represent only the cystatins present in the plant kingdom. A global level analysis of large collection of cystatin superfamily representatives revealed that at the level of particular taxanomic group (in angiosperms, vertebrates and mammals only) as well as at the level of orthologous families great diversification has occurred, namely in kininogen, feutins A and B, HRG, latexins and T1G1, cathelicidins, Spp24, cystatins C, E/M and F, CRES subgroup and stefins A and B (for detail review see, Kordis and Turk, 2009].

1.6. Structure of Cystatins

Amino Acid Composition and Covalent Structure

Most of the members of cystatin superfamily are polypeptides of 98-126 amino acid residues with Mr values in the range of 11-14 kDa. As regards to the amino acid composition of cystatins few distinctive features can be attributed to the subfamilies. Stefins are devoid of disulphide linkages (human cystatin A and rat cystatin α lack cysteine residues while human cystatin B and rat cystatin β have 1 and 2 cysteine residues, respectively) and tryptophan. Turk et al. [1993] however reported the presence of tryptophan in stefin C. A unique feature of stefin B is the conserved QVVAG region in the stefins of mammalian origin, with Val54 replaced by Leu54. Ni et al. [1998] reported the presence of an additional disulphide bridge in cystatin F, for stabilizing the N-terminal part of the molecule in addition to the presence of second tryptophan residue, along with the conserved Trp 106, characteristic of type 2 cystatins. The alignment of sequences of cystatins reveals common features, significant to the structure and activity of the proteins. Four residues are common to all the sequences of cystatins and inhibitory kininogen segments: Gly9, Gln53, Val55 and Gly57. These residues are considered to be of functional importance since they are absent from the non-inhibitory segment D1 of kininogens. Another six conserved residues are Val47, Val55, Ala56, Tyr60, Cys71 and Tyr100. The segment Gln53 to Gly57 is the most highly conserved region. The crystalline form of chicken cystatin reported by Bode et al. [1988] revealed a new fold, the cystatin fold which is, a five stranded anti-parallel β-sheet wrapped around the central N-terminal helix [Figure 2].

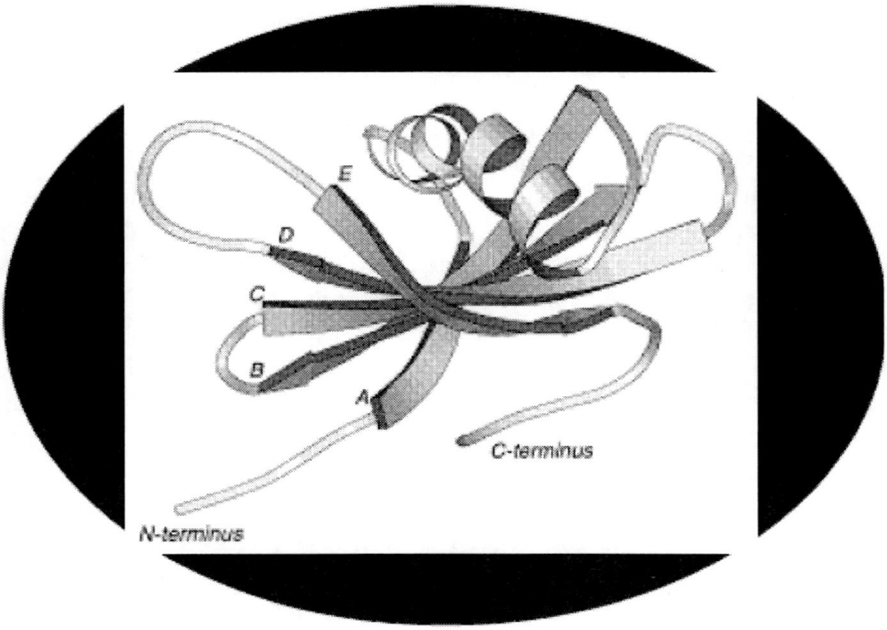

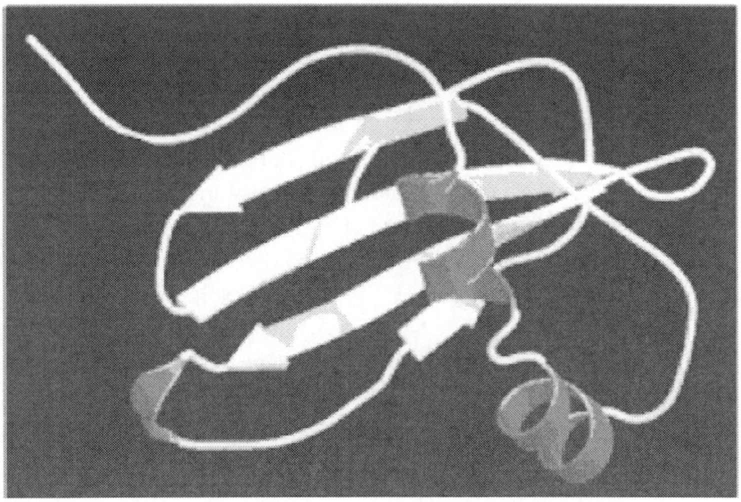

Figure 4. (A) Three dimensional structure of stefin A (where, A, B, C, D, E are the five antiparallel β sheets strands). (B) Three dimensional structure of stefin B (antiparallel β sheets are in white colour and α-helix is in grey colour).

Secondary Structure

The cystatin fold has been shown to exist in HCC, chicken cystatin, cystatin D, as well as in family 1 cystatins A and B [Martin et al., 1994; 1995; Alvarez-Fernandez et al., 2005; Stubbs et al., 1990]. An appending segment of partial α-helical geometry is present in chicken cystatin [Saxena and Tayyab, 1997], but absent in HCC [Bode et al., 1988]. Tryptophan was found only in the second hairpin loop of cystatins [Bode et al., 1988]. A unique feature was observed in crystal structure of cystatin F in its dimeric 'off' state. The two monomers interacted in a fashion not seen before for cystatins or cystatin like proteins, crucially dependent on an unusual intermolecular disulphide bridge. The core sugars for one of the two N-linked glycosylation sites for cystatin F are well ordered and probably their conformation and interactions with the protein modulate its inhibitory properties in particular its reduced affinity toward asparaginyl endopeptidase compared with other cystatins [Schuttelkopf et al., 2006]. There is considerable similarity between the structural features of stefins A and B [Figure 4], but there are also some important differences in the regions which are fundamental to proteinase binding. The difference primarily consists of the two regions of high conformational heterogeneity in free stefin A which correspond in stefin B to two of the components of the tripartite wedge that docks into the active site of target proteinases. These regions which are mobile in solutions are the five N-terminal residues and the second binding loop. In the bound conformation of stefin B they form a turn and a short helix, respectively.

Circular dichroism and computer prediction of secondary structure from the sequence indicates that chicken cystatin has about 20% α-helix, 42% β-structure, 24% β-turn and 12% random coil [Schawbe et al., 1984]. Recombinant human cystatin A in good comparison to cystatin A, in far UV-CD spectrum revealed ~45% β -structure and a low α -helix content (~15%) [Pol et al., 1995].

1.7. Inhibition of Proteases

Specificity

Cystatins are highly specific for CPs except for thyropins which show inhibitory activity against aspartic and metalloproteases [Mihelic and Turk, 2007; Lenarcic and Turk, 1999]. However there are few cystatins capable of inhibiting mammalian legumain [Alvarez-Fernandez et al., 1999] and calpains [Crawford et al., 1987]. To date, none of the cytoplasmic inhibitors have been tested on ubiquitin processing and recycling proteases. Stefin A and B are potent inhibitors of papain, cathepsin L, S and H but have decreased activity against cathepsin B [Musil et al., 1991]. Type 2 cystatins are important endogenous inhibitors of papain like CPs including cathepsins, parasite proteases like cruzipain and mammalian legumain [Turk et al., 2005; Turk and Bode, 1991]. HCC and chicken cystatin inhibit papain, cathepsin L and S [Abrahamson et al., 2003]. HCC shows strong inhibitory capacity for rapid binding thus neutralizing protease activity in an emergency inhibition [Turk et al., 2005]. It also inhibits cruzipain, suggesting its possible defensive role after infection [Stoka et al., 1995]. Cystatin F inhibits cathepsin F, K, V, S, L and H [Langerholc et al., 2005] and weakly legumain [Alvarez-Fernandez et al., 1999].

Cystatin D inhibits cathepsin S, H and L but not cathepsin B or pig legumain [Alvarez-Fernandez et al., 2005]. Human cystatin E/M inhibits papain, cathepsin B, L, V and legumain [Ni et al., 1997; Sotiropoulou et al., 1997; Cheng et al., 2006; Alvarez-Fernandez et al., 1999]. Clostripain (protease not belonging to papain family) is also inhibited by cystatins [Barrett et al., 1986].

Table 2. Equilibrium constants for dissociation (Ki) of complexes between human cystatins and chicken cystatin with lysosomal cysteine proteases (papain, human cathepsins and cruzipain)

CPI	Ki (nM)				
Cystatin	Papain	Cathepsin B	Cathepsin H	Cathepsin L	Cruzipain
Stefin A	0.019	8.2	0.31	1.3	0.0072
Stefin B	0.12	73	0.58	0.23	0.060
Cystatin C	0.00001	0.27	0.28	<0.005	0.014
Cystatin D	1.2	>1000	7.5	18	n.d.
Cystatin E/M	0.39	32	n.d.	n.d.	n.d.
Cystatin F	1.1	>1000	n.d.	0.31	n.d.
Cystatin S	108	n.d.	n.d.	n.d.	n.d
Cystatin SA	0.32	n.d.	n.d.	n.d.	n.d.
Cystatin SN	0.016	19	n.d.	n.d.	n.d.
Chicken cystatin	0.005	1.7	0.06	0.019	0.001
L-kininogen	0.015	600	0.72	0.017	0.041
H-kininogen	0.02	400	1.1	0.109	n.d.

n.d. (not determined), Ki values for human cystatins [Abrahamson et al., 2003], chicken cystatin [Barrett et al., 1986] and cruzipain inhibition by cystatins [Stoka et al., 1995].

Kinetic Behaviour

Cystatins are the first group of protein inhibitors of CPs for which the mechanism of inhibition was investigated. All the cystatins are non-covalent, competitive, reversible, tight binding inhibitors which inhibit the target enzymes in micromolar to picomolar range. They form tight equimolar complexes with CPs [Anastasi et al., 1983]. Some of the reported values of equilibrium constants for dissociation of complexes between human cystatins and lysosomal CPs are summarized in Table 2. The affinity differences can be explained by the differences in the active site regions of endo- and exopeptidases. The access of the inhibitor to the active site of exopeptidases is partially obstructed by occluding loops in cathepsin B [Musil et al., 1991] and cathepsin X [Guncar et al., 2000] and propeptide parts in cathepsin H [Guncar et al., 1998] and cathepsin C [Turk et al., 2001].

Reactive Site and Mechanism of Action

It has been established that no disulphide bond is formed between the active site cysteine residue and the inhibitor because the complexes dissociated when denatured without reduction as was found in chicken cystatin [Nicklin and Barrett, 1984] and kininogens [Gounaris et al., 1984]. The complex formation is accompanied by pronounced spectroscopic changes [Bjork et al., 1989].

On the basis of cystatin domain structure, it was proposed that there are three regions crucial for interaction with proteases: the amino terminus and two β-hairpin loops, one in the middle and one in the C-terminal segment of the protein. The first loop contains a QXVXG sequence conserved in almost all inhibitory members of cystatins, whereas the second loop contains a P-W motif, which is also highly conserved [Table 3]. Both these loops and the amino terminus forms a wedge shaped edge, which is highly complementary to the active site of the enzyme. The N-terminally truncated forms of chicken cystatin confirmed the crucial importance for the binding of the residues preceding the conserved Gly-9 residue [Machleidt et al., 1989]. The essential interactive elements of this hypothetical complex are shown in figure 5. Complex formed on interaction of stefin B with cathepsin H is shown in figure 6. This was supported by Brzin et al. [1984] who demonstrated that the truncated form of HCC starting with Leu-Val before Gly-11 (corresponding to Gly-9 of chicken cystatin) has virtually the same affinity for papain as the full length form whereas the truncated form starting with Gly-12 has been reported to show 1000 fold weaker inhibition [Abrahamson et al., 1987]. However, Nycander and Bjork [1990] emphasized the role of Trp-104 in the inhibition of CP. According to their model, Trp-104 of cystatin interacts primarily with two Trp side chains in the active side cleft of papain, Trp 177 and Trp 181, in such a manner that the indole ring of Trp-104 stacks on the side chain of Trp 177 and the edge lies on the indole ring of Trp 181. A two step mechanism of inhibition of the lysosomal CP cathepsin B by its endogenous inhibitor cystatin C was observed by Nycander et al. [1998]. An initial weak interaction in which N-terminal of the inhibitor binds to the proteinase is followed by a conformational change. Subsequently, the occluding loop of the proteinase that partially obscures the active site is displaced by the inhibitor bringing about another conformational change. The presence of occluding loop of cathepsin B renders it much less susceptible to inhibition by cystatin than other proteinases. A similar two step binding of cystatin A to the CP was suggested by Estrada and Bjork [2000]. The flexible N-terminal region of the cystatin binds independently to the target proteinases after the binding of hairpin loops. It is interesting that the replacement of the three N-terminal residues preceding the conserved Gly of stefin A by the corresponding 10-residues long segment of cystatin C increased affinity of the inhibitor for cathepsin B by about 15-fold [Pavlova and Bjork, 2003], suggesting that the inhibitory potency of cystatin can be substantially improved by protein engineering.

Table 3. Conserved amino-acid residues in binding segments of human cystatins [a, b]

Cystatin	N-terminus	I loop	II loop
A	MIPGG	QVVAG	
B	AcMMCGA	QVVAG	
C	RLVGG	QIVAG	VPWQ
D	TLAGG	QIVAG	VPWE
E	RMVGE	QLVAG	VPWQ
F	VKPGF	QIVKG	VPWL
S	IIPGG	QTFGG	VPWE
SA	IIEGG	QIVGG	VPWE
SN	IIPGG	QTVGG	VPWE
H-kininogen			
1-domain	*(QESQS)*	*(TVGSD)*	*(RSST)*
2-domain	DCLGC	QVVAG	*(DIQL)*
3-domain	ICVGC	QVVAG	VPWE
L-kininogen			
1-domain	*(QESQS)*	*(TVGSD)*	*(RSST)*
2-domain	DCLGC	QVVAG	*(DIQL)*
3-domain	ICVGC	QVVAG	VPWE

[a] Sequences in parenthesis correspond to the appropriate binding sequences of cystatins.

[b] Grzonka et al., 2001.

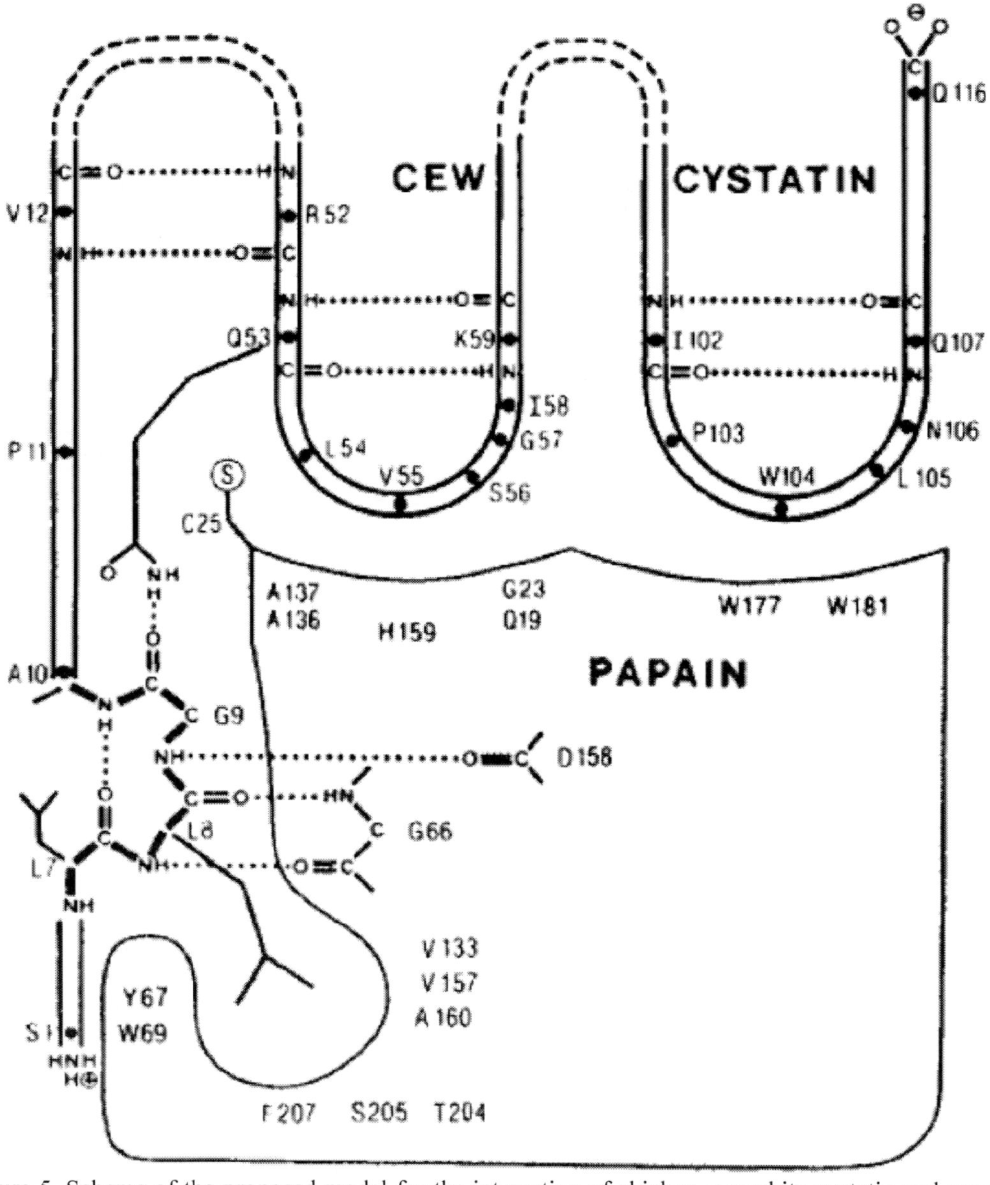

Figure 5. Scheme of the proposed model for the interaction of chicken egg white-cystatin and papain. (Adapted from Turk and Bode, FEBS Lett. 1991, 285, 213-219).

The crystal structure of human stefin A-porcine cathepsin H complex showed small distortion of the structure upon formation of the complex [Jenko et al., 2003]. In addition to the structurally derived data, the contribution of the individual residues within protease binding region of cystatins was additionally investigated by mutational analysis and kinetic studies performed by several different groups [Pol and Bjork, 2001; Auerswal et al., 1994; Estrada et al., 1999; Pavlova et al., 2000].

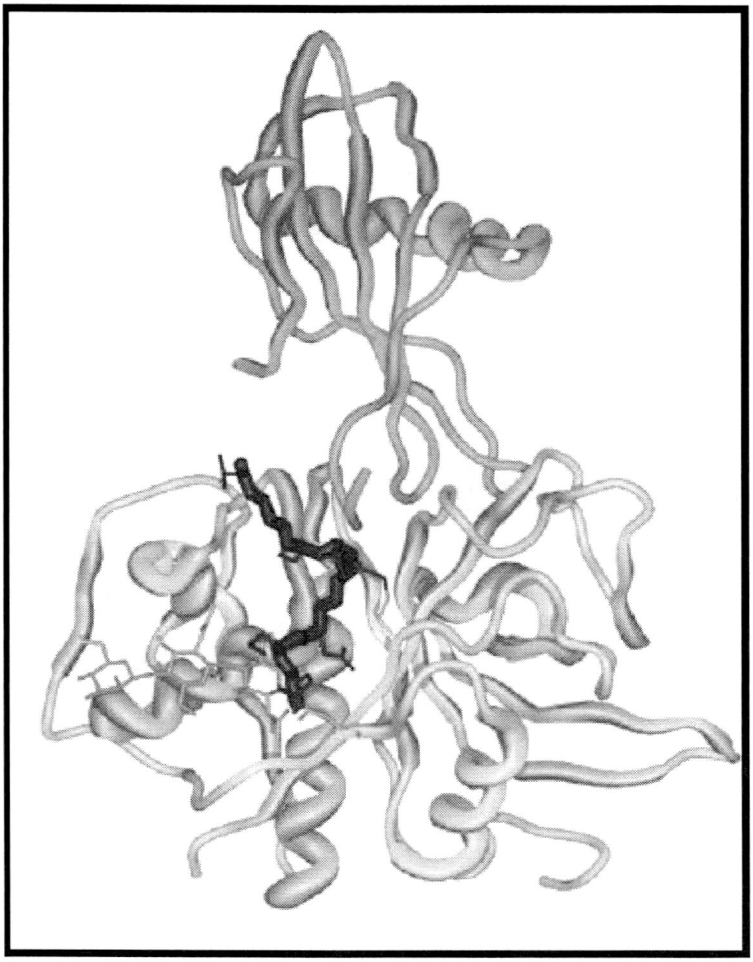

Figure 6. Three dimensional structure of the complex formed between stefin A and cathepsin H. Binding of the stefin A into cathepsin H active site. Stefin A fold is shown as a dark grey chain trace. Whereas cathepsin H fold is shown in light grey. Cathepsin H mini-chain residues are shown as black sticks which are thicker for the main chain. The identified carbohydrate rings are shown as thin sticks. The N-terminus of stefin A displaces the C-terminus of the minichain by pushing its residues outside the binding cleft. Adopted from Turk and Turk, Acta Chim. Slov. 2008, 55, 727–738.

1.8. Biological Aspects and Pathophysiology of Cystatins

Proteases and their natural inhibitors may co-exist at different levels of cellular evolution. Disturbing the harmony of the normal balance of enzymatic activities of proteases and their natural inhibitors may lead to severe biological effects.

Cystatins constitute a powerful regulatory system for endogenous CPs which are often secreted or leaked from the lysosomes of dying and diseased cells [Ekiel et al., 1997]. Besides regulation of the enormous hydrolytic potential of CPs, plethora of actions has now been ascribed to cystatins. They are known to play important roles in various pathophysiologic conditions such as sepsis [Assfalg-Machleidt et al., 1988], cancer [Cox, 2009], rheumatoid arthritis [Trabandt et al., 1991], purulent bronchiectasis [Buttle et al., 1990], multiple sclerosis [Bever and Garver, 1995], muscular dystrophy [Sohar et al., 1988], etc. which indicate that a tight enzyme regulation by cystatin is a necessity in the normal state.

Cystatins and cancer: Cathepsins involved in the degradation of extracellular matrix facilitate the growth, invasion and metastasis of tumour cells and also in tumour angiogenesis [Gocheva and Joyce, 2007; Mohamed and Sloane, 2006; Turk et al., 2004; Vasiljeva et al., 2006]. A broad spectrum of cysteine protease inhibitor was shown to inhibit tumour angiogenesis [Joyce et al., 2004].

Generally, cathepsin to cystatin ratio is found to be increased in most tumour types compared to normal tissues [Paraoan et al., 2009; Rivenbark and Coleman, 2009]. Elevated TPI level in various tumour types have been correlated to better prognosis like, stefin A positive breast cancer patients are less likely to develop distant metastasis [Parker et al., 2008], stefin A and B in non small cell lung cancer [Werle et al., 2006], cystatin SN is upregulated in gastric cancer [Choi et al., 2009], cystatin C [Sokol and Schiemann, 2004], cystatin M [Zhang et al., 2004], cystatin F [Utosunomiya et al., 2002], were found to be expressed in epithelial and mesenchymal tumour cells. Cystatin M is often hailed as tumour suppressor.

Stefin A and cystatin C overexpression has been shown to inhibit cancer cell invasion and metastasis [Li et al., 2005; Kopitz et al., 2005]. Cystatins may also inhibit cell migration by interfering with cell signalling pathways, by direct cathepsin and calpain inhibition [Cox, 2009]. Cathepsins B, L and S promote tumour growth in a murine model of pancreatic tumourigenesis [Gocheva et al., 2006]. The tumour microvascular density declined by about half in pancreatic tumours in cathepsin B or cathepsin S null mice. Significant increases in apoptosis were also noted in cathepsin B, L and S null pancreatic tumours. High activity of cysteine proteases such as legumain and the cathepsins have been shown to facilitate growth and invasion of a variety of tumor types. Briggs et al. [2010]_suggested that the level of cystatin E/M regulates legumain activity and hence the invasive potential of human melanoma cells.

Cystatins and neurodegeneration: Though CPs are implicated in various pathologies of brain, there are only few studies concerning the role of cystatins in pathologies of brain. Only two genetic diseases are known in which mutations in cystatin C and stefin B are associated with disease status, Hereditary cystatin C amyloid angiopathy (HCCAA) the first human disorder known to be caused by deposition of cystatin C amyloid fibrils in walls of brain arteries leading to single to multiple strokes with fatal outcomes. The amyloid deposited is composed mainly of the Leu 68 Gln variant of cystatin C and is associated with mutation in cystatin C gene [Palsdottir et al., 2006]. Normal cystatin C may protect pathogenesis of Alzheimer's disease by binding to soluble amyloid-β-peptide and preventing its deposition [Mi et al., 2007].

Progressive myoclonus epilepsy [EPM1] is exhibited by a group of inherited diseases characterized by myoclonic seizures, generalized epilepsy and progressive neurological degeneration caused by mutations in cystatin B gene (in the conserved QVVAG region) [Pennacchio et al., 1997; Joensuu et al., 2007].

Cystatins and cell death: Cystatins are shown to be involved in normal cell apoptosis, and most dramatically in selective tissue type for example in EPM1 [Lieuallen et al., 2001]. In fibrosarcoma cells, cystatins regulate cell death in response to TNFα [Foghsgaard et al., 2001]. An elevated cystatin C/cathepsin B ratio was found to be associated with chemoresistance in non-small cell lung carcinoma patients [Petty et al., 2006]. Intracellular cystatins normally inhibit low level lysosomal leakage. High cystatin levels are expected to be protective for general cathepsin mediated cell death.

Cystatins and immunomodulation: Cystatins have emerged as effector molecules of immunomodulation [Zavasnik-Bergant, 2008]. They can stimulate nitric oxide release from macrophages [Verdot et al., 1999]; modulate respiratory burst and phagocytosis in neutrophils [Leung-Tack et al., 1990]; and interleukin, cytokine production in T-cells and fibroblasts [Schierack et al., 2003; Kato et al., 2002; 2004]. Most of these functions operate via putative cell surface cystatin-binding molecules or membrane domains [Kato et al., 2002]. Cystatin C has been shown to be a TGFβ receptor antagonist and TGFβ signalling pathway blocker [Sokol and Schiemann, 2004; Sokol et al., 2005]. Type 2 cystatins are also known to increase interleukin-6 expression in fibroblasts and splenocytes [Kato et al., 2000]. Cystatin C is a potent, reversible inhibitor in vitro of the human lysosomal CPs e.g., cathepsin S (Ki = 8 pM), cathepsin L (Ki = 8 pM) and cathepsin H (Ki = 220 pM). These proteases are located all along the endocytic pathway of dendritic cell and are involved in the controlled proteolysis associated with the degradation of antigenic peptides [Pluger et al., 2002]. Cystatin F by targeting cathepsin C is known to regulate diverse immune cell effector functions [Hamilton et al., 2008].

Cystatins as antimicrobial and antiviral agents: Horse-shoe crab hemocyte cystatin has antimicrobial activity against Gram negative bacteria with IC50s against *S. typhimurium*, *E. coli* and *K. pneumoniae* in the 80-100 μg/ml range [Agarwala et al., 1996]. Both chicken and human cystatins were found to inhibit the growth of *P. gingivalis* with an IC50 of 1.1 and 1.2 fM, respectively [Blank et al., 1996]. Cystatin C is also an effective inhibitor of replication of coronavirus [Collins and Grubb, 1998]. Sialostatin L displays anti-inflammatory role and inhibits proliferation of cytotoxic T-lymphocytes [Kotsyfakis et al., 2006]. Ordóñez-Gutiérrez et al. [2009] have found that recombinant barley cystatins, HvCPI5 and HvCPI6 have antiproliferative effect on on promastigotes and intracellular amastigotes of *Leishmania infantum* invoking the interest of this recombinant cystatin in the chemotherapy of leishmaniasis.

Cystatin C in clinical diagnostics: Cystatin C was the first protein to be used in clinical diagnostics. Levels of cystatin C in various body fluids is used as a barometer of disease [Shah and Bano, 2008]. Recent studies indicate that it is a better marker of glomerular filtration rate and is a stronger predictor of cardiovascular disease and mortality than serum creatinine [Fried, 2009]. Cystatin C levels can also be used to reflect the characteristics of peritoneal membrane in dialysis patients [Al-Wakeel et al., 2009]. Korolenko et al. [2008] recently found that serum cystatin C concentration can be used as one of the prognostic criteria in patients with several kinds of hemoblastoses. IL-6 levels along with that of cystatin C may be regarded as markers of increased osteoblastic activity associated to bisphosphate treatments in prostrate cancer patients with bone metastases [Tumminello et al., 2009]. Yang et al. [2009] found that its levels decrease significantly in cerebrospinal fluids of patients with Guillain-Barre syndrome and may be involved in its pathophysiology. Cystatin B was found to be specifically over expressed in most hepatocellular carcinomas and alone or in combination with α-fetoprotein may be a useful marker for diagnosis of the diseases [Lee et al., 2008]. Cystatin C has been proposed as a novel marker of renal function and as a predictor of cardiovascular risk in the elderly [Cepeda et al., 2010]. The prevalence of an elevated cystatin C level in the general population was found to be high and was associated with the presence of classical cardiovascular risk factors such as diabetes, hypertension and chronic renal disease, along with higher levels of C-reactive protein, homocysteine and fibrinogen.

Fibrillogenic cystatins: Cystatins are prone to form amyloids [Morgan et al., 2008; Turk et al., 2008]. Human cystatin C is highly amyloidogenic protein. The fibril formation is also known to occur in chicken cystatin [Staniforth et al., 2001], stefin B under in vitro conditions [Zerovnik et al., 2007], stefin A [Jenko et al., 2004], latexin [Pallares et al., 2007], CRES protein [Horsten et al., 2007]. In the case of HCC the oligomers and fibrils are formed by propagated domain swapping. This model is not compatible with stefin B, in which proline (Pro) isomerization is important in preventing steric clashing [Morgan et al., 2008]. Trans to cis isomerization of Pro 74 is involved in formation of stefin B dimers [Jenko-Kokalj et al., 2007]. Since this Pro is widely conserved in cystatin superfamily its isomerization can play role in amyloidogenesis [Jenko-Kokalj et al., 2007].

Role of cystatins in other diseases: Cystatins are now known to participate in neuronal differentiation [Taupin et al., 2000]. Numerous studies have demonstrated that cathepsin K, L and S are involved in elastic fibre degradation, associated with the development of different pathological conditions of cardiovascular system. Elastinolyitc activites of cathepsin K, L and S can be blocked by cystatins [Novinec et al., 2007]. Equistatin is known to inhibit the growth of the red flour beetle *Triboleum castaneum* [Oppert et al., 2003], suggesting to be promising candidates for the transgenic seed technology to enhance seed resistance to storage pests. Heparin binding and cell-binding domain 5 (light chain) of H-kininogen has antibacterial acitivity against *E. coli*, *P. aeruginosa* and *Enterococcus faecalis* [Andersson et al., 2005]. Bradykinin induces dendritic cell maturation and can modulate innate or adaptive immunity [Aliberti et al., 2003; Scharfstein et al., 2007]. Cystatin C appears to be up-regulated in response to injury in the brain [Shah and Bano, 2008]. Cystatin M/E is a key molecule in a biochemical pathway that controls skin barrier formation by the regulation of both crosslinking and desquamation of stratum corneum [Zeeuwen et al., 2009]. Lower cystatin C level is also implicated in retinal degeneration in (rd1) mouse model of retinitis pigmentation [Ahuja et al., 2008]. An imbalance in cathepsin B/cystatin C level may contribute to the progression of pelvic inflammatory disease [Tsai et al., 2009]. There are many other diseases with decreased cystatin levels, such as inflammatory diseases, osteoporosis, arthritis, and diabetes as well as a number of other neurodegenerative diseases [Turk et al., 2008]. Numerous lines of evidence suggest a role of oxidative stress in initiation and progression of heart failure. Xie et al., [2010] recently found that that H_2O_2 caused an elevated cystatin C protein in the conditioned medium from cardiomyocytes. In myocardial tissue from the ischaemic area, an increase in cystatin C correlates with the inhibition of cathepsin B activity and accumulation of fibronectin and collagen I/III. Overexpressing cystatin C gene or exposing fibroblasts to cystatin C protein results in an inhibition of cathepsin B and accumulation of fibronectin and collagen I/III playing a role in cardiac extracellular matrix remodelling.

CONCLUSION

Cystatins, the long-established endogenous inhibitors of C1 cysteine proteinases, have been extensively studied. The cystatin superfamily has witnessed a baffling growth over the last few decades. Proteins containing cystatin domains but lacking cysteine protease inhibitory activities have been isolated and characterized, and unsurprisingly more will be described in the near future. Equipped with novel attributes these cystatin superfamily members have ventured into tumorigenesis, immunomodulation, cell cycle regulation, apoptosis, etc. The future awaits researches targeted to unmask the another facet of the cystatin superfamily regarding their potential involvement in serious human ailments; development of recombinant potent cysteine protease inhibitors as therapeutic modalities or even vaccines with regards to cysteine protease involvement in microbial invasion and divison inside the host.

REFERENCES

[1] Rawlings, ND; Tolle, DP; Barrett, AJ. MEROPS: the peptidase database. *Nucleic Acids Res.* 2004; 32 (Database issue), D160–D164.

[2] Rawlings, ND; Barrett, AJ. Evolution of proteins of the cystatin superfamily. *J. Mol. Evol.* 1990; 30, 60-71.

[3] Turk, B. Targeting proteases: successes, failures and future prospects. *Nat. Rev. Drug Discov.* 2006; 5, 785-799.

[4] Turk, B; Turk, D; Turk, V. Lysosomal cysteine proteases: more than scavenger. *Biochim. Biophys. Acta.* 2000; 1477, 98-111.

[5] Barrett, AJ; Rawlings, ND; O' Brien, EA. The MEROPS database as a protease information system. *J. Struct. Biol.* 2001; 134, 95-102.

[6] Punturieri, A; Filippov, S; Allen, E; Caras, I; Murray, R; et al. Regulation of elastinolytic cysteine proteinase activity in normal and cathepsin K-deficient human macrophages. *J. Exp. Med.* 2000; 192, 789-799.

[7] Zaidi, M; Troen, B; Moonga, BS; Abe, E. Cathepsin K, osteoclastic resorption, and osteoporosis therapy. *J. Bone Miner. Res.* 2001; 16, 1747-1749.

[8] Rocken, C; Stix, B; Brömme, D; Ansorge, S; Roessner, A; et al. A putative role for cathepsin K in degradation of AA and AL amyloidosis. Cell death inhibition: keeping caspases in check. *Am. J. Pathol.* 2001; 158, 1029-1038.

[9] Turk, V; Turk, B; Guncar, G; Turk, D; Kos, J. Lysosomal cathepsins: structure, role in antigen processing and presentation, and cancer. *Adv. Enzyme Regul.* 2002; 42, 285-303.

[10] Brix, K; Dunkhorst, A; Mayer, K; Jordans, S. Cysteine cathepsins: cellular roadmap to different functions. *Biochimie* 2008; 90, 194-207.

[11] Vasiljeva, O; Reinheckel, T; Peters, C; Turk, D; Turk, V; et al. Emerging roles of cysteine cathepsins in disease and their potential as drug targets. *Curr. Pharm. Des.* 2007; 13, 387-403.

[12] Nakanishi, H. Neuronal and microglial cathepsins in aging and age-related diseases. *Ageing Res. Rev.* 2003; 2, 367-381.

[13] van Acker, GJ; Saluja, AK; Bhagat, L; Singh, VP; Song, AM; et al. Cathepsin B inhibition prevents trypsinogen activation and reduces pancreatitis severity. *Am. J. Physiol. Gastrointest. Liver Physiol.* 2002; 283, G794-800.

[14] Gocheva, V; Joyce, JA. Cysteine cathepsins and the cutting edge of cancer invasion. *Cell Cycle* 2007; 6, 60-64.

[15] Lutgens, SP; Cleutjens, KB; Daemen, MJ; Heeneman, S. Cathepsin cysteine proteases in cardiovascular disease. *FASEB J.* 2007; 21, 3029-3041.

[16] Turk, B; Stoka, V.http://www.ncbi.nlm.nih.gov/pubmed/17544407?ordinalpos=9&itool=EntrezSystem2.PEntrez.Pubmed.Pubmed_ResultsPanel.Pubmed_DefaultReportPanel.Pubmed_RVDocSum Protease signalling in cell death: caspases versus cysteine cathepsins. *FEBS Lett.* 2007; 581, 2761-2767.

[17] Huang, Y; Wang, KK. The calpain family and human disease. *Trends Mol. Med.* 2001; 7, 355-362.

[18] Goyal, L. Cell death inhibition: keeping caspases in check. *Cell* 2001; 104, 805-808.

[19] Manoury, B; Hewitt, EW; Morrice, N. An asparaginyl endopeptidase processes a microbial antigen for class II MHC presentation. *Nature* 1998; 396, 695-699.

[20] Choi, SJ; Reddy, SV; Devlin, RD; Menaa, C; Chung, H; Boyce, BF; Roodman, GD. Identification of human asparaginyl endopeptidase (legumain) as an inhibitor of osteoclast formation and bone resorption. *J. Biol. Chem.* 1999; 274(39), 27747-53.

[21] Alvarez-Fernandez, M; Barrett, AJ; Gerhartz, B; Dando, PM; Ni, J; et al. Inhibition of mammalian legumain by some cystatins is due to a novel second reactive site. *J. Biol. Chem.* 1999; 274, 19195-19203.

[22] Crawford, C. Inhibition of chicken calpain II by proteins of the cystatin superfamily and alpha 2-macroglobulin. *Biochem. J.* 1987; 248, 589-594.

[23] Hayashi, H; Tokuda, A; Udaka, K. Biochemical study of cellular antigen-antibody reaction in tissue culture. I. Activation and release of a protease *J. Exp. Med.* 1960; 112, 237-247.

[24] Fossum, K; Whitaker, JR. Ficin and papain inhibitor from chicken egg white. *Arch. Biochem. Biophys.* 1968; 125, 367-375.

[25] Sen, LC; Whitaker, JR. Some properties of a ficin-papain inhibitor from avian egg white. *Arch. Biochem. Biophys.* 1973; 158, 623-632.

[26] Keilova, H; Tomasek, V. Inhibition of cathepsin C by papain inhibitor from chicken egg white and by complex with cathepsin B1. *Coll. Czech. Chem. Commun.* 1975; 40, 218-224.

[27] Barrett, AJ. Cystatin, the egg white inhibitor of cysteine proteinases. *Methods Enzymol.* 1981; 80, 771-778.

[28] Kopitar, M; Brzin, J; Zvonar, T; Locnikar, P; Kregar, I; et al. Inhibition studies of an intracellular inhibitor on thiol proteinases. *FEBS Lett.* 1978; 91, 355-359.

[29] Turk, V; Brzin, J; Longer, M; Ritonja, A; Eropkin, M; et al. Protein inhibitors of cysteine proteinases. III. Amino-acid sequence of cystatin from chicken egg white. *Hoppe Seylers Z. Physiol. Chem.* 1983; 364, 1487-1496.

[30] Schwabe, C; Anastasi, A; Crow, H; McDonald, JK; Barrett, AJ. Cystatin. Amino acid sequence and possible secondary structure. *Biochem. J.* 1984; 217, 813-817.

[31] Machleidt, W; Borchart, U; Fritz, H; Brzin, J; Ritonja, A; et al. Protein inhibitors of cysteine proteinases. II. Primary structure of stefin, a cytosolic protein inhibitor of

cysteine proteinases from human polymorphonuclear granulocytes. *Hoppe Seylers Z. Physiol. Chem*. 1983; 364, 1481-1486.

[32] Brzin, J; Popovic, T; Turk, V; Borchart, U; Machleidt, W. Human cystatin, a new protein inhibitor of cysteine proteinases. *Biochem. Biophys. Res. Commun*. 1984; 118, 103-109.

[33] Barrett, AJ; Davies, ME; Grubb, A. The place of human gamma-trace (cystatin C) amongst the cysteine proteinase inhibitors. *Biochem. Biophys. Res. Commun*. 1984; 120, 631-636.

[34] Nawa, H; Kitamura, N; Hirose, T; Asai, M; Inayama, S; et al.: Primary structures of bovine liver lFow molecular weight kininogen precursors and their two mRNAs. Proc *Natl. Acad. Sci. USA* 1983; 80, 90-94.

[35] Müller-Esterl, W; Fritz, H; Machleidt, W; Ritonja, A; Brzin, J; et al. Human plasma kininogens are identical with alpha-cysteine proteinase inhibitors. Evidence from immunological, enzymological and sequence data. *FEBS Lett.* 1985a; 182, 310-314.

[36] Ohkubo, I; Kurachi, K; Takasawa, T; Shiokawa, H; Sasaki, M. Isolation of a human cDNA for alpha 2-thiol proteinase inhibitor and its identity with low molecular weight kininogen. *Biochemistry* 1984; 23, 5691-5697.

[37] Barrett, AJ; Rawlings, ND; Davies, ME; Machleidt, W; Salvesen, G; et al. Cysteine proteinase inhibitors of the cystatin superfamily. In: Proteinase inhibitors, Barrett, AJ and Salvesen, G (editors). Amsterdam: Elsevier, 1986; pp 515-569.

[38] Bode, W; Engh, R; Musil, D; Thiele, U; Huber, R; et al. The 2.0 A X-ray crystal structure of chicken egg white cystatin and its possible mode of interaction with cysteine proteinases. *EMBO J.* 1988; 7, 2593-2599.

[39] Brown, WM; Dziegielewska, KM. Friends and relations of the cystatin superfamily-new members and their evolution. *Protein Sci*. 1997; 6, 5-12.

[40] Abrahamson, M; Barrett, AJ; Salvesen, G; Grubb, A. Isolation of six cysteine proteinase inhibitors from human urine. Their physicochemical and enzyme kinetic properties and concentrations in biological fluids. *J. Biol. Chem*. 1986; 261, 11282-11289.

[41] Turk, B; Krizaj, I; Kralj, B; Dolenc, I; Popovic, T; et al. Bovine stefin C, a new member of the stefin family. *J. Biol. Chem*. 1993; 268, 7323-7329.

[42] Lenarcic, B; Krizaj, I; Zunec, P; Turk, V. Differences in specificity for the interactions of stefins A, B and D with cysteine proteinases. *FEBS Lett.* 1996; 395, 113-118.

[43] Tsui, FW; Tsui, HW; Mok, S; Mlinaric, I; Copeland, NG; et al. Molecular characterization and mapping of murine genes encoding three members of the stefin family of cysteine proteinase inhibitors. *Genomics* 1993; 15, 507-514.

[44] Fraki, JE. Human skin proteases. Separation and characterization of two acid proteases resembling cathepsin B1 and cathepsin D and of an inhibitor of cathepsin B1. *Arch. Dermatol. Res*. 1976; 255, 317-330.

[45] Jarvinen, M. Purification and some characteristics of the human epidermal SH-protease inhibitor. *J. Invest. Dermatol.* 1978; 71, 114-118.

[46] Brzin, J; Kopitar, M; Turk, V; Machleidt, W. Protein inhibitors of cysteine proteinases. I. Isolation and characterization of stefin, a cytosolic protein inhibitor of cysteine proteinases from human polymorphonuclear granulocytes. *Hoppe-Seylers Z. Physiol. Chem*. 1983; 364, 1475-1480.

[47] Green, GDJ; Kembhavi, AA; Davies, ME; Barrett, AJ. Cystatin-like cysteine proteinase inhibitors from human liver. *Biochem. J.* 1984; 218, 939-946.

[48] Hopsu-Havu, VK; Joronen, I; Jarvinen, M; Rinnie, A. Separation of cysteine proteinase inhibitors from psoriatic scale. *Br. J. Dermatol.* 1983a; 109, 77-85.

[49] Rinnie, A; Jarvinen, M; Rasanen, O. A protein reminiscent of the epidermal SH-protease inhibitor occurs in squamous epithelia of man and rat. *Acta Histochem.* 1978; 63, 183-192.

[50] Rinnie, A; Alavaikko, M; Jarvinen, M; Martikinen, J; Kartunen, T; et al. Demonstration of immunoreactive acid cysteine-proteinase inhibitor in reticulum cells of lymph node germinal centres.*Virchows Arch. Cell Pathol.* 1983; 43, 121-126.

[51] Minakata, K; Asano, M. Acidic cysteine proteinase inhibitor in seminal plasma. *Biol. Chem. Hoppe-Seyler.* 1985; 366, 15-18.

[52] Turk, B; Ritonja, A; Bjork, I; Stoka, V; Dolenc, I; et al. Identification of bovine stefin A, a novel protein inhibitor of cysteine proteinases. *FEBS Lett.* 1995; 360, 101-105.

[53] Rinnie, A; Rasanen, O; Jarvinen, M; Dammert, K; Kallioinen, M; et al. Occurrence of acid and neutral cysteine proteinase inhibitors in epidermal malignancies: immunohistochemical study. *Acta Histochem.* 1984; 74, 75-79.

[54] Jarvinen, M. Purification and properties of two protease inhibitors from rat skin inhibiting papain and other SH-proteases. *Acta Chem. Scand.* B. 1976; 30(10), 933-40.

[55] Jarvinen, M; Rasanen, O; Rinnie, A. The low-molecular-weight SH-protease inhibitor in rat skin is epidermal. *J. Invest. Dermatol.* 1978; 71, 119-121.

[56] Mihelic, M; Teuscher, C; Turk, V; Turk, D. Mouse stefins A1 and A2 (Stfa1 and Stfa2) differentiate between papain-like endo- and exopeptidases. *FEBS Lett.* 2006; 580, 4195-4199.

[57] Lenney, JF; Tolan, JR; Sugai, WJ; Lee, AG. Thermostable endogenous inhibitors of cathepsins B and H. *Eur. J. Biochem.* 1979; 101:153-161.

[58] Jarvinen, M; Rinnie, A. Human spleen cysteineproteinase inhibitor. Purification, fractionation into isoelectric variants and some properties of the variants. *Biochim. Biophys. Acta.* 1982; 708, 210-217.

[59] Finkenstaedt, JT. Intracellular distribution of proteolytic enzymes in rat liver tissue *Proc. Soc. Exp. Biol. Med.* 1957; 95, 302-304.

[60] Kominami, E; Wakamatsu, N; Katunuma, N. Endogenous thiol protease inhibitor from rat liver *Biochem. Biophys. Res. Commun.* 1981; 99, 568-575.

[61] Grubb, A; Lofberg, H; Barrett, AJ. . The disulphide bridges of human cystatin C (g-trace) and chicken cystatin. *FEBS Lett.* 1984; 170, 370-374.

[62] Ni, J; Abrahamson, M; Zhang, M; Alvarez-Fernandez, M; Grubb, A; et al. Cystatin E is a novel human cysteine proteinase inhibitor with structural resemblance to family 2 cystatins. *J. Biol. Chem.* 1997; 272, 10853-10858.

[63] Sotiropoulou, G; Anisowicz, A; Sager, R. Identification, cloning, and characterization of cystatin M, a novel cysteine proteinase inhibitor, down-regulated in breast cancer. *J. Biol. Chem.* 1997; 272, 903-910.

[64] Ni, J; Fernandez, MA; Danielsson, L. Cystatin F is a glycosylated human low molecular weight cysteine proteinase inhibitor. *J. Biol. Chem.* 1998; 273, 24797-24804.

[65] Esnard, F; Esnard, A; Faucher, D; Capony, JP; Derancourt, J; et al. Rat cystatin C: the complete amino acid sequence reveals a site for N-glycosylation. *Biol. Chem. Hoppe Seyler.* 1990; 371, S161-S166.

[66] Turk, V; Bode, W. The cystatins: protein inhibitors of cysteine proteinases. *FEBS Lett.* 1991; 285, 213-219.

[67] Balbin, M; Hall, A; Grubb, A; Mason, RW; Lopez-Otin, C; Abrahamson M. Structural and functional characterization of two allelic variants of human cystatin D sharing a characteristic inhibition spectrum against mammalian cysteine proteinases. . *J. Biol. Chem.* 1994; 269, 23156-23162.

[68] Butler, EA; Flynn, FV. The occurrence of post-gamma protein in urine: a new protein abnormality. *J. Clin. Pathol.* 1961; 14, 172-178.

[69] Hochwald, GM; Thorbecke, GJ. Use of an antiserum against cerebrospinal fluid in demonstration of trace proteins in biological fluids. *Proc. Soc. Exp. Biol. Med.* 1962; 109, 91-95.

[70] Cejka, J; Fleischmann, LE. Post-γ-globulin: isolation and physicochemical characterization. *Arch Biochem. Biophys.* 1973; 157, 168-176.

[71] Colle, A; Guinet, R; Leclercq, M; Manuel, Y. Occurrence of beta2-microglobulin and post-gamma globulin in human semen. *Clin. Chim. Acta.* 1976; 67, 93-97.

[72] Lofberg, H; Grubb, A; Davidsson, J; Kjellander, B; Stromblad, LG; et al. Occurrence of gamma-trace in the calcitonin-producing C-cells of simian thyroid gland and human medullary thyroid carcinoma *Acat. Endocrinol.* 1983; 104, 69-76.

[73] Lofberg, H; Grubb, AO; Jornvau, H; Moller, CA; Stromblad, LG; et al. Demonstration of gamma-trace in normal endocrine cells of the adrenal medulla and in phaeochromocytoma. An immunohistochemical study in monkey, dog and man. *Acat. Endocrinol.* 1982; 100, 595-598.

[74] Freiji, JP; Abrahamson, M; Olafsson, I; Velasco, G; Grubb, A; et al. Structure and expression of the gene encoding cystatin D, a novel human cysteine proteinase inhibitor *J. Biol. Chem.* 1991; 266, 20538-20543.

[75] Isemura, S; Saitoh, E; Ito, S; Isemura, M; Sanada, K. Cystatin S: a cysteine proteinase inhibitor of human saliva. *J. Biochem.* 1984b; 96, 1311-1314.

[76] Isemura, S; Saitoh, E; Sanada, K; Isemura, M; Ito, S. In: Cysteine Proteinase and their inhibitors, Turk V (editor). Berlin, Walter de Gruyter: 1986.

[77] Isemura, S; Saitoh, E; Sanada, K. Isolation and amino acid sequence of SAP-1, an acidic protein of human whole saliva, and sequence homology with human gamma-trace. *J. Biochem.* 1984a; 96, 489-498.

[78] Isemura, S; Saitoh, E; Ito, S; Sanada, K; Minakata, K. Identification of full-sized forms of salivary (S-type) cystatins (cystatin SN, cystatin SA, cystatin S, and two phosphorylated forms of cystatin S) in human whole saliva and determination of phosphorylation sites of cystatin S. *J. Biochem.* 1991; 110, 648-654.

[79] Cheng, T; Hitomi, K; van Vlijmen-Willems, IMJJ; de Jongh, GJ; Yamamoto, K; et al. Cystatin M/E is a high affinity inhibitor of cathepsin V and cathepsin L by a reactive site that is distinct from the legumain-binding site: a novel clue for the role of cystatin M/E in epidermal cornification. *J. Biol. Chem.* 2006; 281, 15893-15899.

[80] Halfon, S; Ford, J; Foster, J; Dowling, L; Lucian, L; et al. Leukocystatin, a new Class II cystatin expressed selectively by hematopoietic cells. *J. Biol. Chem.* 1998; 273, 16400-16408.

[81] Schuttelkopf, AW; Hamilton, G; Watts, C; van Aalten, DMF. Structural Basis of Reduction dependent Activation of Human Cystatin F. *J. Biol. Chem.* 2006; 281, 16570-16575.

[82] DeLa Cadena, RA; Colman, RW. Structure and functions of human kininogens. *Trends Pharmacol. Sci.* 1991; 12, 272-275.

[83] Müller-Esterl, W. Novel functions of the kininogens. *Semin. Thromb Hemost* 1987; 13, 115-126.

[84] Salvesen, G; Parkes, C; Abrahamson, M; Grubb, A; Barrett, AJ. Human low-Mr kininogen contains three copies of a cystatin sequence that are divergent in structure and in inhibitory activity for cysteine proteinases. *Biochem. J.* 1986a; 234, 429-434.

[85] Salvesen, G; Parkes, C; Rawlings, ND; Brown, MA; Barrett, AJ; et al. In: Cysteine Proteinases and their Inhibitors, Turk V (editor). Berlin and New York: Walter de Gruyter, 1986b; pp 413-428.

[86] Turk, V; Turk, B. Lysosomal Cysteine Proteases and Their Protein Inhibitors: Recent Developments. *Acta Chim. Slov.* 2008; 55, 727-738.

[87] Sutton, HG; Fusco, A; Cornwall, GA. Cystatin related epididymal spermatogenic protein colocalizes with luteinizing hormone-beta protein in mouse anterior pituitary gonadotropes. *Endocrinology* 1999; 140, 2721-2732.

[88] Tohonen, V; Osterlund, C; Nordqvist, K. Testatin: a cystatin-related gene expressed during early testis development. *Proc. Natl. Acad. Sci. USA* 1998; 95, 14208-14213.

[89] Li, Y; Friel, PJ; Robinson, MO; McLean, DJ; Griswold, MD. Identification and characterization of testis- and epididymis-specific genes: cystatin SC and cystatin TE-1. *Biol. Reprod*. 2002; 67, 1872-1880.

[90] Hamil, KG; Liu, Q; Sivashanmugam, P; Yenugu, S; Soundararajan, R; et al. Cystatin 11: a new member of the cystatin type 2 family. *Endocrinology* 2002; 143, 2787-2796.

[91] Xiang, Y; Nie, DS; Wang, J; Tan, XJ; Deng, Y; et al. Cloning, characterization and primary function study of a novel gene, Cymg1, related to family 2 cystatins. *Acta Biochim. Biophys. Sin.* (Shanghai) 2005; 37, 11-18.

[92] Shoemaker, K; Holloway, JL; Whitmore, TE; Maurer, M; Feldhaus, AL. Molecular cloning, chromosome mapping and characterization of a testisspecific cystatin-like cDNA, cystatin T. *Gene* 2000; 245,103-108.

[93] Cornwall, GA; Hsia, N. A new subgroup of the family 2 cystatins. *Mol. Cell Endocrinol.* 2003; 200, 1-8.

[94] Sutton-Walsh, HG; Whelly, S; Cornwall, GA. Differential effects of GnRH and androgens on Cres mRNA and protein in male mouse anterior pituitary gonadotropes. *J. Androl.* 2006; 27, 802-815.

[95] von Horsten, HH; Johnson, SS; San Francisco, SK; Hastert, MC; Whelly, SM; et al. Oligomerization and transglutaminase cross-linking of the cystatin CRES in the mouse epididymal lumen: potential mechanism of extracellular quality control. *J. Biol. Chem.* 2007; 282, 32912-32923.

[96] Janowski, R; Kozak, M; Jankowska, E; Grzonka, Z; Grubb, A; et al. Human cystatin C, an amyloidogenic protein, dimerizes through threedimensional domain swapping. *Nat. Struct. Biol.* 2001; 8, 316-320.

[97] Wahlbom, M; Wang, X; Lindstrom, V; Carlemalm, E; Jaskolski, M; et al. Fibrillogenic oligomers of human cystatin C are formed by propagated domain swapping. *J. Biol. Chem*. 2007; 282, 18318-18326.

[98] Zerovnik, E; Pompe-Novak, M; Skarabot, M; Ravnikar, M; Musevic, I; et al. Human stefin B readily forms amyloid fibrils in vitro.*Biochim. Biophys. Acta* 2002a; 1594, 1-5

[99] Jenko Kokalj, S; Guncar, G; Stern, I; Morgan, G; Rabzelj, S; et al. Essential role of praline isomerization in stefin B tetramer formation. *J. Mol. Biol.* 2007; 366, 1569-1579.

[100] Koshizuka, Y; Yamada, T; Hoshi, K; Ogasawara, T; Chung, UI; et al. Cystatin 10, a novel chondrocyte-specific protein, may promote the last steps of the chondrocyte differentiation pathway. *J. Biol. Chem.* 2003; 278, 48259-48266.

[101] Sun, H; Li, N; Wang, X; Liu, S; Chen, T; et al. Molecular cloning and characterization of a novel cystatin-like molecule, CLM, from human bone marrow stromal cells. *Biochem. Biophys. Res. Commun.* 2003; 301, 176-182.

[102] Elzanowski, A; Barker, WC; Hunt, LT; Seibel-Ross, E. Cystatin domains in alpha-2-HS-glycoprotein and fetuin. *FEBS Lett.* 1988; 227, 167-170.

[103] Dziegielewska, KM; Brown, WM; Casey, SJ; Christie, DL; Foreman, RC; et al. The complete cDNA and amino acid sequence of bovine fetuin. Its homology with alpha 2HS glycoprotein and relation to other members of the cystatin superfamily. *J. Biol. Chem.* 1990; 265, 4354-4357.

[104] Dziegielewska, KM; Mollgard, K; Reynolds, ML; Saunders, NR. A fetuin- related glycoprotein (azHS) in human embryonic and fetal development. *Cell Tissue Res.* 1987; 248, 33-41.

[105] Dziegielewska, KM; Brown, WM. Fetuin. Molecular Biology Intelligence Unit, Landes RG Co., Texas, Int. Austin: Springer-Verlag, 1995.

[106] Leung, L. Histidine-rich glycoprotein: An abundant plasma protein in search of a function *J. Lab. Clin. Med.* 1993; 121, 630-631.

[107] Koide, T; Foster, D; Yoshitake, S; Davie, EW. Amino acid sequence of human histidine-rich glycoprotein derived from the nucleotide sequence of its cDNA. *Biochemistry* 1986; 25, 2220-2225.

[108] Agarwala, KL; Kawabata, S; Hirata, M; Miyagi, M; Tsunasawa, S; Iwanaga, S. A cysteine protease inhibitor stored in the large granules of horseshoe crab hemocytes: purification, characterization, cDNA cloning and tissue localization. *J. Biochem. Tokyo* 1996; 119, 85-94.

[109] Lenarcic, B; Bevec, T. Thyropins--new structurally related proteinase inhibitors. *Biol. Chem.* 1998; 379, 105-111.

[110] Lenarcic, B; Ritonja, A; Strukelj, B; Turk, B; Turk, V. Equistatin, a new inhibitor of cysteine proteinases from Actinia equina, is structurally related to thyroglobulin type-1 domain. *J. Biol. Chem.* 1997; 272, 13899-13903.

[111] Mihelic, M; Turk, D. Two decades of thyroglobulin type-1 domain research. Biol. *Chem.* 2007; 388:1123-1130.

[112] Lenarcic, B; Turk, V. Thyroglobulin type-1 domains in equistatin inhibit both papain-like cysteine proteinases and cathepsin D. *J. Biol. Chem.* 1999; 274, 563-566.

[113] Kotsyfakis, M; Karim, S; Andersen, JF; Mather, TN; Ribeiro, JMC. Selective cysteine protease inhibition contributes to blood-feeding success of the tick Ixodes scapularis. *J. Biol. Chem.* 2007; 282, 29256-29263.

[114] Filipek, R; Rzychon, M; Oleksy, A; Gruca, M; Dubin, A; et al. The Staphostatinsta-phopain complex: a forward binding inhibitor in complex with its target cysteine protease. *J. Biol. Chem.* 2003; 278, 40959-40966.

[115] Brzin, J; Rogelj, B; Popovic, T; Strukelj, B; Ritonja, A. Clitocypin, a new type of cysteine proteinase inhibitor from fruit bodies of mushroom clitocybe nebularis. *J. Biol. Chem.* 2000; 275, 20104-20109.

[116] Monteiro, ACS; Abrahamson, M; Lima, APCA; Vannier-Santos, MA; Scharfstein, J. Identification, characterization and localization of chagasin, a tight-binding cysteine protease inhibitor in Trypanosoma cruzi *J. Cell Sci.* 2001; 114, 3933-3942.

[117] Arai, S; Matsumoto, I; Emori, Y; Abe, K. Plant seed cystatins and their target enzymes of endogenous and exogenous origin. *J. Agric. Food Chem.* 2002; 50, 6612-6617.

[118] Nagata, K; Kudo, N; Abe, K; Arai, S; Tanokura, M. Three-dimensional solution structure of oryzacystatin-I, a cysteine proteinase inhibitor of the rice, Oryza sativa L. japonica. *Biochemistry* 2000 Dec 5;39(48), 14753-60.

[119] Abe, M; Abe, K; Kuroda, M; Arai, S. Corn kernel cysteine proteinase inhibitor as a novel cystatin superfamily member of plant origin. Molecular cloning and expression studies. *Eur. J. Biochem.* 1992; 209, 933-937.

[120] Chen, MS; Johnson, B; Wen, L; Muthukrishnan, S; Kramer, KJ; et al. Rice cystatin: bacterial expression, purification, cysteine proteinase inhibitory activity, and insect growth suppressing activity of a truncated form of the protein. *Protein Expr. Purif.* 1992; 3, 41- 49.

[121] Lalitha, S; Shade, RE; Murdock, LM; Hasegawa, PM; Bressan, RA; et al. Comparison of chemical characteristics of three soybean cysteine proteinase inhibitors. *J. Agric. Food Chem.* 2005; 53, 1591-1597.

[122] Oliva, ML; Carmona, AK; Andrade, SS; Cotrin, SS; Soares-Costa, A; et al. Inhibitory selectivity of canecystatin: a recombinant cysteine peptidase inhibitor from sugarcane. *Biochem. Biophys. Res. Commun.* 2004; 320, 1082-1086.

[123] Martinez, M; Diaz-Mendoza, M; Carrillo, L; Diaz, I. Carboxy terminal extended phytocystatins are bifunctional inhibitors of papain and legumain cysteine proteinases. *FEBS Lett.* 2007; 581, 2914-2918.

[124] Diop, NN; Kidric, M; Repellin, A; Gareil, M; d'Arcy-Lameta, A; et al. A multicystatin is induced by drought-stress in cowpea (Vigna unguiculata (L.) Walp.) Leaves. *FEBS Lett.* 2004; 577, 545-550.

[125] Wu, J; Haard, NF. Purification and characterization of a cystatin from the leaves of methyl jasmonate treated tomato plants. *Comp. Biochem. Physiol. C Toxicol. Pharmacol.* 2000; 127, 209-220.

[126] Grzonka, Z; Jankowska, E; Kasprzykowski, F; Kasprzykowska, R; Lankiewicz, L; et al. http://www.ncbi.nlm.nih.gov/pubmed/11440158?ordinalpos=1&itool=EntrezSystem2.PEntrez.Pubmed.Pubmed_ResultsPanel.Pubmed_DefaultReportPanel.Pubmed_RVDocSum Structural studies of cysteine proteases and their inhibitors. *Acta Biochim. Pol.* 2001; 48, 1-20.

[127] Koiwa, H; Bressan, RA; Hasegawa, PM. Regulation of protease inhibitors and plant defense. *Trends Plant Sci.* 1997; 21, 379-384.

[128] Brzin, J; Kidric, M. Proteinases and their inhibitors in plants: role in normal growth and in response to various stress conditions. *Biotechnol. Genet. Eng. Rev.* 1995; 13, 420-467.

[129] Aguiar, JM; Franco, OL; Rigden, DJ; Bloch, C Jr; Monteiro, AC; et al. Molecular modeling and inhibitory activity of cowpea cystatin against bean bruchid pests. *Proteins* 2006; 63, 662-670.

[130] Evans HJ; Barrett, AJ. A cystatin-like cysteine proteinase inhibitor from venom of the African puff adder (Bitis arietans). *Biochem. J.* 1987; 246, 795-797.

[131] Delbridge, ML; Kelly, LE. Sequence analysis, and chromosomal localization of a gene encoding a cystatin-like protein from Drosophila melanogaster. *FEBS Lett.* 1990; 274, 141-145.

[132] Brzin, J; Kopitar, M; Locnikar, P; Turk, V. An endogenous inhibitor of cysteine and serine proteinases from spleen. *FEBS Lett.* 1982; 138, 193-197.

[133] Aghajanyan, HG; Arzumanyan, AM; Arutunyan, AA; Akopyan, TN. Cystatins from bovine brain: purification, some properties, and action on substance P degrading activity. *Neurochem. Res.* 1988; 13, 721-727.

[134] Tsushima, H; Higashiyama, K; Mine, H. Isolation and amino acid sequence of a cystatin-type cysteine proteinase inhibitor from bovine hoof. *Arch Dermatol. Res.* 1996; 288, 484-488.

[135] Hirado, M; Tsunasawa, S; Sakiyama, F; Niiobe, M; Fujii, S. Complete amino acid sequence of bovine colostrum low-Mr cysteine proteinase inhibitor. *FEBS Lett.* 1985; 186, 41-45.

[136] Poulik, MD; Shinnick, CS; Smithies, O. Partial amino acid sequences of human and dog post-gamma globulins. *Mol. Immunol.* 1981; 18, 569-572.

[137] Sekine, T; Poulik, MD. Post-gamma globulin: tissue distribution and physicochemical characteristics of dog post-gamma globulin. *Clin. Chim. Acta* 1982; 120, 225-235.

[138] Rashid, F; Sharma, S; Bano, B. Detailed biochemical characterization of human placental cystatins (HPC). *Placenta* 2006a; 2, 822-831.

[139] Pontremoli, S; Melloni, E; Salamino, F; Sparatore, B; Michetti, M; Horecker, BL. Endogenous inhibitors of lysosomal proteinases. *Proc. Natl. Acad. Sci. USA* 1983; 80, 1261-1264.

[140] Kopitar, M; Stern, F; Marks, N. Cerebrocystatin suppresses degradation of myelin basic protein by purified brain cysteine proteinase. *Biochem. Biophys. Res. Commun.* 1983; 112, 1000-1006.

[141] Aoki, N; Deshimaru, M; Kihara, K; Terada, S. Snake fetuin: isolation and structural analysis of new fetuin family proteins from the sera of venomous snakes. *Toxicon.* 2009 Sep 15;54(4), 481-90.

[142] Tarasuk, M; Vichasri, S; Viyanant, V; Grams, R. Type I cystatin (stefin) is a major component of Fasciola gigantica excretion/secretion product. *Mol. Biochem. Parasitol.* 2009 Sep;167(1), 60-71.

[143] Kotsyfakis, M; Sa-Nunes, A; Francischetti, IMB; Mather, TN; Andersen, JF, Ribeiro, JMC. Antiinflammatory and immunosuppressive activity of sialostatin L, a salivary cystatin from the tick Ixodes scapularis. *J. Biol. Chem.* 2006; 281, 26298-26307.

[144] Li, F; Gai, X; Wang, L; Song, L; Zhang, H; Qiu, L; Wang, M; Siva, VS. Identification and characterization of a Cystatin gene from Chinese mitten crab Eriocheir sinensis. *Fish Shellfish Immunol.* 2010 Sep;29(3), 521-9.

[145] Liu, YH; Han, YP; Li, ZY; Wei, J; He, HJ; Xu, CZ; Zheng, HQ; Zhan, XM; Wu, ZD; Lv, ZY. Molecular cloning and characterization of cystatin, a cysteine protease inhibitor, from Angiostrongylus cantonensis. *Parasitol. Res.* 2010 Jun 22.

[146] Salát, J; Paesen, GC; Rezácová, P; Kotsyfakis, M; Kovárová, Z; Sanda, M; Majtán, J; Grunclová, L; Horká, H; Andersen, JF; Brynda, J; Horn, M; Nunn, MA; Kopácek, P;

Kopecký, J; Mares, M. Crystal structure and functional characterization of an immunomodulatory salivary cystatin from the soft tick Ornithodoros moubata. *Biochem. J.* 2010 Jun 1;429(1), 103-12.

[147] Khaznadji, E; Collins, P; Dalton, JP; Bigot, Y; Moire, N. A new multi-domain member of the cystatin superfamily expressed by Fasciola hepatica. *Int. J. Parasitol.* 2005; 35, 1115-1125.

[148] Zehra, S; Shahid, PB; Bano, B. Isolation, characterization and kinetics of goat cystatins. *Comp. Biochem. Physiol-Part B* 2005; 142, 361-368.

[149] Sumbul, S; Bano, B. Purification and characterization of high molecular mass and low molecular mass cystatins from goat brain. *Neurochem. Res*. 2006; 31, 1327-1336.

[150] Khan, MS; Bano, B. Purification, characterization and kinetics of thiol protease inhibitor from goat (Capra hircus) lung. *Biochemistry* (Mosc). 2009 Jul; 74(7), 781-8.

[151] Priyadarshini, M; Bano, B. Cystatin like thiol proteinase inhibitor from pancreas of Capra hircus: purification and detailed biochemical characterization. *Amino Acids*. 2010 Apr;38(4), 1001-10

[152] Chung, YB; Yang, HJ. Partial purification and characterization of a cysteine protease inhibitor from the plerocercoid of Spirometra erinacei. *Korean J. Parasitol.* 2008; 46, 183-186.

[153] Müller-Esterl, W; Fritz, H; Kellermann, J; Lottspeich, F; Machleidt, W; Turk, V. Genealogy of mammalian cysteine proteinase inhibitors. Common evolutionary origin of stefins, cystatins and kininogens. F*EBS Lett.* 1985b; 191, 221-226.

[154] Lee, C; Bongcam-Rudloff, E; Sollner, C; Jahnen-Dechent, W; Claesson-Welsh, L. Type 3 cystatins; fetuins, kininogen and histidine-rich glycoprotein. *Front Biosci.* 2009; 14, 2911-2922.

[155] Kordis, D; Turk, V. Phylogenomic analysis of the cystatin superfamily in eukaryotes and prokaryotes. *BMC Evol. Biol.* 2009 Nov 18;9, 266.

[156] Alvarez-Fernandez, M; Liang, YH; Abrahamson, M; Su, XD. Crystal structure of human cystatin D, a cysteine peptidase inhibitor with restricted inhibition profile. *J. Biol. Chem*. 2005; 280, 18221-18228.

[157] Martin, JR; Craven, CJ; Jerala, R; Kroon-Zitko, L; Zerovnik, E; et al. The three-dimensional solution structure of human stefin A. *J. Mol. Biol.* 1995; 246, 331-343.

[158] Martin, JR; Jerala, R; Kroon-Zitko, L; Zerovnik, E; Turk, V; et al. Structural characterisation of human stefin A in solution and implications for binding to cysteine proteinases. *Eur. J. Biochem.* 1994; 225, 1181-1194.

[159] Saxena, I; Tayyab, S. Protein proteinase inhibitors from avian egg whites *Cell Mol. Life Sci.* 1997; 53, 13-23.

[160] Schwabe, C; Anastasi, A; Crow, H; McDonald, JK; Barrett, AJ. Cystatin. Amino acid sequence and possible secondary structure. *Biochem. J.* 1984; 217, 813-817.

[161] Pol, E; Olsson, SL; Estrada, S; Prasthofer, TW; Bjork, I. Characterization by spectroscopic, kinetic and equilibrium methods of the interaction between recombinant human cystatin A (stefin A) and cysteine proteinases. *Biochem. J.* 1995; 311, 275-282.

[162] Musil, D; Zucic, D; Turk, D; Engh, RA; Mayr, I; et al. The refined 2.15 A X-ray crystal structure of human liver cathepsin B: the structural basis for its specificity. *EMBO J.* 1991; 10, 2321-2330.

[163] Turk, B; Turk, D; Salvesen, GS. Regulating cysteine protease activity: essential role of protease inhibitors as guardians and regulators. *Medicinal Chem. Rev. Online* 2005; 2, 283-297.

[164] Abrahamson, M; Alvarez-Fernandez, M; Nathanson, CM. Cystatins. *Biochem. Soc. Symp.* 2003; 70, 179-199.

[165] Langerholc, T; Zavasnik-Bergant, V; Turk, B; Turk, V; Abrahamson, M; Kos, J. Inhibitory properties of cystatin F and its localization in U937 promonocyte cells. *FEBS J.* 2005; 272, 1535-1545.

[166] Anastasi, A; Brown, MA; Kembhavi, AA; Nicklin, MJH; Sayers, CA; et al. Cystatin, a protein inhibitor of cysteine proteinases. Improved purification from egg white, characterization, and detection in chicken serum. *Biochem. J.* 1983; 211, 129-138.

[167] Guncar, G; Podobnik, M; Pungercar, J; Strukelj, B; Turk, V; et al. Crystal structure of porcine cathepsin H determined at 2.1 A resolution: location of the mini-chain C-terminal carboxyl group defines cathepsin H aminopeptidase function. *Structure* 1998; 6, 51-61.

[168] Guncar, G; Klemencic, I; Turk, B; Turk, V; Karaoglanovic-Carmona, A; et al. Crystal structure of cathepsin X: a flip-flop of the ring of His23 allows carboxy-monopeptidase and carboxy-dipeptidase activity of the protease. *Structure* 2000; 8, 305-313.

[169] Turk, D; Janjic, V; Stern, I; Podobnik, M; Lamba, D; et al. Structure of human dipeptidyl peptidase I (cathepsin C): exclusion domain added to an endopeptidase framework creates the machine for activation of granular serine proteases. *EMBO J.* 2001; 20, 6570-6582.

[170] Nicklin, MJH; Barrett, AJ. Inhibition of cysteine proteinases and dipeptidyl peptidase I by egg-white cystatin. *Biochem. J.* 1984; 223, 245-253.

[171] Gounaris, AD; Brown, MA; Barrett, AJ. Human plasma alpha-cysteine proteinase inhibitor. Purification by affinity chromatography, characterization and isolation of an active fragment *Biochem. J.* 1984; 221(2), 445-52.

[172] Bjork, I; Alriksson, E; Ylinenjarvi, K. Kinetics of binding of chicken cystatin to papain. *Biochemistry* 1989; 28, 1568-1573.

[173] Machleidt, W; Thiele, U; Laber, B; Assfalg-Machleidt, I; Esterl, A; Wiegand, G; Kos, J; Turk, V; Bode, W. Mechanism of inhibition of papain by chicken egg white cystatin. Inhibition constants of N-terminally truncated forms and cyanogen bromide fragments of the inhibitor. *FEBS Lett.* 1989; 243, 234-238

[174] Abrahamson, M; Ritonja, A; Brown, MA; Grubb, A; Machleidt, W; et al. Identification of the probable inhibitory reactive sites of the cysteine proteinase inhibitors human cystatin C and chicken cystatin. *J. Biol. Chem.* 1987b; 262, 9688-9694.

[175] Nycander, M; Bjork, I. Evidence by chemical modification that tryptophan-104 of the cysteine-proteinase inhibitor chicken cystatin is located in or near the proteinase-binding site. *Biochem. J.* 1990; 271, 281-284.

[176] Nycander, M; Estrada, S; Mort, JS; Abrahamson, M; Bjork, I. Two-step mechanism of inhibition of cathepsin B by cystatin C due to displacement of the proteinase occluding loop *FEBS Lett.* 1998; 422, 61-64.

[177] Estrada, S; Olson, ST; Raub-Segall, E; Bjork, I. The N-terminal region of cystatin A (stefin A) binds to papain subsequent to the two hairpin loops of the inhibitor.

Demonstration of two-step binding by rapid-kinetic studies of cystatin A labeled at the N-terminus with a fluorescent reporter group. *Protein Sci.* 2000; 9, 2218-2224.
[178] Pavlova, A; Bjork, I. Grafting of features of cystatins C or B into the N-terminal region or second binding loop of cystatin A (stefin A) substantially enhances inhibition of cysteine proteinases. *Biochemistry* 2003; 42, 11326-11333.
[179] Jenko, S; Dolenc, I; Guncar, G; Dobersek, A; Podobnik, M. Crystal structure of Stefin A in complex with cathepsin H: N-terminal residues of inhibitors can adapt to the active sites of endo- and exopeptidases. *J. Mol. Biol.* 2003; 326, 875-885.
[180] Pol, E; Bjork, I. Role of the single cysteine residue, Cys 3, of human and bovine cystatin B (stefin B) in the inhibition of cysteine proteinases. *Protein Sci.* 2001; 10, 1729-1738.
[181] Auerswald, EA; Nagler, DK; Schulze, AJ; Engh, RA; Genenger, G; Machleidt, W; Fritz, H. Production, inhibitory activity, folding and conformational analysis of an N-terminal and an internal deletion variant of chicken cystatin. *Eur. J. Biochem.* 1994; 224, 407-415.
[182] Estrada, S; Pavlova, A; Bjork, I. The contribution of N-terminal region residues of cystatin A (stefin A) to the affinity and kinetics of inhibition of papain, cathepsin B, and cathepsin L. *Biochemistry* 1999; 38, 7339-7345.
[183] Pavlova, A; Krupa, JC; Mort, JS; Abrahamson, M; Björk, I. Cystatin inhibition of cathepsin B requires dislocation of the proteinase occluding loop. Demonstration By release of loop anchoring through mutation of his110. *FEBS Lett.* 2000; 487, 156-160.
[184] Ekiel, I; Abrahamson, M; Fulton, DB; Lindahl, P. (NMR) Structural studies of human cystatin C dimers and monomers. *J. Mol. Biol.* 1997; 271, 266-271.
[185] Assfalg-Machleidt, I; Jochun, M; Klaubert, W; Inthorn, D; Machleidt, W. Enzymatically active cathepsin B dissociating from its inhibitor complexes is elevated in blood plasma of patients with septic shock and some malignant tumors. *Biol. Chem. Hoppe-Seyler* 1988; 369, 263-269.
[186] Cox, JL. Cystatins and cancer. *Front Biosci.* 2009; 14, 463-474.
[187] Trabandt, A; Gay, RE; Fassbender, HG; Gay, S. Cathepsin B in synovial cells at the site of joint destruction in rheumatoid arthritis. Trabandt A, Gay RE, Fassbender HG, Gay S. *Arthritis Rheum.* 1991 Nov;34(11):1444-51.
[188] Sohar, I; Laszlo, A; Gaal, K; Mechler, F. Cysteine and metalloproteinase activities in serum of Duchenne muscular dystrophic genotypes. *Biol. Chem. Hoppe-Seyler* 1988; 369, 277-279.
[189] Buttle, DJ; Burnett, D; Abrahamson, M. Levels of neutrophil elastase and cathepsin B activities, and cystatins in human sputum: relationship to inflammation. *Scand. J. Clin. Lab. Invest* 1990; 50:509-516.
[190] Bever, CT; Garver, DW. Increased cathepsin B activity in multiple sclerosis brain. *J. Neurol. Sci.* 1995; 131, 71-73.
[191] Mohamed, MM; Sloane, BF. Cysteine cathepsins: multifunctional enzymes in cancer. *Nat. Rev. Cancer* 2006; 6, 764-775.
[192] Turk, V; Kos, J; Turk, B. Cysteine cathepsins (proteases)--on the main stage of cancer? *Cancer Cell* 2004; 5, 409-410.
[193] Vasiljeva, O; Papazoglou, A; Kruger, A; Brodoefel, H; Korovin, M; Deussing, J; Augustin, N; Nielsen, BS; Almholt, K; Bogyo, M; Peters, C; Reinheckel, T. Tumor

cell-derived and macrophage-derived cathepsin B promotes progression and lung metastasis of mammary cancer. *Cancer Res.* 2006; 66, 5242-5250.

[194] Joyce, JA; Baruch, A; Chehade, K; Meyer-Morse, N; Giraudo, E; Tsai, FY; Greenbaum, DC; Hager, JH; Bogyo, M; Hanahan, D. Cathepsin cysteine proteases are effectors of invasive growth and angiogenesis during multistage tumorigenesis. *Cancer Cell* 2004; 5, 443-453.

[195] Paraoan, L; Gray, D; Hiscott, P; Garcia-Finana, M; Lane, B; Damato, B; Grierson, I. Cathepsin S and its inhibitor cystatin C: imbalance in uveal melanoma. *Front Biosci.* 2009; 14, 2504-2513.

[196] Rivenbark, AG; Coleman, WB. Epigenetic regulation of cystatins in cancer. *Front Biosci.* 2009; 14, 453-462.

[197] Werle, B; Schanzenbacher, U; Lah, TT; Ebert, E; Julke, B; et al. Cystatins in non-small cell lung cancer: tissue levels, localization and relation to prognosis. *Oncol. Rep.* 2006; 16, 647-655.

[198] Parker, B; Ciocca, D; Bidwell, B; Gago, F; Fanelli, M; et al. Primary tumour expression of the cysteine cathepsin inhibitor Stefin A inhibits distant metastasis in breast cancer. *J. Pathol.* 2008; 214, 337-346.

[199] Choi, EH; Kim, JT; Kim, JH; Kim, SY; Song, EY; Kim, JW; Kim, SY; Yeom, YI; Kim, IH; Lee, HG. Upregulation of the cysteine protease inhibitor, cystatin SN, contributes to cell proliferation and cathepsin inhibition in gastric cancer. *Clin. Chim. Acta* 2009; 406:45- 51.

[200] Sokol, JP; Schiemann, WP. Cystatin C antagonizes transforming growth factor beta signaling in normal and cancer cells. *Mol. Cancer Res.* 2004; 2, 183-195.

[201] Zhang, J; Shridhar, R; Dai, Q; Song, J; Barlow, SC; Yin, L; Sloane, BF; Miller, FR; Meschonat, C; Li, BDL; Abreo, F; Keppler, D. Cystatin M: a novel candidate tumor suppressor gene for breast cancer. *Cancer Res.* 2004; 64, 6957-6964.

[202] Utsunomiya, T; Hara, Y; Kataoka, A; Morita, M; Arakawa, H; Mori, M; Nishimura, S. Cystatin-like metastasis-associated protein mRNA expression in human colorectal cancer is associated with both liver metastasis and patient survival. *Clin. Cancer Res.* 2002; 8, 2591-2594.

[203] Li, W; Ding, F; Zhang, L; Liu, Z; Wu, Y; Luo, A; Wu, M; Wang, M; Zhan, Q; Liu, Z. Overexpression of stefin A in human esophageal squamous cell carcinoma cells inhibits tumor cell growth, angiogenesis, invasion, and metastasis. *Clin. Cancer Res.* 2005; 11, 8753-8762.

[204] Kopitz, C; Anton, M; Gansbacher, B; Kruger, A. Reduction of experimental human fibrosarcoma lung metastasis in mice by adenovirus-mediated cystatin C overexpression in the host. *Cancer Res.* 2005; 65, 8608-8612.

[205] Gocheva, V; Zeng, W; Ke, D; Klimstra, D; Reinheckel, T; Peters, C; Hanahan, D; Joyce, JA. Distinct roles for cysteine cathepsin genes in multistage tumorigenesis. *Genes Dev.* 2006; 20, 543-556.

[206] Briggs, JJ; Haugen, MH; Johansen, HT; Riker, AI; Abrahamson, M; Fodstad, Ø; Maelandsmo, GM; Solberg, R. Cystatin E/M suppresses legumain activity and invasion of human melanoma. *BMC Cancer*. 2010 Jan 15; 10, 17.

[207] Palsdottir, A; Snorradottir, AO; Thorsteinsson, L. Hereditary cystatin C amyloid angiopathy: genetic, clinical, and pathological aspects. *Brain Pathol.* 2006; 16, 55-59.

[208] Mi, W; Pawlik, M; Sastre, M; Jung, SS; Radvinsky, DS; et al. Cystatin C inhibits amyloid-beta deposition in Alzheimer's disease mouse models. *Nat. Genet.* 2007; 39, 1440-1442.

[209] Pennacchio, LA; Lehesjoki, AE; Stone, NE; Willour, VL; Virtaneva, K; et al. Mutations in the gene encoding cystatin B in progressive myoclonus epilepsy (EPM1). *Science* 1996; 271, 1731-1734.

[210] Joensuu, T; Kuronen, M; Alakurtti, K; Tegelberg, S; Hakala, P; et al. Cystatin B: mutation detection, alternative splicing and expression in progressive myclonus epilepsy of Unverricht-Lundborg type (EPM1) patients. *Eur. J. Hum. Genet.* 2007; 15, 185-193.

[211] Lieuallen, K; Pennacchio, LA; Park, M; Myers, RM; Lennon, GG. Cystatin B-deficient mice have increased expression of apoptosis and glial activation genes. *Hum. Mol. Genet.* 2001; 10, 1867-1871.

[212] Foghsgaard, L; Wissing, D; Mauch, D; Lademann, U; Bastholm, L; et al. Cathepsin B acts as a dominant execution protease in tumor cell apoptosis induced by tumor necrosis factor. *J. Cell Biol.* 2001; 153, 999-1010.

[213] Petty, RD; Kerr, KM; Murray, GI; Nicolson, MC; Rooney, PH; et al. Tumor transcriptome reveals the predictive and prognostic impact of lysosomal protease inhibitors in non-small-cell lung cancer. *J. Clin. Oncol.* 2006; 24, 1729-1744.

[214] Zavasnik-Bergant, T. Cystatin protease inhibitors and immune functions. *Front Biosci.* 2008; 13, 4625-4637.

[215] Verdot, L; Lalmanach, G; Vercruysse, V; Hoebeke, J; Gauthier, F; et al. Chicken cystatin stimulates nitric oxide release from interferon-gamma-activated mouse peritoneal macrophages via cytokine synthesis. *Eur. J. Biochem.* 1999; 266, 1111-1117.

[216] Leung-Tack, J; Tavera, C; Gensac, MC; Martinez, J; Colle, A. Modulation of phagocytosis-associated respiratory burst by human cystatin C: role of the N-terminal tetrapeptide Lys–Pro–Pro–Arg. *Exp. Cell Res.* 1990; 188, 16-22.

[217] Kato, T; Imatani, T; Minaguchi, K; Saitoh, E; Okuda, K. Salivary cystatins induce interleukin-6 expression via cell surface molecules in human gingival fibroblasts. Mol. *Immunol.* 2002; 39, 423-430.

[218] Kato, T; Ito, T; Imatani, T; Minaguchi, K; Saitoh, E; Okuda, K. Cystatin SA, a cysteine proteinase inhibitor, induces interferon-gamma expression in CD4-positive T cells. *Biol. Chem.* 2004; 385, 419-422.

[219] Schierack, P; Lucius, R; Sonnenburg, B; Schilling, K; Hartmann, S. Parasite-specific immunomodulatory functions of filarial cystatin *Infect. Immun.* 2003; 71, 2422-2429.

[220] Sokol, JP; Neil, JR; Schiemann, BJ; Schiemann, WP. The use of cystatin C to inhibit epithelialmesenchymal transition and morphological transformation stimulated by transforming growth factor-beta. *Breast Cancer Res.* 2005; 7, R844-853.

[221] Kato, T; Imatani, T; Miura, T; Minaguchi, K; Saitoh, E; et al. Cytokine-inducing activity of family 2 cystatins. *Biol. Chem.* 2000; 381, 1143-1147.

[222] Pluger, EB; Boes, M; Alfonso, C; Schröter, CJ; Kalbacher, H; et al. Specific role for cathepsin S in the generation of antigenic peptides in vivo. *Eur. J. Immunol.* 2002; 32, 467-476.

[223] Hamilton, G; Colbert, JD; Schuettelkopf, AW; Watts, C. Cystatin F is a cathepsin C-directed protease inhibitor regulated by proteolysis. *EMBO J.* 2008; 27, 499-508.

[224] Blank, EMF; Henskens, YM; Van't, H; Veerman, EC; Nieuw, AV. Inhibition of the growth and cysteine proteinase activity of Porphyromonos gingivalis by human salivary cystatin S and chicken cystatin. *Biol. Chem.* 1996; 377, 847-850.

[225] Collins, AR; Grubb, A. Cystatin D, a natural salivary cysteine protease inhibitor, inhibits coronavirus replication at its physiologic concentration. *Oral. Microbial. Immunol.* 1998; 13, 59-61

[226] Ordóñez-Gutiérrez, L; Martínez, M; Rubio-Somoza, I; Díaz, I; Mendez, S; Alunda, JM. Leishmania infantum: antiproliferative effect of recombinant plant cystatins on promastigotes and intracellular amastigotes estimated by direct counting and real-time PCR. *Exp. Parasitol.* 2009 Dec;123(4), 341-6.

[227] Shah, A; Bano, B. Cystatins in health and disease. *Int. J. Pept. Res. Ther.* 2009; 15, 43-48.

[228] Fried, LF. Creatinine and cystatin C: what are the values? *Kidney Int.* 2009; 75, 578-580.

[229] Al-Wakeel, JS; Hammad, D; Memon, NA; Tarif, N; Shah, I; et al. Serum cystatin C: a surrogate marker for the characteristics of peritoneal membrane in dialysis patients. *Saudi J. Kidney Dis. Transpl.* 2009; 20, 227-231.

[230] Korolenko, TA; Filatova, TG; Cherkanova, MS; Khalikova, TA; Bravve, IIu. Cystatins: cysteine proteases regulation and disturbances in tumors and inflammation. *Biomed. Khim.* 2008; 54, 210-217.

[231] Tumminello, FM; Badalamenti, G; Incorvaia, L; Fulfaro, F; D'Amico, C; et al. Serum interleukin-6 in patients with metastatic bone disease: correlation with cystatin C. *Med. Oncol.* 2009; 26, 10-15.

[232] Cepeda, J; Tranche-Iparraguirre, S; Marín-Iranzo, R; Fernández-Rodríguez, E; Riesgo-García, A; García-Casas, J; Hevia-Rodríguez, E. Cystatin C and cardiovascular risk in the general population. *Rev. Esp. Cardiol.* 2010 Apr;63(4), 415-22.

[233] Morgan, GJ; Giannini, S; Hounslow, AM; Craven, CJ; Zerovnik, E; et al. Exclusion of the native alpha-helix from the amyloid fibrils of a mixed alpha/beta protein. *J. Mol. Biol.* 2008; 375, 487-498.

[234] Turk, V; Stoka, V; Turk, D. Cystatins: Biochemical and structural properties, and medical relevance. *Front Biosci.* 2008; 13, 5406-5420.

[235] Staniforth, RA; Giannini, S; Higgins, LD; Conroy, MJ; Hounslow, AM; et al. Three-dimensional domain swapping in the folded and molten-globule states of cystatins, an amyloid-forming structural superfamily. *EMBO J.* 2001; 20, 4774-4781.

[236] Zerovnik, E; Skarabot, M; Skerget, K; Giannini, S; Stoka, V; et al.: Amyloid fibril formation by human stefin B: influence of pH and TFE on fibril growth and morphology. *Amyloid.* 2007; 14, 237-247.

[237] Pallares, I; Berenguer, C; Aviles, FX; Vendrell, J; Ventura, S. Self-assembly of human latexin into amyloid-like oligomers. *BMC Struct. Biol.* 2007; 7, 75.

[238] Zerovnik, E. Amyloid-fibril formation. Proposed mechanisms and relevance to conformational disease. *Eur. J. Biochem.* 2002; 269, 3362-3371.

[239] Soto, C. Protein misfolding and disease; protein refolding and therapy. *FEBS Lett.* 2001; 498, 204-207.

[240] Ceru, S; Kokalj, SJ; Rabzelj, S; Skarabot, M; Gutierrez-Aguirre, I; et al. Size and morphology of toxic oligomers of amyloidogenic proteins: a case study of human stefin B. *Amyloid.* 2008; 15, 147-159.

[241] Rabzelj, S; Viero, G; Gutiérrez-Aguirre, I; Turk, V; Dalla Serra, M; et al. Interaction with model membranes and pore formation by human stefin B: studying the native and prefibrillar states. *FEBS J.* 2008; 275, 2455-2466.

[242] Bucciantini, M; Calloni, G; Chiti, F; Formigli, L; Nosi, D; et al. , Prefibrillar amyloid protein aggregates share common features of cytotoxicity. *J. Biol. Chem.* 2004; 279, 31374-31382.

[243] Taupin, P; Ray, J; Fischer, WH; Suhr, ST; Hakansson, K; et al. FGF-2-responsive neural stem cell proliferation requires CCg, a novel autocrine/paracrine cofactor. *Neuron* 2000; 28, 385-397.

[244] Novinec, M; Grass, RN; Stark, WJ; Turk, V; Baici, A; et al. Interaction between human cathepsins K, L, and S and elastins: mechanism of elastinolysis and inhibition by macromolecular inhibitors. *J. Biolm. Chem.* 2007; 282, 7893–7902.

[245] Oppert, B; Morgan, TD; Hartzer, K; Lenarcic, B; Galesa, K; et al. Effects of proteinase inhibitors on digestive proteinases and growth of the red flour beetle, Tribolium castaneum (Herbst) (Coleoptera: Tenebrionidae). *Comp. Biochem. Physiol. C Toxicol. Pharmacol.* 2003; 134, 481-490.

[246] Andersson, NE; Rydengard, V; Mörgelin, M; Schmidtchen, A. Domain 5 of high molecular weight kininogen is antibacterial. *J. Biol. Chem.* 2005; 280, 34832- 34839.

[247] Aliberti, J; Viola, JP; Vieira-de-Abreu, A; Bozza, PT; Sher, A; et al. Cutting edge: bradykinin induces IL-12 production by dendritic cells: a danger signal that drives Th1 polarization. J. *Immunol.* 2003; 170, 5349-5353.

[248] Scharfstein, J; Schmitz, V; Svensjö, E; Granato, A; Monteiro, AC. Kininogens coordinate adaptive immunity through the proteolytic release of bradykinin, an endogenous danger signal driving dendritic cell maturation. *Scand. J. Immunol.* 2007; 66, 128-136.

[249] Zeeuwen, PL; Cheng, T; Schalkwijk, J. The biology of cystatin M/E and its cognate target proteases. *J. Invest. Dermatol.* 2009; 129, 1327-1338.

[250] Ahuja, S; Ahuja-Jensen, P; Johnson, LE; Caffe, AR; Abrahamson, M; et al. rd1 Mouse retina shows an imbalance in the activity of cysteine protease cathepsins and their endogenous inhibitor cystatin C. *Invest. Ophthalmol. Vis. Sci.* 2008; 49, 1089-1096.

[251] Tsai, HT; Wang, PH; Tee, YT; Lin, LY; Hsieh, YS; et al. Imbalanced serum concentration between cathepsin B and cystatin C in patients with pelvic inflammatory disease. *Fertil. Steril.* 2009; 91, 549-555.

[252] Xie, L; Terrand, J; Xu, B; Tsaprailis, G; Boyer, J; Chen, QM. Cystatin C increases in cardiac injury: a role in extracellular matrix protein modulation. *Cardiovasc. Res.* 2010; 87(4), 628-635.

In: Cystatins: Protease Inhibitors ...
Editors: John B. Cohen and Linda P. Ryseck
ISBN: 978-1-61209-343-7

Chapter 2

ANTI-TRYPANOSOMATID PROPERTIES OF CYSTATIN SUPERFAMILY: IMPLICATIONS ON PARASITE DEVELOPMENT AND VIRULENCE

André Luis Souza dos Santos, Claudia Masini d'Avila-Levy and Marta Helena Branquinha

Departamento de Microbiologia Geral, Instituto de Microbiologia Prof. Paulo de Góes (IMPPG), Centro de Ciências da Saúde (CCS), Universidade Federal do Rio de Janeiro (UFRJ), Ilha do Fundão,Rio de Janeiro, RJ, Brazil
Laboratório de Biologia Molecular e Doenças Endêmicas, Instituto Oswaldo Cruz, Fundação Oswaldo Cruz, Rio de Janeiro, RJ, Brazil

ABSTRACT

The Trypanosomatidae family, assorted on the Kinetoplastida order, consists of distinct genera of eukaryotic monoflagellated protozoa. Among the trypanosomatids with a digenetic life cycle, some species stand out: *Trypanosoma cruzi*, *Trypanosoma brucei* and several *Leishmania* species. These parasites are the causative agents of Chagas' disease, African sleeping sickness and leishmaniasis, respectively. Likewise, some species belonging to the *Phytomonas* genus can induce serious diseases in plants, which indicates the economical importance of these trypanosomatids, a problem especially affecting developing countries. Trypanosomatid cysteine proteases have been implicated in several processes including proliferation, differentiation, nutrition, host cell infection, and evasion of the host immune responses. For instance, *Leishmania* spp. possess three major cysteine proteases of the papain family (designated clan CA, family C1), namely, the cathepsin L-like CPA and CPB and the cathepsin B-like CPC, which are directly linked to the parasite survival inside macrophage cells and modulation of host immune response. In addition, cruzipain is a major cysteine protease, expressed in all developmental forms of *T. cruzi*, being highly immunogenic in patients with chronic Chagas' disease. Since cysteine proteases are present in trypanosomatids and their catalytic properties can vary considerably from those of the host enzymes they are considered important targets for new chemotherapeutical intervention. Typical cysteine protease inhibitors like E-64, leupeptin and K777 are able to inhibit both the cell-associated and released cysteine proteases produced by trypanosomatid cells in different extensions. Moreover, the superfamily of cystatins (stefins, cystatins and kininogens), which are endogenous proteins and tight-binding reversible competitive inhibitors of clan CA, family C1 papain-like cysteine inhibitors, interfere with different physiological aspects of trypanosomatid cells. In this sense, the present chapter will summarize the knowledge about the inhibitory effects of cystatins directly on cysteine proteases produced by pathogenic trypanosomatids or indirectly by potentiating the host immune cells.

Keywords: Cystatin superfamily, cysteine proteases, Leishmania, Trypanosoma, Phytomonas, development

TRYPANOSOMATIDAE FAMILY

The Trypanosomatidae family, which belongs to the Kinetoplastida order, is composed by a diverse group of exclusively parasitic protozoa. Trypanosomatids are distinguishable from other protozoa by distinctive organizational features such as the presence of (i) a single flagellum, (ii) a sub-pellicle microtubule cytoskeleton, (iii) a relatively small kinetoplast containing densely packed DNA that is organized in mini- and maxi-circles and that corresponds to the parasite mitochondrial genome, localized near the basal body of the flagellum, (iv) isolation of glycolytic enzymes in a special organelle, the glycosome, and (v) use of the flagellar pocket for molecular traffic into and out of the cell (Figure 1) [1–3]. In addition, these microorganisms exhibit unusual molecular phenomena such as antigenic variation, *trans*-splicing, RNA editing and peculiar nuclear organization. The trypanosomatids seem to be able to adapt with ease their energy metabolism to the availability of substrates and oxygen, and this may give them the ability to institute new life cycles if host behavior patterns allow [1–3].

Trypanosomatids are cosmopolitan group of unicellular eukaryotes capable in infecting a broad range of organisms such as mammals, fishes, plants and invertebrates, the latter group preferentially insects from the orders Diptera and Hemiptera. Trypanosomatids are able to cause a wide spectrum of diseases of social, economical and medical relevance, such as trypanosomiasis and leishmaniasis, especially in developing countries. Invertebrates act as hosts of monoxenous parasites, such as *Blastocrithidia*, *Crithidia*, *Herpetomonas* and *Leptomonas*, or serve as vectors of heteroxenous genera *Trypanosoma*, *Endotrypanum*, *Leishmania* and bug-transmitted parasites of the genus *Phytomonas*, found in plants. So, these trypanosomatids parasites have complex life cycles that involve multiple hosts. Inside the insect vector or mammalian/plant host, these parasites undergo differentiation and multiplication phases [1–3].

In humans and domestic animals, *Trypanosoma* and *Leishmania* cause severe diseases including sleeping sickness, Chagas' disease and kala-azar [3], while *Phytomonas* are pathogens of different families of plants including coconut, oil palm, tomato and corn [4]. Conversely, insect trypanosomatids are used routinely as safe models for initial biochemical and molecular studies on conserved features of the Trypanosomatidae family because they are easily cultured under axenic conditions and are traditionally nonpathogenic to humans [2, 3].

However, some possible mammalian infections due to insect flagellates have been described in immunosupressed patients, mainly in human immunodeficiency virus (HIV)-infected individuals [2, 5–11]. Corroborating these findings, some of the plant and insect trypanosomatids (collectively designated as lower trypanosomatids) contain homologues of virulence factors from the classic human trypanosomatid pathogens. For instance, two distinct protease classes were described in lower trypanosomatids: metalloproteases that have similarity with gp63 (synonymous leishmanolysin) of *Leishmania* spp., and cysteine proteases that present similar antigenic properties with the major cysteine protease (named cruzipain) produced by *T. cruzi* [12–30].

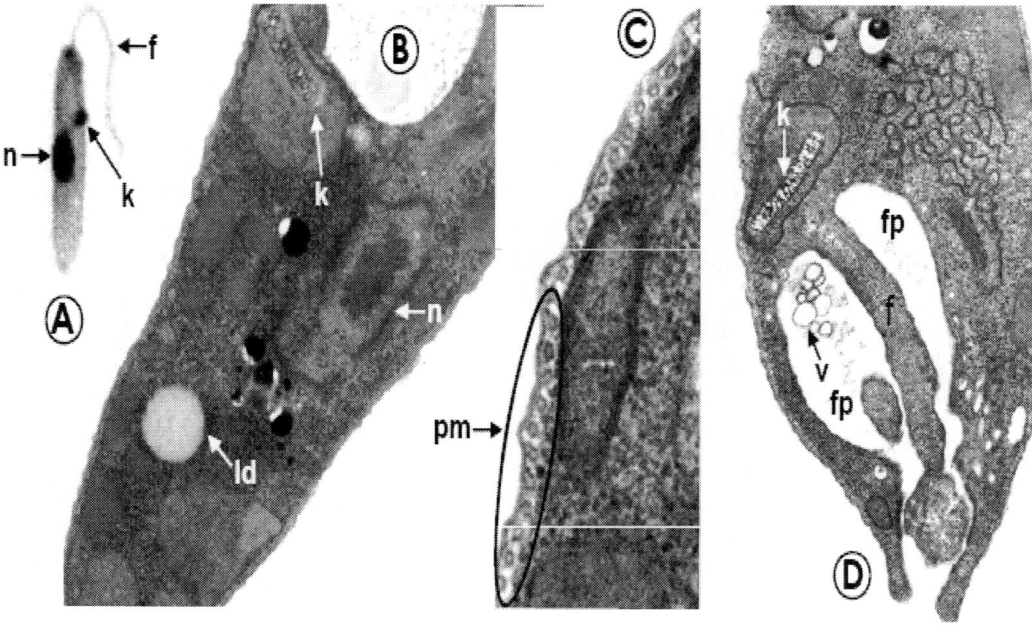

Figure 1. Representative morphology of a trypanosomatid cell. (A) Optical microscopy evidencing the Giemsa-stained parasite. This staining shows the position of the nucleus (n) in relation to the kinetoplast (k) in trypanosomatids. Note the presence of only one flagellum (f). (B–D) Transmission electron micrografies showing the different sections of a trypanosomatid cell. Note in (B) the presence of typical structures like nucleus, kinetoplast and lipid droplets (ld). In (C) observe the subpelicular microtubules along all extension of the plasmatic membrane (pm) and in (D) the flagellum, the flagellar pocket (fp) and the secretory vesicles (v).

These trypanosomiases are neglected tropical diseases, since they are largely ignored by medical science, first-world public opinion and pharmaceutical companies. The effective treatment of diseases caused by trypanosomatids is still an open issue, nevertheless of paramount importance. Current therapy, based on drugs developed decades ago, is suboptimal due to the low efficacy and high toxicity of the available therapeutic agents and the emergence of drug resistance [31, 32]. On the other hand, new treatments based on the use of lipid formulations of amphotericin B or miltefosine, are too expensive considering the target population [33–37]. Compounding these problems is the increase in the number of cases of trypanosomiases-HIV coinfection, due to the overlap between the acquired immunodeficiency syndrome (AIDS) epidemic and leishmaniasis/trypanosomiasis [38, 39]. A promising line of research targets the proteases produced by these parasites aiming the establishment of novel, effective and selective chemotherapies, since proteases participate in several phases of trypanosomatids development and virulence, including proliferation, nutrition, differentiation, interaction with both invertebrate and vertebrate/plant hosts, and evasion of host immune responses. For these reasons, this class of hydrolytic enzymes has attracted the attention of the scientific community to exploitation as targets for rational chemotherapy against leishmaniasis and trypanosomiasis. Moreover, proteases produced by trypanosomatids are highly immunogenic and have been exploited for serodiagnosis and as vaccine targets, in special cysteine-type proteases [40–45].

Cysteine Proteases and Clan CA, Family C1 Inhibitors

Proteases are involved in virtually all biological functions, which are explained by their participation in controlling protein synthesis and degradation through the hydrolysis of specific amide bonds in proteins and peptides [46, 47]. A comprehensive classification of proteases and their inhibitors has been developed in the MEROPS database (http://merops.sanger.ac.uk), and it is employed in this chapter. The organizational principle of the database is a hierarchical classification in which homologous sets of proteins of interest are grouped in families and the homologous families, principally thought to have common evolutionary origins from tertiary structure comparisons, are grouped in clans [48].

Cysteine proteases are widespread in nature and can be found in all organisms, representing one of the seven mechanistic classes of proteases. Their common feature is their catalytic mechanism: all cysteine proteases utilize a cysteine residue as the nucleophile and a histidine residue as the general base for proton shuttling. Among the nine clans already characterized as representative of cysteine proteases, the most abundant one is clan CA, which contains all the families of proteases that are known to have structural similarity to that of papain, the clan-type protease. For instance, clan CA includes the papain family C1 (Figure 2) and the calpain family C2, the former is the major scope of this chapter and the latter is the focus of the next book chapter. Other 22 families are assigned to clan CA on the basis of sequence motifs. In addition to the residues Cys^{158} and His^{292} of the catalytic dyad, two other functionally important residues are commonly present in papain and its relatives: these are Gln^{152} that helps in the formation of the 'oxyanion hole', an electrophilic center that stabilizes the tetrahedral intermediate, and Asn^{308}, which is thought to orientate the imidazolium ring of the catalytic His [47, 49].

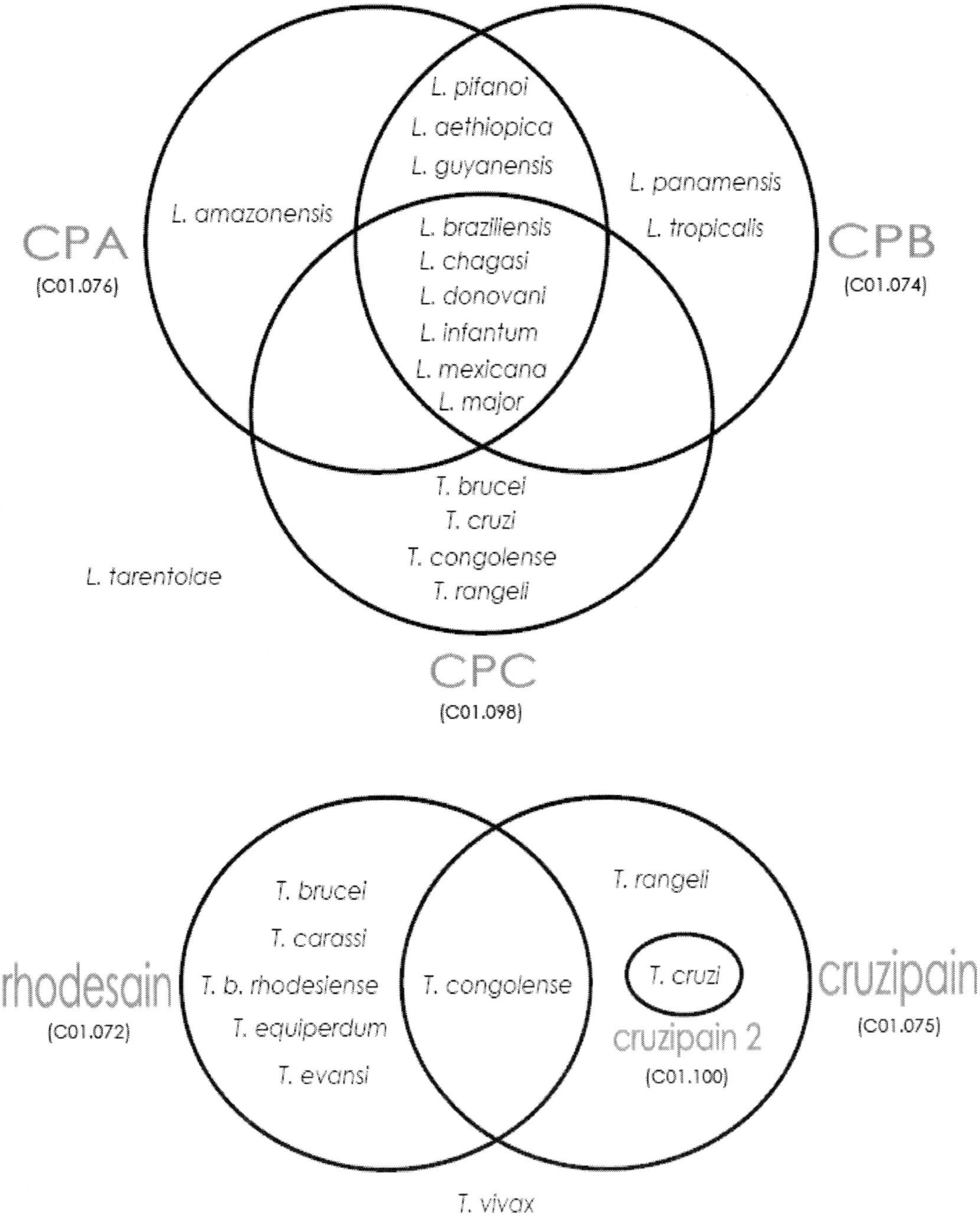

Figure 2. Major cysteine proteases, belonging to the papain-like family (cpa, cpb, cpc, rhodesain, cruzipain and cruzipain 2), produced by different trypanosomatids. The numbers inside the parenthesis represent the MEROPS database codes.

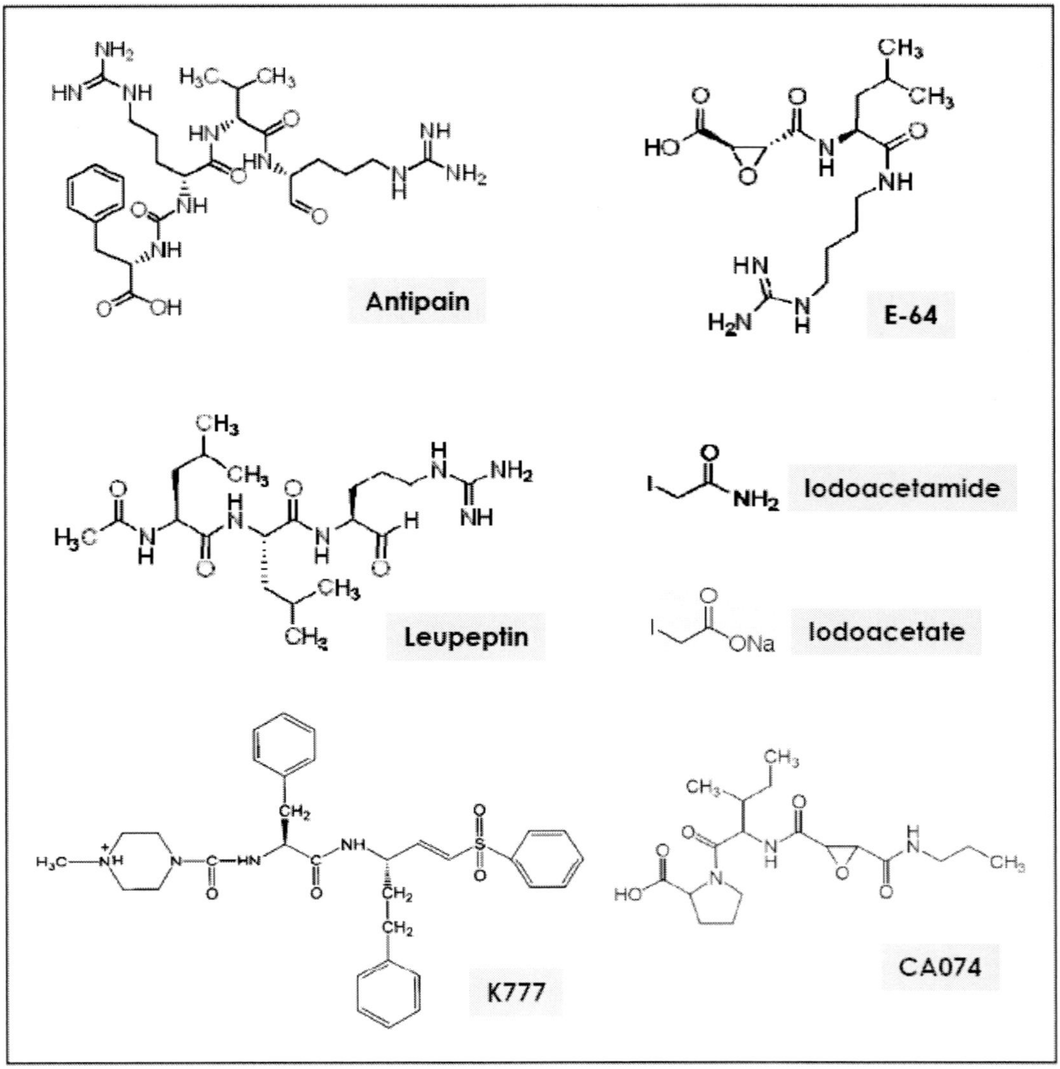

Figure 3. Typical cysteine protease inhibitors of the papain-like family.

Family C1 comprises papain, related plant proteases, many human lysosomal cathepsins such as B, C, F, H, K, L, O, S, V, W and X, and cysteine proteases of parasitic origin, including many found in trypanosomatids (Figure 2). The enzymes share similar sequences and tertiary structures, are optimally active in slightly acidic, reducing environments and are mostly endoproteases, sensible to typical cysteine protease inhibitors including antipain ([(S)-1-carboxy-2-phenylethyl]-carbamoyl-L-arginyl-L-valyl-argininal), E-64 (L-*trans*-epoxysuccinyll-leucyl-amido-(4-guanidino) butane), iodoacetamide, iodoacetate, leupeptin (*N*-acetyl-L-leucyl-L-leucyl-D,L-argininaldehyde), CA074 (*N*-(L-3-*trans*-propylcarbamoyloxirane-2-carbonyl)-L-isoleucyl-L-proline) and K777 (N-methyl-piperazine-Phe-homoPhe-vinylsulfone-phenyl) (Figure 3). The papain-like cathepsins are long believed to be primarily responsible for intralysosomal protein degradation, but gene-knockout studies revealed that this function is not exclusively dependent on any single cathepsin. Some of the lysosomal cathepsins have specific and individual functions, and these proteases have been associated with a number of pathologies due to their presence in the extracellular and extralysosomal environments [50]. The better understanding of the biological roles and pathological consequences of uncontrolled cysteine protease activity in humans have emerged, and these enzymes have been implicated in distinct disease processes ranging from cardiovascular, inflammatory, neurological, respiratory, musculoskeletal, immunological to cancer [51–54].

Once cysteine proteases control a great variety of physiological processes, their activity must be highly controlled. Cysteine proteases have their activity regulated at either the transcriptional level, the rate of synthesis/degradation or by specific endogenous inhibitors, such as cystatins, in which this review is focused. Protease inhibitors act both intra- or extracellularly; complexes are formed with their target enzymes in order to maintain an appropriate equilibrium of free enzyme and their complexes. As a matter of fact, proteases and their natural inhibitors may co-exist at different levels of cellular evolution, so that disturbances on the harmony of the normal balance of enzymatic activities of proteases and their natural inhibitors can lead to severe biological effects [55]. Consequently, an understanding of the cellular processes involving control of protein degradation may enable enhancement of the efficiency of treatment for many pathological diseases [56–59]. Besides regulating endogenous protease activity, these natural inhibitors must have a general protective role, since many pathogens use cysteine proteases to evade target organism defense mechanisms, as well as the pathogens themselves may use protease inhibitors to evade the host defense system [55].

Studies on the mechanisms of action and the binding of natural protease inhibitors would provide valuable information for developing drugs to treat distinct diseases [60, 61]. In this sense, the development of cysteine protease inhibitors as potential drug candidates for the treatment of human diseases has considerably increased [54, 62, 63]. The chemical modification of the cysteine thiol group in the active site usually results in a protein with drastically different properties, and the enzyme is essentially inhibited or inactivated. Peptides have been traditionally used to probe specific binding to cysteine proteases, which can then be applied to the design and development of new inhibitors with high membrane permeability, long plasma half-lives, high selectivity for the target protease and good oral availability, among other aspects. Nevertheless, peptidyl inhibitors have their use hampered by undesirable properties, such as susceptibility to protease cleavage and poor pharmacological properties [64, 65]. The use of combinatorial chemistry is an emerging approach to lead discovery that exploits the biological target molecule as a template for the construction of its own inhibitor through self-assembly within the target active site [reviewed in 54, 65].

Papain-like cysteine proteases were shown to be inhibited by different proteins, including propeptide-like inhibitors, thyropins and the general protease inhibitor α_2-macroglobulin [reviewed in 50]. Although usually not considered as typical inhibitors, the N-terminal proregions of the lysosomal cysteine proteases, which are removed during the final maturation step, act as potent reversible inhibitors of the mature enzymes in order to prevent inappropriate activation. Propeptides are competitive inhibitors and possess high selectivity for the enzymes from which they originate. Thyropins contain a cysteine-rich motif and share considerable sequence homology with thyroglobulin type-1 domain found in 11 copies in the prohormone thyroglobulin and in a variety of other proteins from various organisms, including the major histocompatibility complex class II-associated p41 invariant chain, equistatin from the sea anemone *Actinia equina*, the chum salmon egg cysteine protease inhibitor and saxiphilin from the bullfrog *Rana castebiana* [55]. The best known endogenous class of family C1 inhibitors, though, is formed by cystatins, described below.

CYSTATIN SUPERFAMILY

Cystatins (clan IH, family I25 – MEROPS database identifiers) are endogenous protein and tight-binding reversible competitive inhibitors of clan CA, family C1 papain-like cysteine proteases found in a wide range of metazoan and plant taxa, and they are also reported in some lower eukaryotes [66–69]. In this sense, 12 functional cystatins from humans have already been described. Cystatins usually function as a protection against lysosomal proteases (mainly cathepsins B, H and L) released either occasionally at normal cell death or phagocyte degranulation, or by proliferating cancer cells or by invading organisms, such as parasites [55, 69, 70].

On the basis of primary sequence homology, distribution in the body and physiological roles, the cystatin superfamily is subdivided into three families: stefins (family 1), cystatins (family 2) and kininogens (family 3) (Figure 4) [69–71]. Cystatins are all similarly folded: the elucidation of the tertiary structure of several representative members of the family defined a characteristic "cystatin fold", which suggests that the interactions of all cystatins are the same and it may involve three conserved regions that block the active site of C1 cysteine proteases: the N-terminal glycine-containing region; a central pentamer (Gln^{55}-Xaa-Val-Xaa-Gly^{59}) that stabilizes the complex cystatin-protease; and the C-terminal dimer Pro^{105}-Trp^{106} (or His/Gly^{106} in family 1) [69–71].

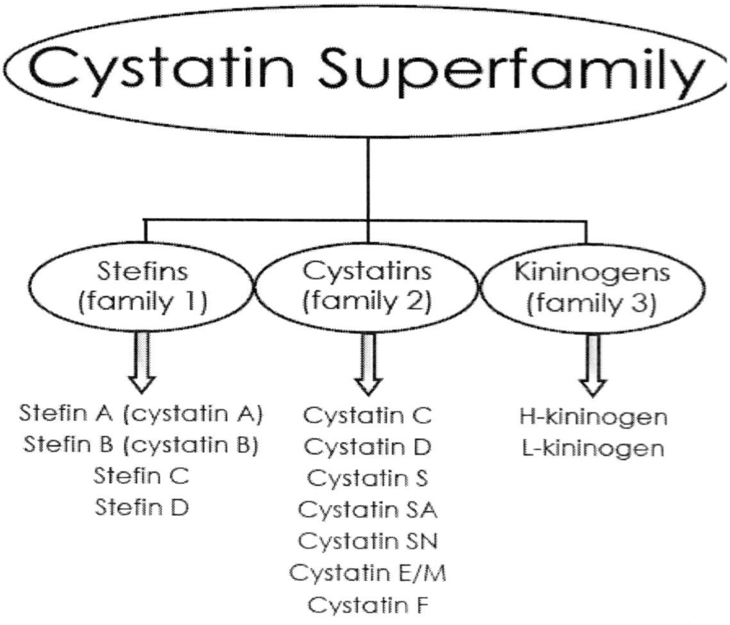

Figure 4. Cystatin superfamily.

Stefins, belonging to family 1, are predominantly intracellular non-glycosylated proteins that lack disulphide bonds. Representatives of this family are mammalian stefin A (cystatin A) and mammalian stefin B (cystatin B), showing extensive sequence identity (>50%), but stefin variants have been already detected, including stefin C in cows and stefin D in pigs. The expression pattern of stefin A is restricted to the skin and some blood cells in mammals, suggesting that its main role is to protect the body against cysteine proteases from pathogens [72]. The bacteriostatic properties of the skin are associated to cystatin A, among other components [73]. Stefin B is widely distributed in the cytosol of almost all cells and tissues in mammals, and probably protects the cells against uncontrolled proteolysis by cathepsins, such as lysosomal cathepsins B, H, L and S. However, alterations in the gene encoding this protein lead to disorders only in a very selective population of cells, suggesting that genetic redundancy of cystatins may largely act to compensate for its function [55]. In this sense, an imbalance between cathepsins and stefins in favor of proteolysis contributes to the development of the invasive cell phenotype and to tumorigenicity or metastasis [74]. In addition, the role of cystatin B in the maintenance of normal neuron structure is well documented: cystatin B-deficient mice present an apoptotic loss of cerebellar granule cells [75].

Family 2 contains cystatins, which are usually found in extracellular spaces at significant high concentrations, may be glycosylated and possess two disulphide bonds at the C-terminus. The family is represented by the non-glycosylated cystatins C, D, S, SA and SN, with more than 50% amino acid sequence identity, as well as the glycosylated cystatins E/M and F. Cystatin C is one of the best characterized members of this family; this protein is produced by a wide variety of tissues, and it is ubiquitous in all extracellular fluids, being the most abundant extracellular cysteine protease inhibitor. As a consequence, cystatin C is described as being able to scavenge any cysteine protease released to body fluids [50]. Nevertheless, cystatin C-deficient mice show no gross pathological abnormalities [55]. A point mutation in the cystatin C gene is responsible for human hereditary cystatin C amyloid angiopathy, which results in abnormal protein aggregation responsible for the pathology; the inhibitory activity of the mutant form remains unaffected, though plasma levels are decreased [76]. Further concerning the cystatin C-related abnormalities, it was found that there was an imbalance between the activity levels of cathepsins B, H, K, L and S and their inhibitors, cystatins B and C, in favor of the former, in a baboon model of bronchopulmonary dysplasia, which may be an alternative to increase protease activity in the development of an inflammatory disease [77]. In addition, decreased inhibitor levels were demonstrated in arteriosclerosis and abdominal aortic aneurysm, inflammatory diseases that involve extensive extracellular matrix degradation and vascular wall remodeling [78]. Cystatin C is also involved in the modulation of tumorigenicity: some studies have demonstrated a direct correlation between high cystatin C levels and improved tumor prognosis [79]. Cathepsin B plays a key role in tumor cell invasion, and although cathepsins are regarded as intracellular proteases in normal cells, they are expressed extracellularly on the surface of tumor cells, where the proteolytic activity may be inhibited by cystatins, particularly cystatin C [79, 80]. Nevertheless, the presence of high levels of cystatin C in the sera of patients with melanoma and colorectal cancers is not explained [81, 82].

Unlike cystatin C, the remaining non-glycosylated cystatins are mainly found in saliva, although they are also found in tears and, for cystatins S, SA and SN, in urine and seminal plasma as well [55, 83]. All these proteins most probably play a protective role against endogenous and/or pathogen-derived proteases: for instance, the level of salivary cystatins is elevated during experimental *Trypanosoma cruzi* infection in rats [84], and they also inhibit cysteine protease activity and growth of *Porphyromonas gingivalis* [85]. In addition, a dramatic increase in cystatin S levels is found when papain is administered to the oral cavity of rats [86].

Cystatins E/M and F have low sequence homology with the remaining family 2 members [87, 88]. Cystatin E/M was identified in the skin epithelium and cystatin F was found in hematopoietic cells and blood cells of the immune system [70, 88]. It is proposed that cystatin E/M may suppress breast cancer tumors, as it is down-regulated in mammary tumor cells [89]. High levels of cystatin F are found intracellularly, probably in lysosomes, in contrast to the lower amounts detected in blood plasma, which may suggest that cystatin F acts as a possible regulator of antigen processing and/or presentation [90]. In addition, cystatin F overexpression is associated to metastasis [55].

While both families 1 and 2 are very stable molecules even at extreme pH and high temperature and are low molecular mass inhibitors (10–15 kDa), kininogens, members of family 3, are larger blood plasma proteins expressed in the liver. There are two glycosylated forms of human kininogens that differ in the length of their C-terminal regions: H-kininogen (120 kDa) and L-kininogen (50–80 kDa), thus designated due to their higher and lower molecular masses, respectively [91]. Both kininogens harbor three tandemly repeated cystatin-like domains in their N-terminus (D1–D3), two of which (D2 and D3) have inhibitory activities. In addition, domain 2 is able to inhibit calpain, which is considered to serve as a protection against this proteolytic activity at accidental intravascular release in pathological states [55]. Together with α_2-macroglobulin, they are the major inhibitors of plasma cysteine proteases belonging to the cathepsin family. Smaller amounts are found in other body fluids as well [92].

Kininogens are also studied due to their correlation to the kallikrein-kinin system [93]. Kallikrein is released by blood coagulation activated factor XII, and it cleaves kininogen to bradykinin. The latter mediates distinct processes, including smooth muscle contraction, vascular permeability and pain production [94]. Nevertheless, the deficiencies of kininogens in humans do not produce pronounced phenotypic effects [55].

Besides these three families described above, proteins containing cystatin domains but lacking cysteine protease inhibitory activity have been identified and thus represent new families [80]. These include histidine-rich glycoprotein (HRG), of unknown physiological function; fetuin-A, which has been shown to be involved in a number of physiological processes; cystatin-related protein (CRP), an androgen-regulated protein expressed in prostate; and Spp24, a bone phosphoprotein. On the other hand, cystatin-related epididymal spermatogenic (CRES) proteins contain only the C-terminal Pro^{105}-Trp^{106} sequence and lack the central Gln^{55}-Xaa-Val-Xaa-Gly^{59} pentamer. It is supposed that more proteins with cystatin-like domains will be identified and their physiological roles described in the future.

Cystatin and Immunity

In general, cystatins function as a protection against lysosomal proteases released occasionally at normal cell death or phagocyte degranulation, or 'intentionally' by proliferating cancer cells or by invading organisms, such as protozoan parasites. Accordingly, cystatins not only have capacity to regulate normal body processes and perhaps cause disease when down-regulated, but may also participate in the defense against microbial infections [70]. Furthermore, Verdot and co-workers [95] reported that cystatins (human stefin B, chicken cystatin and kininogen) potentiated the nitric oxide (NO) production from interferon-gama (IFN-γ)-activated mouse peritoneal macrophages in a concentration-dependent manner. For instance, macrophages activated with IFN-γ and then stimulated for 48 h with IFN-γ (100 IU/ml) plus chicken cystatin (0.1 μM) generated increased amounts of NO (about 8-fold) in comparison with macrophages only activated with IFN-γ. NO induction was due to increased inducible NO synthase (iNOS) protein synthesis, which generates NO via L-arginine pathway, since the potent iNOS inhibitor N^G-monomethyl-L-arginine (L-NMMA) completely abolished the NO production by macrophage cells. NO is important in various physiological processes (e.g., vasodilatation, smooth muscle regulation and neurotransmission) and particularly in the immune response. NO is also involved in several inflammatory diseases and is cytotoxic or cytostatic in a wide range of infections [96, 97]. In turn, NO has inhibitory activity on cysteine proteases, especially those from parasitic protozoa [98–100]. Interestingly, the complexation of the inhibitory site of the cystatins (chicken cystatin, human stefin B and rat kininogen) with the carboxymethylated papain induced an increase of the NO levels identical to that of free cystatins after 48 h of incubation, providing evidence that NO enhancement was unrelated to the inhibitory function of cystatins. A subsequent study of Verdot and co-workers [101] proposed a new relationship between cystatins, cytokines, inflammation and the immune response. In that work, macrophages incubated with chicken cystatin alone or with IFN-γ plus chicken cystatin produced increased amounts of both tumor necrosis factor-alpha (TNF-α) and interleukin (IL)-10. The addition of recombinant murine TNF-α alone or in combination with recombinant murine IL-10 mimicked the effect of chicken cystatin, and the addition of neutralizing anti-TNF-α antibody reduced sharply NO production by chicken cystatin/IFN-γ-activated mouse peritoneal macrophages. However, cystatins did not stimulate NO production by unactivated macrophage, indicating that they did not induce the NO synthesis pathway, as did IFN-γ. In fact, they seem to modulate NO production, as do many cytokines produced during infection or inflammation, such as TNF-α [102], IL-10 [103, 104], granulocyte macrophage-colony stimulating factor (GM-CSF) [105], IFN-α/β [106] or bacterial products like lipopolysaccharide (LPS) [107]. Additionally, cystatins modulate cathepsin activities and antigen presentation. On the whole, cystatins and cystatin-like molecules belong to a new category of immunomodulatory molecules [108, 109].

Effects of Cystatin on Trypanosomatids

Cysteine proteases have been incriminated in the growth, replication and virulence of protozoa, including *Trypanosoma* spp. and *Leishmania* spp. [41, 44, 45, 110, 111]. Since cysteine proteases are present in all trypanosomatid parasites (Figure 5) and their catalytic properties can vary considerably from those of the host enzymes they are considered important chemotherapeutic targets [44, 112]. Cysteine proteases are synthesized as proenzymes and during the final maturation step the N-terminal proregion is removed. Once the cysteine proteases are in their mature state, their activity is regulated by different means but the major one is their interaction with inhibitors [50, 113]. In this section, we have aimed to summarize our present knowledge about the inhibitory effects of cystatins directly on cysteine proteases produced by pathogenic trypanosomatids or indirectly benefit by potentiating the host immune cells. In this line of thinking, appropriate T cell-mediated responses are of primary importance in an effective host defense against both leishmaniasis and trypanosomiasis [114, 115].

Leishmania

Infections caused by *Leishmania* are accompanied by parasite-specific immune depression mediated by T cells and macrophages, thereby preventing spontaneous cure and the development of protective immunity in humans [116, 117] and in experimental animal models [118, 119]. Therefore, immunostimulation of the infected host is an effective strategy for circumventing immunosuppression. Studies suggest that the *Leishmania* cysteine proteases may themselves help to ensure a T_H2-like response in BALB/c mice that leads to parasite proliferation [41, 44, 45, 110, 111]. To date, the majority of studies on the cysteine protease activities of *Leishmania* have concerned just a few enzymes, designated CPA and CPB, which are both cathepsin L-like in terms of primary amino acid sequences, and CPC, which is cathepsin B-like. All of these are papain-like and belong to the same group of cysteine proteases (designated Clan CA, Family C1) [110].

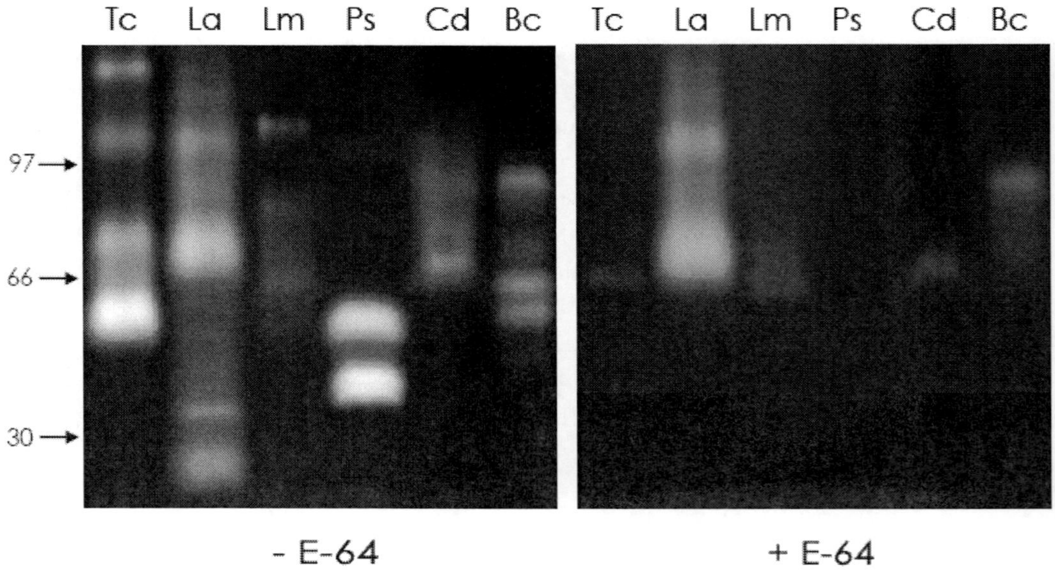

Figure 5. SDS-PAGE containing gelatin as the co-polymerized protein substrate showing the major acidic proteases produced by different trypanosomatids: *Trypanosoma cruzi* (Tc), *Leishmania amazonensis* (La), *Leishmania major* (Lm), *Phytomonas serpens* (Ps), *Crithidia deanei* (Cd) and *Blastocrithidia culicis* (Bc). The gels were incubated in phosphate buffer supplemented with the reducing agent dithiotreitol in the absence or in the presence of E-64, a powerful cysteine protease inhibitor. The numbers on the left represent molecular masses of the protein standards expressed in kilodaltons.

Thus, inhibition of such cysteine proteases might slow or even prevent parasite proliferation and allow the host immune system to function effectively and confer protective immunity by effecting a switch in $CD4^+$ T cell differentiation from T_H2 to T_H1 [120]. Information about the roles and importance of these enzymes in host–parasite interactions was also obtained by the generation of cysteine protease-deficient mutants [121–123] or by comparison between avirulent and virulent strains of the same species, for example as described by *L. amazonensis* [124] and *L. braziliensis* [125].

The pioneer work published by Das and co-workers [126] tested the capacity of cystatin to elicit a T_H1-mediated adaptive response and to prevent and treat infections with *L. donovani*, a lethal T_H2-mediated disease, in BALB/c mice. The authors observed a 6-fold increase in NO production in IFN-γ-activated macrophages when simultaneously were treated with 0.5 μM chicken cystatin in comparison with IFN-γ alone. Moreover, although *L. donovani* infection caused a suppression of NO production in IFN-γ-activated macrophages (1.12 ± 0.17 nmol nitrite/10^6 cells), combined treatment of infected macrophages with cystatin and IFN-γ for 48 h produced 12.18 ± 1.24 nmol nitrite/10^6 cells, up regulating the effector molecule (NO) responsible for antileishmanial activity. In a 45-day BALB/c mouse model of visceral leishmaniasis, complete elimination of spleen parasite burden was achieved by cystatin in synergistic activation with a suboptimal dose of IFN-γ. In contrast to the case with promastigote forms, cystatin and IFN-γ inhibited the growth of amastigotes in macrophages. Although *in vitro* cystatin treatment of macrophages did not induce any NO generation, significantly enhanced amounts of NO were generated by macrophages of cystatin-treated animals. Their splenocytes secreted soluble factors required for the induction of NO biosynthesis, and the increased NO production was paralleled by a concomitant increase in antileishmanial activity. Moreover, splenocyte supernatants treated with anti-IFN-γ or anti-TNF-α antibodies suppressed inducible NO generation, whereas intravenous administration of these anti-cytokine antibodies along with combined therapy reversed protection against infection. mRNA expression and flow cytometric analysis of infected spleen cells suggested that cystatin and IFN-γ treatment, in addition to greatly reducing parasite numbers, resulted in reduced levels of IL-4 but increased levels of IL-12 and iNOS. Not only was this treatment curative when administered 15 days postinfection, but it also imparted resistance to reinfection, since the combination regimen conferred long-standing immunity. In this sense, infected BALB/c mice treated with cystatin and IFN-γ were later reinfected intravenously 60 days after primary infection. Spleen parasite burden in the reinfected animals progressed prominently in PBS-treated BALB/c mice, whereas cystatin- and IFN-γ-treated mice were largely resistant. Thus, infected BALB/c mice subjected to combination chemotherapy with cystatin (5 mg/kg/day) and a suboptimal dose of IFN-γ (10^4 U/mouse) acquired protective immunity [126].

Mukherjee and co-workers [127] characterized the NO-stimulatory domain in cystatin with a view to decipher the minimal peptide sequence involved in NO generation to develop an immunopotent biopeptide, since cysteine protease-inhibitory activity and NO-up-regulatory activity correspond to different regions. The authors used three nonoverlapping recombinant proteins spanning the sequences of the N-terminal region (amino acid residues 1 to 28), the intermediate region (amino acid residues 29 to 72, comprising the conserved QLSVG segment, known to play a central role in the cysteine protease-cystatin interaction) and the C-terminal region (amino acid residues 73 to 116) to demonstrate the presence of the up-regulatory activity in the N-terminal end. Using overlapping synthetic peptides derived from this potent region, it was also demonstrated that apart from being able to generate NO with subthreshold amounts of IFN-γ, a 10-mer peptide (amino acid residues 11 to 20 – PVPVDENDEG) was capable in inducing a protective T_H1 response to *L. donovani* infection in susceptible mice, with increased levels of IL-12 and TNF-α, and also resulted in resistance to reinfection. The generation of a natural peptide from cystatin with robust immunomodulatory potential may therefore provide a promising therapeutic agent for macrophage-associated diseases.

The parasitic protozoa *L. donovani* induces an immune-silencing mechanism for its intracellular survival inside the hostile environment of macrophages. Impairment of mitogen-activated protein kinase (MAPK) signaling is one of the major manipulative strategies through which *Leishmania* can promote numerous phagocyte dysfunctions, such as failure to respond to IFN-γ and inhibition of iNOS gene expression [128]. Infection by *L. amazonensis* amastigotes altered phosphorylation of extracellular signal-regulated kinases-1/2 (ERK1/2) in response to LPS [129] and *L. donovani* infection caused inactivation of ERK1/2, which was accompanied by the inhibition of transcription factors elk-1 and c-fos [130]. In both studies, it was suggested that protein tyrosine phosphatases (PTPs) are responsible for the dephosphorylation of ERK1/2. Conversely, for establishing infection, *Leishmania* inhibits CD40-triggered p38 MAPK signaling [131]. Gene expression profiling revealed that *L. donovani* infection markedly up-regulated the expression of three phosphatases: MAPK phosphatase 1 (MKP1), MKP3 and protein phosphatase 2A (PP2A). Inhibition of these phosphatases prior to infection points toward preferential induction of the T_H2 response through deactivation of p38 by MKP1. On the other hand, MKP3 and PP2A might play significant roles in the inhibition of iNOS expression through deactivation of ERK1/2 [132]. As previously reported, cystatin acts as an immunomodulator in activating macrophages for the release of NO and T_H1 cytokines, which play a central role in curing visceral leishmaniasis in an experimental animal model [126]. Kar and co-workers [132] also demonstrated that cystatin switches the PTP/PTK balance in infected macrophages toward PTK with a parallel activation of ERK and down-regulation of MKP1, MKP3 and PP2A expression. Furthermore, inhibition of phosphatases in *L. donovani*-infected BALB/c mice by cystatin shifted the cytokine balance in favor of the host by inducing TNF-α and iNOS expression (T_H1 response) and subsequent elimination of organ parasite burden of infected mice.

Understanding the molecular mechanisms through which cystatin modulates macrophage effector responses will contribute to better define its potential for macrophage-associated diseases, in general. In this sense, curative effect of cystatin on experimental visceral leishmaniasis was clearly associated with strong upregulation of iNOS and was almost completely abrogated in the presence of L-NMMA [132]. The LPS and/or IFN-γ-dependent iNOS induction is mostly controlled by two regulatory regions present in the iNOS promoter, which contains binding sequences for two transcription factors, NF-κB and IFN regulatory factor 1 (IRF-1) [133]. The synergistic role of IFN-γ is due to its ability to induce the expression of IRF-1 and signal transducer and activator of transcription 1 (STAT1), transcriptional complexes that can also bind to the IFN-γ activating sequence and IRF response element sequences, respectively, at the 50-flanking region of the iNOS promoter. Of all the signaling pathways involved in the production of NO in response to IFN-γ, the Janus kinase family (JAK)/STAT and NF-κB signaling pathways seem most important [134]. Moreover, MAPK signaling pathways have been identified as the upstream kinases that induce NF-κB activation through the phosphorylation of its inhibitor, IκBα [135]. The transductional mechanisms underlying this cellular response was investigated in the murine macrophage cell line RAW 264.7 and in the BALB/c mouse model of visceral leishmaniasis [136]. Cystatin synergizes with IFN-γ in inducing ERK1/2 phosphorylation and NF-κB DNA-binding activity. Pretreatment of cells with specific inhibitors of NF-κB or ERK1/2 pathway blocked the cystatin plus IFN-γ-inducible NF-κB activity and markedly reduced the expression of iNOS at both mRNA and protein levels. Silencing of mitogen- and stress-activated protein kinase 1 (MSK1) significantly reduced cystatin-mediated NF-κB-dependent iNOS gene transcription suggesting the involvement of MSK1 activation in ERK1/2 signaling. DNA binding as well as silencing experiments revealed the requirement of IFN-γ-mediated JAK-STAT activation even though cystatin did not modulate this signaling cascade by itself. In the *in vivo* situation, key steps in the activation cascade of NF-κB, including nuclear translocation of NF-κB subunits, IκB phosphorylation and IκB kinase, are all remarkably enhanced in *Leishmania*-infected mice by cystatin. Taken together, these findings suggest that the sequential activation of MAPK-, ERK-, MSK1- and NF-κB-dependent signaling events play pivotal roles in cystatin plus IFN-γ-inducible macrophage NO generation and that STAT-1a is also involved in the regulation of iNOS gene expression [132, 136].

Two recombinant barley cystatins, HvCPI5 and HvCPI6, were tested *in vitro* against promastigotes and intracellular amastigotes of *L. infantum* in the J774 monocytic cell line. Low concentrations (2 μM) from both cystatins were unable to inhibit promastigote replication [137]. Similarly, chicken cystatin at 10 μg/mL (0.06 μM) did not show any antiproliferative effect on *L. donovani* promastigotes [126]. The phytocystatin HvCPI6 did not exhibit any toxicity for J774 cells up to 6 μM, but inhibited the intracellular *L. infantum* amastigote multiplication in a dose-dependent manner, presenting an IC_{50} value of 1.5 μM [137]. Interestingly, in the Ordóñez-Gutiérrez and co-workers study, no exogenous IFN-γ was added to the macrophage cultures and approximately 5 μM almost completely cleared parasite infection in J774 cells, suggesting that phytocystatin may possess distinct ways to potentiate macrophage killing strategies.

Kinins is a general designation for a group of potent vasoactive peptides proteolytically liberated from an internal segment of their plasma protein precursors, high and/or low molecular mass kininogens, that includes bradykinin (BK), lysyl-BK and their metabolites. Kinins were recently recognized as signals alerting the innate immune system [138]. Although classically associated with acute inflammatory responses (e.g. increased blood flow, edema formation, vasodilatation and pain sensations), BK was recently identified as potent stimulators of dendritic cell maturation, an effect coupled to IL-12-driven polarization of T_H1-type responses [139]. Once liberated, kinins act in a paracrine mode, activating a broad range of host cell types (e.g. endothelial cells, epithelial cells, neurons and dendritic cells) through the constitutively expressed B_2 bradykinin receptor (B_2R) [140, 141]. The duration of the effects of kinins on their G-protein coupled receptors is normally limited by the degradative action of several metalloproteases, particularly, angiotensin-converting enzyme (ACE), a di-peptidylpeptidase expressed at high levels on the vascular endothelium. Svenjö and co-workers [142] showed that *L. donovani* and *L. chagasi*, two etiological agents of visceral leishmaniasis, were able to activate the kinin system. Intravital microscopy in the hamster cheek pouch showed that topically applied promastigotes induced macromolecular leakage through postcapillary venules. The parasite-induced leakage through postcapillary venules was drastically enhanced by captopril an inhibitor of (ACE). In parallel, the enhanced microvascular responses were cancelled by HOE-140, an antagonist of the B_2R, or by pre-treatment of promastigotes with the irreversible cysteine protease inhibitor K777 [143], suggested that promastigote-derived cysteine protease participate in the molecular mechanisms underlying kinin system. Analysis of parasite-induced breakdown of high molecular mass kininogens, combined with active site-affinity-labeling with biotin-N-Pip-hF-VSPh, identified 35-40 kDa proteins as kinin-releasing cysteine proteases. It was also revealed that full-fledged B_2R engagement resulted in vigorous increase of *L. chagasi* uptake by resident macrophages. Evidence that inflammatory macrophages treated with HOE-140 became highly susceptible to amastigote outgrowth, assessed 72 h after initial macrophage interaction, further suggests that the kinin/B_2R activation pathway may critically modulate inflammation and innate immunity in visceral leishmaniasis [142].

Trypanosoma cruzi

T. cruzi also expresses a potent papain-like cysteine protease, called cruzipain, cruzain or gp57/51, which is essential for its life cycle as well as acts as a virulence factor when the parasite infects the human host, causing the devastating Chagas' disease [144, 145]. Of relevance, cruzipain is the main antigen recognized by the immune system at infection [146–150]. Because of these properties, cruzipain is a promising target for therapy and it has been demonstrated that synthetic cysteine protease inhibitors strongly limit the parasitic replication both in cell culture, mice and dogs [151–155].

Natural cruzipain is a complex of isoforms, as judged from ionic exchange chromatography, isoelectrofocusing, reversed-phase HPLC and SDS–PAGE in substrate containing gels [150]. This heterogeneity is probably due to both the simultaneous expression of several genes encoding amino acid substitutions and to the presence, in different cruzipain molecules, of high mannose-type, hybrid monoantennary-type, or complex biantennary-type oligosaccharide chains at the only *N*-glycosylation site in the C-terminal [156]. Besides, the presence of non-conservative amino acids substitutions at the C-terminal, change the predicted isoelectric point of the protein and are likely to result in structural variants at this level [157]. This C-terminal domain contains a number of post-translational modifications and is responsible for the immunodominant antigenic character of cruzipain in natural human infections. In addition, this domain is probably the cause of most of the microheterogeneities found in natural cruzipain [158]. It can be assumed that, as in other unicellular and multicellular organisms, a balance exists between cysteine protease(s) and specific inhibitor(s). Chagasin, a tight-binding inhibitor of cruzipain found in *T. cruzi* [159], exhibits no sequence similarity with cystatins (GenBank/EMBL accession number AJ299433) [160], despite its similar size (110 residues). Molecular modeling studies predicted an immunoglobulin-like fold for chagasin, the first protease inhibitor to do so [161], which was essentially confirmed by subsequent nuclear magnetic resonance [162] and crystallographic studies [163, 164]. A hypothetical model of chagasin-cruzipain interaction suggests that chagasin may dock to the cruzipain active site in a similar manner with the conserved NPTTG motif of chagasin forming a loop that is similar to the wedge structures formed at the active sites of papain and cathepsin L by two unrelated inhibitors, stefin and p41, that belong to the cystatin and thyroglobulin families, respectively [165]. Studies of the functional role of chagasin demonstrated that it modulates the endogenous activity of cruzipain, thus indirectly interfering with *T. cruzi* metacyclogenesis (a complex differentiation process from *T. cruzi* epimastigotes, proliferative and non-infectious forms, to metacyclic trypomastigotes, nonproliferative and infective forms) and/or to invade mammalian host cells [166]. The intersection of chagasin and cruzipain trafficking pathways may represent a checkpoint for downstream regulation of proteolysis in *T. cruzi* [159]. Along similar lines, experimental results suggest that chagasin homologues in *L. mexicana* modulate the outcome of host-parasite interactions [167].

The ability of *T. cruzi* proteases to bind to cystatin was employed in order to purify by affinity chromatography distinct cysteine proteases produced by *T. cruzi* cells, including cruzipain. For instance, Duschak and co-workers [168] reported the partial characterization of a subset of atypical cruzipain molecules which did not bind to Concanavalin A (Con A)-Sepharose column, which were presented in different strains of epimastigote forms of *T. cruzi* and represented a 2-4% of total cruzipain. Duschak and co-workers [169] purified a novel cysteine protease, named TcCPmet, released into the culture medium during *T. cruzi* metacyclogenesis induced by stimulation of epimastigotes with *Triatoma infestans* hindgut homogenate. TcCPmet had apparent molecular mass range between 97 and 116 kDa and was not related to cruzipain, presenting distinct different elution pattern on Con A-Sepharose in comparison to cruzipain and, in addition, it was not recognized by anti-cruzipain serum.

The inhibitory properties of two representative members of the cystatin family (chicken cystatin and rat kininogen) [170] with respect to purified cruzipain were assayed using experimental conditions such that neither temperature nor substrate concentration were limiting [171]. Rat kininogen and chicken cystatin both inhibited cruzipain with K_i values lower than 1 nM. Unlike mammalian cathepsins B, H and L and papain, which are all preferentially inhibited by chicken cystatin [170, 172], cruzipain was more strongly inhibited by kininogen, indicating some differences in the mechanism of inhibition. Encouraged by the observation that cruzipain is strongly inhibited by several members of the cystatin superfamily [173, 174], Lalmanac and co-workers [175] designed peptide substrates with intramolecularly quenched fluorescence and peptidyl diazomethane inhibitors from the N-terminal substrate-like segment of cystatin C, and from the conserved QxVxG sequence in cystatins [176–178]. The inhibition of cruzipain by Biot-LVG-CHN_2 (where Biot represents biotinyl) and Biot-Ahx-LVG-CHN_2 (where Ahx represents 6-aminohexanoic acid) was similar to that of unlabeled inhibitor. Biotin labelling of the inhibitor slowed the inhibition of both cathepsin B and cathepsin L. Analysis of the blotted purified proteases revealed marked differences in the accessibility of extravidin-peroxidase conjugate to the protease-bound biotinylated inhibitor. Cruzipain molecules exposed to Biot-LVG-CHN_2 or Biot-Ahx-LVG-CHN_2 were readily identified, but the reaction was much stronger when the enzyme was treated with the spacer-containing inhibitor. In contrast with the parasite enzyme, rat cathepsin B and cathepsin L treated with either Biot-LVG-CHN_2 or Biot-Ahx-LVG-CHN_2 produced no detectable bands. Papain, the archetype of this family of cysteine proteases, was poorly labeled with Biot-LVG-CHN_2, but strong staining was obtained with Biot-Ahx-LVG-CHN_2. The ability of biotinylated peptidyl diazomethane inhibitors, derived from the cystatin sequence, to specifically label cruzipain makes them versatile tools for targeting the parasite protease in mammalian cells infected by *T. cruzi.*

Plasma extravasation is a common endothelium response to tissue injury provoked by pathogens and host protease inhibitors (e.g., α_2-macroglobulins and kininogens) interact with protozoan cysteine proteases in extravascular infection sites, linking inflammation to innate immunity by different mechanisms [179]. For instance, α_2-macroglobulin entrapment of *T. cruzi* cruzipain reduced the activation threshold of cruzipain-specific $CD4^+$ T cells due to facilitated uptake of α_2-macroglobulins–cruzipain complexes by the multiscavenger receptor (CD91) from human monocytes [180].

At first sight, the notion that kininogens would serve as substrates for cruzipain seemed paradoxical, because kininogens bind and inactivate cruzipain [173, 181]. Interestingly, addition of kininogen to the *T. cruzi* lysates before electrophoresis resulted in the appearance of additional bands of proteolytic activity in the 160–190 kDa molecular mass range as previously reported for other trypanosomatids [182]. In that study, the authors described a novel 30 kDa kininogen-binding protease from *T. cruzi.* The mechanism of activity enhancement may involve a structural change in the cysteine protease to a metastable state at low temperatures [182].

Studies of the mechanisms underlying inflammation elicited by *T. cruzi* revealed that kininogens, once bound to glycosaminoglycans, are not able to efficiently inactivate cruzipain via their inhibitory cystatin-like domains. Instead, cruzipain readily processes surface-bound kininogens, liberating bioactive kinins that act as paracrine hormones, which vigorously activate host cells through bradykinin receptors, thus stimulating endocytic uptake of the pathogen [183]. Rather than unilaterally enhancing parasite infectivity, the liberated kinins activate innate immunity by potently stimulating dendritic cell maturation via the B_2R [184]. As an attempt to understand this intriguing findings, Scharfstein and co-workers [184] tested the hypothesis that orientation of cystatin-like domains on cell-bound forms of kininogens may somewhat hinder their ability to optimally bind and inactivate cruzipain [184]. This possibility was considered because one of the cell surface binding sites of the kininogens was mapped to the D3 domain, i.e., it overlaps with the cystatin-like domain involved in cysteine protease inhibition [185, 186]. Some of the kinin parental molecules bind to the extracellular matrix in orientations that hamper the cysteine protease inhibitory activity of the cystatin-like domains. This conformation switch favors kinin excision because surface-bound kininogens become vulnerable to proteolytic attack by cruzipain [187].

Long-range effects of kinins on B_2R is attenuated by the swift action of metalloproteases collectively designated as ''kininases'', such as ACE/kininase II [188]. Invasion of human primary umbilical vein endothelial cells (HUVECs) or Chinese hamster ovary (CHO) cells overexpressing the B_2 type of bradykinin receptor (CHO-B_2R) by tissue culture trypomastigotes of *T. cruzi* is subtly modulated by the combined activities of kininogens, kininogenases and kinin-degrading proteases [184]. The presence of captopril, an inhibitor of bradykinin degradation by kininase II, drastically potentiated parasitic invasion of HUVECs and CHO-B_2R, but not of mock-transfected CHO cells, whereas the B_2R antagonist HOE 140 or monoclonal antibody MBK3 to bradykinin blocked these effects. Invasion competence correlated with the parasites' ability to liberate the short-lived kinins from cell-bound kininogen and to elicit vigorous intracellular free calcium ($[Ca^{2+}]_i$) transients through B_2R. Invasion was impaired by membrane-permeable cysteine protease inhibitors such as Z-

(SBz)Cys-Phe-CHN_2 but not by the hydrophilic inhibitor E-64 or cystatin C, suggesting that kinin release is confined to secluded spaces formed by juxtaposition of host cell and parasite plasma membranes [184]. Experimental assays confirmed that purified cruzipain elicits strong $[Ca^{2+}]_i$ responses by generating kinin agonists for B_2R, which is highly expressed in such cells [184]. Importantly, the robust $[Ca^{2+}]_i$ responses which the trypomastigotes elicit through the released kinin agonists stimulated the endocytic uptake of the pathogen [179, 184, 189]. Of note, parasite infectivity was potentiated in cultures supplemented with a well-known ACE inhibitor (captopril), thus suggesting that ACE down-modulates the effects of kinins liberated by cruzipain. Although limited to *in vitro* settings, these results suggested that cruzipain and ACE play opposing roles in the host-parasite balance [179, 184, 189].

Analysis of the dynamics of *T. cruzi*-evoked inflammation revealed that activation of TLR2/neutrophils drives the influx of plasma proteins, including kininogens, into peripheral tissues. Once associated to cell surfaces and / or extracellular matrices, the surface-bound kininogens are cleaved by *T. cruzi* cruzipain. Acting as short-lived 'danger' signals, kinins activate dendritic cells via B_2R, converting them into T_H1 inducers [190]. Using the subcutaneous model of *T. cruzi* infection and intravital microscopy, Schmitz and co-workers [191] described the functional interplay of TLR2, CXCR2 and B_2R in edema development during animal model of Chagas' disease. In this context, repertaxin (CXCR2 antagonist) blocked tissue-culture trypomastigotes (TCT)-induced plasma leakage and leukocyte accumulation in the hamster cheek pouch topically exposed to TCT. Furthermore, the TCT-evoked paw edema in BALB/c mice was blocked by repertaxin or HOE-140 (B_2R antagonist), suggesting that CXCR2 propels the extravascular activation of the kinin/B_2R pathway. *In vitro* studies revealed that TCT induce robust secretion of CXC chemokines by resident macrophages in a TLR2-dependent manner. In contrast, $TLR2^{+/+}$ macrophages stimulated with insect-derived metacyclic trypomastigotes or epimastigotes, which lack the developmentally regulated TLR2 agonist displayed by TCT, failed to secrete keratinocyte-derived chemokine/MIP-2, suggesting that edematogenic inflammation evoked by *T. cruzi* is influenced by parasite developmental stage. Collectively, these results suggest that secretion of CXC chemokines by innate sentinel cells links TLR2-dependent recognition of TCT to the kinin system, a proteolytic web that potently amplifies vascular inflammation and innate immunity through the extravascular release of BK [191]. Furthermore, systemic injection of TCT induced IL-12 production by $CD11c^+$ dendritic cells isolated from $B_2R^{+/+}$ spleen, but not by DCs from $B_2R^{-/-}$ mice. Notably, adoptive transfer of $B_2R^{+/+}$ $CD11c^+$ dendritic cells (intravenously) into $B_2R^{-/-}$ mice rendered them resistant to acute challenge, rescued development of type-1 immunity, and repressed T_H17 responses. Collectively, these results demonstrate that activation of B_2R, a dendritic cell sensor of endogenous maturation signals, is critically required for development of acquired resistance to *T. cruzi* infection [192].

African Trypanosomes

African trypanosomes also contain cysteine protease activities (called as trypanopains), which are lysosomally located and, as such, are likely to be centrally involved in intracellular digestive and catabolic proteolysis [193]. Additionally, these enzymes are proposed to help the parasite escape opsonisation by degrading internalized antibody-variant surface glycoprotein complexes [194]. Trypanopain released into the host bloodstream has also been proposed to contribute to pathogenesis more directly by degrading various host proteins. Also, since trypanotolerant cattle infected with *T. congolense* produce antibodies to trypanopain-Tc, while susceptible cattle do not, immune targeting of the enzyme may help protect infected hosts [195]. The apparent molecular masses of the cysteine protease from *T. brucei* (trypanopain-Tb) and *T. congolense* (trypanopain-Tc) are 28 and 32 kDa, respectively, on fibrinogen-containing sodium dodecyl sulfate-polyacrylamide gels (SDS-PAGE). However, the molecular masses of crude trypanopain preparations have proved to be variable depending on the methodology employed [196, 197]. Some of this variability may be caused by the presence of molecules derived from host plasma or serum in the parasite preparations. Indeed, the pattern obtained varies according to the animal species from which the serum is obtained [198]. For instance, two additional bands of trypanopain activity (87 and 105 kDa) were observed when serum was added to the whole parasite lysate before fibrinogen-SDS-PAGE [199]. Formation of the 87 and 105 kDa bands was frequently accompanied by a reduction in the intensity of the 28 kDa activity, suggesting that the extra proteolytic bands were complexes of the 28 kDa trypanopain-Tb and a molecule from rat serum called rat trypanopain modulator (rTM). The physical properties of rTM resemble those of the kininogen family of cysteine protease inhibitors. This resemblance between rTM and kininogens was confirmed by the positive immunoreactivity between anti-(human low-molecular-mass kininogen) antibody and rTM as well as anti-rTM antibody and human low-molecular-mass kininogen. Furthermore, commercial preparations of human-low-molecular-mass kininogen and chicken egg white cystatin mimicked rTM by forming extra bands of proteolytic activity in the presence of trypanopain-Tb [199]. Interaction between rTM and trypanopains appears to reduce the pH optimum of the enzyme activity from about pH 5.3 to about pH 4.8 [197]. This lower value is close to the intralysosomal pH (4.2) reported for an African trypanosome [200] and may result in enhanced lysosomal proteolytic activity *in vivo*. Equally, the formation of similar additional bands of high molecular mass activity was observed when rTM was added to lysates of bloodstream forms of *T. congolense* as well as metacyclic promastigote forms of either *L. donovani* or *L. major* [199]. The activation or stabilization of the trypanosomatid cysteine proteases by mammalian kininogen-like cysteine protease inhibitors may have profound implications with respect to disease pathology as a consequence of the inappropriate control of these proteases when they are released into the hot bloodstream and tissues [201]. In a posterior work, Troeberg and co-workers [202] purified the trypanopain-Tb from *T. brucei brucei* parasite lysate and then studied its interaction with mammalian cysteine protease inhibitors. The authors described that trypanopain-Tb was powerfully inhibited by sheep stefin B (K_i= 0.004 μM), human stefin A (K_i= 0.045 μM), human cystatin C (K_i= 0.001 μM) and human L-kininogen (K_i= 0.0035 μM), predicting that trypanopain-Tb is likely to be effectively controlled by these inhibitors if released into the host bloodstream. Similarly, purified trypanopain-Tc from *T. congolence* lysate was fully inhibited by cystatin at 40 μM [200]. Pike and co-workers [203] reported a similar apparent contradiction in the interaction between a cysteine protease and its putative inhibitor. While purified stefin B inhibited purified cathepsin L as expected, active covalent complexes of the enzyme and inhibitor were shown to form under certain circumstances. An unidentified factor, which may be absent *in vitro* studies using purified components, may contribute to the formation of these active complexes *in vivo*. It is possible that this putative factor modifies the interaction between trypanopain-Tb and L-kininogen *in vivo*, resulting in the formation of active complexes as reported by Lonsdale-Eccles and co-workers [199]. Such an interaction may therefore interfere with effective control of trypanopain in the host bloodstream [202].

Phytomonas

The *Phytomonas* genus includes parasites that circulate in two distinct environments: plants and insects. Phytomonads are found in plants of great economical importance including cashew, coffee, cassava, coconut and oil palms, infecting also edible fruits such as tomato, orange, guava, grape, star fruit and many others [4, 204]. *P. serpens*, specifically, is a parasite able to infect tomatoes, an edible fruit regularly consumed in natura by humans. *P. serpens* represents a biological model, since these trypanosomatids present metabolic routes similar to those of other trypanosomatids pathogenic to humans and animals. Indeed, *P. serpens* parasites share antigens with pathogenic trypanosomatid species, such as *Leishmania* spp and *T. cruzi* [16, 20–22, 24, 25]. In this sense, it was demonstrated that this tomato parasite has cell-associated antigens recognized by human sera from chagasic patients and is able to confer protective immunity in susceptible BALB/c mice [25, 205]. Our research group reported that *P. serpens* produces two major cell-associated cysteine proteases of 38 (exclusively located at the cytoplasmic region) and 40 kDa (located at the cytoplasm, surface and extracellular medium) [20] with biochemical and immunological features similar to cruzipain [20, 22, 24, 25]. Together, these results point to the use of *P. serpens* antigens as an alternative vaccination approach to *T. cruzi* infection, in particular, the cysteine proteases (cruzipain-like molecules) produced by this phytoflagellate trypanosomatid. Cystatin was added to replicating *P. serpens* promastigote forms as a single dose (10 μM), promoting a significant reduction on the cellular growth rate by approximately 50% after 48-96 h of incubation [20]. Cystatin had a reversible effect on *P. serpens* growth since parasites that were transferred to a drug-free fresh medium were still capable of normal development. Alterations on the cell morphology, such as the flagellates becoming short and round, were observed after treatment with different cysteine protease inhibitors including cystatin [20]. Ashall and co-workers [206] and Troeberg and co-workers [207] reported a similar phenomenon and postulated that it indicates osmotic stress caused by protease inhibition, although such morphological changes could also be the consequence of disruptions of the intracellular scaffolding of the target proteins [208]. Therefore, the killing activity of the cysteine protease inhibitors suggests that cysteine proteases are required for the viability of plant trypanosomatid flagellates. Moreover, cystatin and anti-cruzipain antibodies inhibited the interaction of *P. serpens* with explanted salivary glands of the phytophagous insect *Oncopeltus fasciatus*, suggesting that cysteine proteases are relevant to this crucial process along of the parasite life cycle [20].

Conclusion

Trypanosomatid parasites transmitted by insect bites, such as *T. cruzi*, *T. brucei* and *Leishmania* spp. produce large quantities of papain-like (family C1) lysosomal proteases.

These cysteine-type proteases continue to amaze, both with their diversity of structure and range of biological functions. Studies on them have already revealed exciting new insights into the biology of parasitic protozoa. This multiplicity of functions also reflects the great versatility of cysteine as an active site nucleophile of a protease. It can be predicted that the completion of more parasite genome projects and the application of new sophisticated bioinformatic techniques will result in the identification of many more cysteine proteases. This will provide a wealth of exciting new opportunities for elucidating how the enzymes perform their functions in the parasites and designing ways of exploiting the enzymes with new antiparasite therapies. In this sense, cystatin superfamily-derived inhibitors can also bring us with novel and potent pharmacological compounds able to arrest vital physiological processes of trypanosomatid cells as well as their interaction with host structures.

Acknowledgments

This study was supported by MCT/CNPq (Conselho Nacional de Desenvolvimento Científico e Tecnológico), FAPERJ (Fundação Carlos Chagas Filho de Amparo à Pesquisa do Estado do Rio de Janeiro), CAPES (Coordenação de Aperfeiçoamento de Pessoal de Nível Superior) and Fundação Oswaldo Cruz (FIOCRUZ). We would like to thank Leandro Stefano Sangenito and Livia de Oliveira Santos for helping with the experiments presented in this chapter.

References

[1] Wallace, F. G. (1966). The trypanosomatid parasites of insects and arachnids. *Exp. Parasitol. 18*, 124–193.

[2] McGhee, R. B. and Cosgrove, W. B. (1980). Biology and physiology of the lower Trypanosomatidae. *Microbiol. Rev. 44*, 140–173.

[3] Vickerman, K. (1994). The evolutionary expansion of the trypanosomatid flagellates. *Int. J. Parasitol. 24*, 1317–1331.

[4] Camargo, E. P. (1999). *Phytomonas* and other trypanosomatid parasites of plants and fruit. *Adv. Parasitol. 42*, 29–112.

[5] Laveran, A. and Franchini, G. (1913). Infection expérimentales de mammifères par des flagellés du tube digestif de *Ctenocephalus canis* et a'*Anopheles maculipennis*. *C. R. Acad. Sci. 157*, 744–747.

[6] Dedet, J. P.; Roche, B.; Pratlong, F.; Cales-Quist, D.; Jouannelle, J.; Benichou, J. C. and Huerre, M. (1995). Diffuse cutaneous infection caused by a presumed monoxenous trypanosomatid in a patient infected with HIV. *Trans. R. Soc. Trop. Med. Hyg. 89*, 644–646.

[7] Jiménez, M. I.; López-Vélez, R.; Molina, R.; Cañavate, C. and Alvar, J. (1996). HIV co-infection with a currently non-pathogenic flagellate. *Lancet 347*, 264–265.

[8] Pacheco, R. S.; Marzochi, M. C. A.; Pires, M. Q.; Brito, C. M. M.; Madeira, M. F. and Barbosa-Santos, E. G. O. (1998). Parasite genotypically related to a monoxenous

trypanosomatid of dog's flea causing opportunistic infection in an HIV positive patient. *Mem. Inst. Oswaldo Cruz 93*, 531–537.

[9] Dedet, J. P. and Pratlong, F. (2000). *Leishmania*, *Trypanosoma* and monoxenous trypanosomatids as emerging opportunistic agents. *J. Eukaryot. Microbiol. 47*, 37–39.

[10] Chicharro, C. and Alvar, J. (2003). Lower trypanosomatids in HIV/AIDS patients. *Ann. Trop. Med. Parasitol. 97*, 75–78.

[11] Morio, F.; Reynes, J.; Dollet, M.; Pratlong, F.; Dedet, J. P. and Ravel, C. (2008). Isolation of a protozoan parasite genetically related to the insect trypanosomatid *Herpetomonas samuelpessoai* from a human immunodeficiency virus-positive patient. *J. Clin. Microbiol. 46*, 3845–3847.

[12] Inverso, J. A.; Medina-Acosta, E.; O'Connor, J.; Russell, D. G. and Cross, G. A. (1993). *Crithida fasciculata* contains a transcribed leishmanial surface peptidase (gp63) gene homologue. *Mol. Biochem. Parasitol. 57*, 47–54.

[13] Jaffe, C. L. and Dwyer, D. M. (2003). Extracellular release of the surface metalloprotease, gp63, from *Leishmania* and insect trypanosomatids. *Parasitol. Res. 91*, 229–237.

[14] d'Avila-Levy, C. M.; Souza, R. F.; Gomes, R. C.; Vermelho, A. B. and Branquinha, M. H. (2003). A novel extracellular cysteine proteinase from *Crithidia deanei*. *Arch. Biochem. Biophys. 420*, 1–8.

[15] d'Avila-Levy, C. M.; Araújo, F. M.; Vermelho, A. B.; Soares, R. M. A.; Santos, A. L. S. and Branquinha, M. H. (2005). Proteolytic expression in *Blastocrithidia culicis*: influence of the endosymbiont and similarities with virulence factors of pathogenic trypanosomatids. *Parasitology 130*, 413–420.

[16] d'Avila-Levy, C. M.; Santos, L. O.; Marinho, F. A.; Dias, F. A.; Lopes, A. H. C. S.; Santos, A. L. S. and Branquinha, M. H. (2006). Gp63-like molecules in *Phytomonas serpens*: possible role on the insect interaction. *Curr. Microbiol. 52*, 439–444.

[17] d'Avila-Levy, C. M.; Dias, F. A.; Melo, A. C. N.; Martins, J. L.; Lopes, A. H. C. S.; Santos, A. L. S.; Vermelho, A. B. and Branquinha, M. H. (2006). Insights into the role of gp63-like proteins in lower trypanosomatids. *FEMS Microbiol. Lett. 254*, 149–156.

[18] Nogueira de Melo, A. C.; d'Avila-Levy, C. M.; Dias, F. A.; Armada, J. L. A.; Silva, H. D.; Lopes, A. H. C. S.; Santos, A. L. S.; Branquinha, M. H. and Vermelho, A. B. (2006). Peptidases and gp63-like proteins in *Herpetomonas megaseliae*: possible involvement in the adhesion to the invertebrate host. *Int. J. Parasitol. 36*, 415–422.

[19] Elias, C. G. R.; Pereira, F. M.; Silva, B. A.; Alviano, C. S.; Soares, R. M. A. and Santos, A. L. S. (2006). Leishmanolysin (gp63 metallopeptidase)-like activity extracellularly released by *Herpetomonas samuelpessoai*. *Parasitology 132*, 37–47.

[20] Santos, A. L. S.; d'Avila-Levy, C. M.; Dias, F. A.; Ribeiro, R. O.; Pereira, F. M.; Elias, C. G. R.; Souto-Padrón, T.; Lopes, A. H. C. S.; Alviano, C. S.; Branquinha, M. H. and Soares, R. M. A. (2006). *Phytomonas serpens*: cysteine peptidase inhibitors interfere with growth, ultrastructure and host adhesion. *Int. J. Parasitol. 36*, 47–56.

[21] Santos, A. L. S.; Branquinha, M. H. and d'Avila-Levy, C. M. (2006). The ubiquitous gp63-like metalloprotease from lower trypanosomatids: in the search for a function. *An. Acad. Bras. Ciênc. 78*, 687–714.

[22] Santos, A. L. S.; d'Avila-Levy, C. M.; Elias, C. G. R.; Vermelho, A. B. and Branquinha, M. H. (2007). *Phytomonas serpens* immunological similarities with the human trypanosomatid pathogens. *Microbes Infect. 9*, 915–921.

[23] d'Avila-Levy, C. M.; Santos, L. O.; Marinho, F. A.; Matteoli, F. P.; Lopes, A. H. C. S.; Motta, M. C. M.; Santos, A. L. S. and Branquinha, M. H. (2008). *Crithidia deanei*: influence of parasite gp63 homologue on the interaction of endosymbiont-harboring and aposymbiotic strains with *Aedes aegypti* midgut. *Exp. Parasitol. 118*, 345–353.
[24] Elias, C. G. R.; Pereira, F. M.; Dias, F. A.; Alves e Silva, T. L.; Lopes, A. H. C. S.; d'Avila-Levy, C. M.; Branquinha, M. H. and Santos, A. L. S. (2008). Cysteine peptidases in the tomato trypanosomatid *Phytomonas serpens*: influence of growth conditions, similarities with cruzipain and secretion to the extracellular environment. *Exp. Parasitol. 120*, 343–352.
[25] Elias, C. G. R.; Aor, A. C.; Valle, R. S.; d'Avila-Levy, C. M.; Branquinha, M. H. and Santos, A. L. S. (2009). Cysteine peptidases from *Phytomonas serpens*: biochemical and immunological approaches. *FEMS Immunol. Med. Microbiol. 57*, 247–256.
[26] Matteoli, F. P.; d'Avila-Levy, C. M.; Santos, L. O.; Barbosa, G. M.; Holandino, C.; Branquinha, M. H. and Santos, A. L. S. (2009). Roles of the endosymbiont and leishmanolysin-like molecules expressed by *Crithidia deanei* in the interaction with mammalian fibroblasts. *Exp. Parasitol. 121*, 246–253.
[27] Pereira, F. M.; Bernardo, P. S.; Dias Jr, P. F. F.; Silva, B. A.; Romanos, M. T. V.; d'Avila-Levy, C. M.; Branquinha, M. H. and Santos, A. L. S. (2009). Differential influence of gp63-like molecules in three distinct *Leptomonas* species on the adhesion to insect cells. *Parasitol. Res. 104*, 347–353.
[28] Pereira, F. M.; Elias, C. G. R.; d'Avila-Levy, C. M.; Branquinha, M. H. and Santos, A. L. S. (2009). Cysteine peptidases in *Herpetomonas samuelpessoai* are modulated by temperature and dimethylsulfoxide-triggered differentiation. *Parasitology 136*, 45–54.
[29] Pereira, F. M.; Dias, F. A.; Elias, C. G. R.; d'Avila-Levy, C. M.; Silva, C. S.; Santos-Mallet, J. R.; Branquinha, M. H. and Santos, A. L. S. (2010). Leishmanolysin-like molecules in *Herpetomonas samuelpessoai* mediate hydrolysis of protein substrates and interaction with insect. *Protist 161*, 589–602.
[30] Pereira, F. M.; Santos-Mallet, J. R.; Branquinha, M. H.; d'Avila-Levy, C. M. and Santos, A. L. S. (2010). Influence of leishmanolysin-like molecules of *Herpetomonas samuelpessoai* on the interaction with macrophages. *Microbes Infect. 12*, 1061–1070.
[31] Coura, J. R. and de Castro, S. L. (2002). A critical review on Chagas' disease chemotherapy. *Mem. Inst. Oswaldo Cruz 97*, 3–24
[32] Croft, S. L. and Yardley, V. (2002). Chemotherapy of leishmaniasis. *Curr. Pharmaceut. Des. 8*, 319–342.
[33] Urbina, J. A. (1997) Lipid biosynthesis pathways as chemotherapeutic targets in kinetoplastid parasites. *Parasitology 114*, S91–S99.
[34] Harder, A.; Greif, G. and Haberkorn, A. (2001). Chemotherapeutic approaches to protozoa: kinetoplastida--current level of knowledge and outlook. *Parasitol. Res. 87*, 778–780.
[35] Croft, S. L.; Seifert, K. and Duchêne, M. (2003). Antiprotozoal activities of phospholipid analogues. *Mol. Biochem. Parasitol. 126*, 165–172.
[36] Moore, E. M. and Lockwood, D. N. (2010). Treatment of visceral leishmaniasis. *J. Glob. Infect. Dis. 2*, 151–158.
[37] Shukla, A. K.; Singh, B. K.; Patra, S. and Dubey, V. K. (2010). Rational approaches for drug designing against leishmaniasis. *Appl. Biochem. Biotechnol. 160*, 2208–2218.

[38] Alvar, J.; Aparicio, P.; Aseffa, A.; Den Boer, M.; Cañavate, C.; Dedet, J. P.; Gradoni, L.; Ter Horst, R.; López-Vélez, R. and Moreno, J. (2008). The relationship between leishmaniasis and AIDS: the second 10 years. *Clin. Microbiol. Rev. 21*, 334–359.

[39] Diazgranados, C. A.; Saavedra-Trujillo, C. H.; Mantilla, M.; Valderrama, S. L.; Alquichire, C. and Franco-Paredes, C. (2009). Chagasic encephalitis in HIV patients: common presentation of an evolving epidemiological and clinical association. *Lancet Infect. Dis. 9*, 324–330.

[40] McKerrow, J. H.; Sun, E.; Rosenthal, P. J. and Bouvier, J. (1993). The proteases and pathogenicity of parasitic protozoa. *Annu. Rev. Microbiol. 47*, 821–853.

[41] Mottram, J. C.; Brooks, D. R. and Coombs, G. H. (1998). Roles of cysteine proteinases of trypanosomes and *Leishmania* in host–parasite interactions. *Curr. Opin. Microbiol. 1*, 455–460.

[42] Cazzulo, J. J. (2002). Proteinases of *Trypanosoma cruzi:* patential targets for the chemotherapy of Chagas disease. *Curr. Top. Med. Chem. 2*, 1261–1271.

[43] Yao, C. (2010). Major surface protease of trypanosomatids: one size fits all? *Infect. Immun. 78*, 22-31.

[44] Vermelho, A. B.; Giovanni-de-Simone, S.; d'Avila-Levy, C. M.; Santos, A. L. S.; Melo, A. C. N.; Silva Jr, F. P.; Bon, E. P. S. and Branquinha, M. H. (2007). Trypanosomatidae peptidases: a target for drugs development. *Curr. Enz. Inhib. 3*, 19–48.

[45] Vermelho, A. B.; Branquinha, M. H.; d'Ávila-Levy, C. M.; Santos, A. L. S.; Paraguai de Souza, E. and Nogueira de Melo, A. C. (2010). Biological roles of peptidases in trypanosomatids. *Open Parasitol. J. 4*, 5–23.

[46] Barrett, A.J.; Rawlings, N. D. and O'Brien, E. A. (2001). The MEROPS database as a protease information system. *J. Struct. Biol. 134*, 95–102.

[47] Rawlings, N. D.; Morton, F. R.; Kok, C. Y.; Kong, J. and Barrett, A. J. (2008). MEROPS: the peptidase database. *Nucl. Acids Res. 36*, D320–325.

[48] Rawlings, N. D.; Barrett, A. J. and Bateman, A. (2010). MEROPS: the peptidase database. *Nucl. Acids Res. 38*, D227–D233.

[49] Polgar, L. (2004). Catalytic mechanisms of cysteine peptidases. In *Handbook of Proteolytic Enzymes*, 2 edn (Barrett, A.J., Rawlings, N.D. and Woessner, J.F. eds), p.1072–1079. Elsevier, London.

[50] Turk, B.; Turk, D. and Salvesen, G. S. (2002). Regulating cysteine protease activity: essential role of protease inhibitors as guardians and regulators. *Curr. Pharm. Des. 8*, 1623–1637.

[51] Donkor, I. O. (2000). A survey of calpain inhibitors. *Curr. Med. Chem. 7*, 1171–1188.

[52] Dickinson, D. P. (2002). Cysteine peptidases of mammals: their biological roles and potential effects in the oral cavity and other tissues in health and disease. *Crit. Rev. Oral Biol. Med. 13*, 238–275.

[53] Ray, S. K. and Banik, N. L. (2003). Calpain and its involvement in the pathophysiology of CNS injuries and diseases: therapeutic potential of calpain inhibitors for prevention of neurodegeneration. *Curr. Drug Targets CNS Neurol. Disord. 2*, 173–189.

[54] Leung-Toung, R.; Zhao, Y.; Li, W.; Tam, T. F.; Karimiam, K. and Spino, M. (2006). Thiol proteases: inhibitors and potential therapeutic targets. *Curr. Med. Chem. 13*, 547–581.

[55] Dubin, G. (2005). Proteinaceous cysteine protease inhibitors. *Cell. Mol. Life Sci. 62*, 653–669.

[56] Palsdottir, A.; Abrahamson, M.; Thorsteinsson, L.; Amason, A.; Olafsson, I.; Grubb, A. and Jensson, O. (1988). Mutation in cystatin C gene causes hereditary brain haemorrhage. *Lancet 2*, 603–604.

[57] Pennacchio, L. A.; Lehesjoki, A. E.; Stone, N. E.; Willour, V. L.; Virtaneva, K.; Miao, J.; D´Amato, E.; Ramirez, L.; Faham, M.; Koshiniemi, M.; Warrington, J. A.; Norio, R.; de la Chapelle, A.; Cox, D. R. and Myers, R. M. (1996). Mutations in the gene encoding cystatin B in progressive myoclonus epilepsy (EPM1). *Science 271*, 1731–1734.

[58] Goll, D. E.; Thompson, V. F.; Li, H.; Wei, W. and Cong, J. (2003). The calpain system. *Physiol. Rev. 83*, 731–801.

[59] Carragher, N. O. (2006). Calpain inhibitors: a therapeutic strategy targeting multiple disease states. *Curr. Pharmac. Des. 12*, 615–638.

[60] Bode, W. and Huber, R. (2000). Structural basis of the endoproteinase-protein inhibitor interaction. *Biochim. Biochim. Acta 1477*, 241–252.

[61] Abbenante, G. and Fairlie, D. P. (2005). Protease inhibitors in the clinic. *Med. Chem. 1*, 71–104.

[62] Schirmeister, T. and Kaeppler, U. (2003). Non-peptidic inhibitors of cysteine proteases. *Mini Rev. Med. Chem. 3*, 361–373.

[63] Turk, B. and Fritz, H. (2003). Vito Turk – 30 years of research on cysteine proteases and their inhibitors. *Biol. Chem. 384*, 833–836.

[64] Leung, D.; Abbenante, G. and Fairlie, D. P. (2000). Protease inhibitors: current status and future prospects. *J. Med. Chem. 43*, 305–341.

[65] Powers, J. C.; Asgian, J. L.; Ekici, O. D. and James, K. E. (2002). Irreversible inhibitors of serine, cysteine, and threonine proteases. *Chem. Rev. 102*, 4639–750.

[66] Margis, R.; Reis, E. M. and Villeret, V. (1998). Structural and phylogenetic relationships among plant and animal cystatins. *Arch. Biochem. Biophys. 359*, 24–30.

[67] El-Halawany, M. S.; Ohkouchi, S.; Shibata, H.; Hitomi, K. and Maki, M. (2004). Identification of cysteine protease inhibitors that belong to cystatin family 1 in the cellular slime mold Dictyostelium discoideum. *Biol. Chem. 385*, 547–550.

[68] Gregory, W. F. and Maizels, R. M. (2008). Cystatins from filarial parasites: evolution, adaptation and function in the host-parasite relationship. *Int. J. Biochem. Cell Biol. 40*, 1389–1398.

[69] Turk, V. and Bode, W. (1991). The cystatins: protein inhibitors of cysteine proteinases. *FEBS Lett. 285*, 213–219.

[70] Abrahamson, M.; Alvarez-Fernandez, M. and Nathanson, C. M. (2003). Cystatins. *Biochem. Soc. Symp. 3*, 179–199.

[71] Rawlings, N. D. and Barrett, A. J. (1990). Evolution of proteins of the cystatin superfamily. *J. Mol. Evol. 30*, 60–71.

[72] Henskens, Y. M.; Veerman, E. C. and Nieuw Amerongen, A. V. (1996). Cystatins in health and disease. *Biol. Chem. 377*, 71–86.

[73] Takahashi, M.; Tezuka, T. and Katunuma, N. (1994). Inhibition of growth and cysteine proteinase activity of *Staphylococcus aureus* V8 by phosphorylated cystatin α in skin cornified envelope. *FEBS Lett. 355*, 275–278.

[74] Zajc, I.; Sever, N.; Bervar, A. and Lah, T. T. (2002). Expression of cysteine peptidase cathepsin L and its inhibitors stefins A and B in relation to tumorigenicity of breast cancer cell lines. *Cancer Lett. 187*, 185–190.

[75] Pennachio, L. A.; Bouley, D. M.; Higgins, K. M.; Scott, M. P.; Noebels, J. L. and Myers, R. M. (1998). Proggressive ataxia, myoclonic epilepsy and cerebellar apoptosis in cystatin B-deficient mice. *Nat. Genet. 20*, 251–258.

[76] Abrahamson, M.; Jonsdottir, S.; Olafsson, L.; Jensson, O. and Grubb, A. (1992). Hereditary cystatin C amyloid angiopathy: identification of the disease-causing mutation and specific diagnosis by polymerase chain reaction based analysis. *Hum. Genet. 89*, 377–380.

[77] Altiok, O.; Yasumatsu, R.; Bingol-Karakoc, G.; Rise, R. J.; Stahlman, M. T.; Swyer, W.; Pierce, R. A.; Bromme, D.; Weber, E. and Cataltepe, S. (2006). Imbalance between cysteine proteases and inhibitors in a baboon model of bronchopulmonary dysplasia. *Am. J. Respir. Crit. Care Med. 173*, 318–326.

[78] Shi, G. P.; Sukhova, G. K.; Grubb, A.; Ducharme, A.; Rhode, L. H.; Lee, R. T., Ridker, P. M.; Libby, P. and Chapman, H. A. (1999). Cystatin C deficiency in human atherosclerosis and aortic aneurysms. *J. Clin. Invest. 104*, 1191–1197.

[79] Kos, J.; Werle, B.; Lah, T. and Brünner, N. (2000). Cysteine proteinases and their inhibitors in extracellular fluids: markers for diagnosis and prognosis in cancer. *Int. J. Biol. Markers 15*, 84–89.

[80] Ochieng, J. and Chaudhuri, G. (2010). Cystatin superfamily. *J. Health Care Poor Underserved 21*, 51–70.

[81] Kos, J.; Stabuc, B.; Schweiger, A.; Krasovec, M.; Cimerman, N.; Kopitar-Jerala ,N. and Vrhovec, I. (1997). Cathepsins B, H and L and their inhibitors stefin A and cystatin C in sera of melanoma patients. *Clin. Cancer Res. 3*, 1815–1822.

[82] Kos, J.; Krasovec, M.; Cimerman, N.; Nielsen, H. J.; Christensen, I. J. and Brünner, N. (2000). Cysteine proteinase inhibitors stefin A, stefin B and cystatin C in sera from patients with colorectal cancer: relation to prognosis. *Clin. Cancer Res. 6*, 505–511.

[83] Freije, J. P.; Balbin, M.; Abrahamson, M.; Velasco, G.; Dalboge, H.; Grubb, A. and López-Otin, C. (1993). Human cystatin D. cDNA cloning, characterization of the Escherichia coli expressed inhibitor, and identification of the native protein in saliva. *J. Biol. Chem. 268*, 15737–15744.

[84] Alves, J. B.; Alves, M. S. and Naito, Y. (1994). Induction of synthesis of the rat cystatin S protein by the submandibular gland during the acute phase of experimental Chagas disease. *Mem. Inst. Oswaldo Cruz 89*, 81–85.

[85] Blankenvoorde, M. F.; Henskens, Y. M.; van't Hof, W.; Veerman, E. C. and Nieuw Amerongen, A. V. (1996). Inhibition of growth and cysteine proteinase activity of *Porphyromonas gingivalis* by human salivary cystatin S and chicken cystatin. *Biol. Chem. 377*, 847–850.

[86] Naito, Y.; Suzuki, I. and Hasegawa, S. (1992). Induction of cystatin S in rat submandibular glands by papain. *Comp. Biochem. Physiol. 102*, 861–865.

[87] Sotiropoulou, G.; Anisowicz, A. and Sager, R. (1997). Identification, cloning and characterization of cystatin M, a novel cysteine proteinase inhibitor. *J. Biol. Chem. 272*, 903–910.

[88] Ni, J.; Fernandez, M. A.; Danielsson, L.; Chillakuru, R. A.; Zhang, J.; Grubb, A.; Su, J.; Gentz, R. and Abrahamson, M. (1998). Cystatin F is a glycosylated human low molecular weight cysteine proteinase inhibitor. *J. Biol. Chem. 273*, 24797–24804.
[89] Zhang J.; Shridhar, R.; Dai, Q.; Song, J.; Barlow, S. C.; Yin, L.; Sloane, B. F.; Miller, F. R.; Meschonat, C.; Li, B. D.; Abreo, F. and Keppler, D. (2004). Cystatin M: a novel candidate tumor suppressor gene for breast cancer. *Cancer Res. 64*, 6957–6964.
[90] Nathanson, C. M.; Wasselius, J.; Wallin, H. and Abrahamson, M. (2002). Regulated expression and intracellular localization of cystatin F in human U937 cell. *Eur. J. Biochem. 269*, 5502–5511.
[91] DeLa Cadena, R. A. and Colman, R. W. (1991). Structure and functions of human kininogens. *Trends Pharmacol. Sci. 12*, 272–275.
[92] Salvesen, G.; Parkes, C.; Abrahamson, M.; Grubb, A. and Barrett, A. J. (1986). Human low-Mr kininogen contains three copies of a cystatin sequence that are divergent in structure and in inhibitory activity for cysteine proteinases. *Biochem. J. 234*, 429–434.
[93] Schmaier, A. H. (2002). The plasma kallikrein-kinin system counterbalances the rennin-angiotensin system. *J. Clin. Invest. 109*, 1007–1009.
[94] Ueno, A. and Oh-ishi, S. (2003). Roles for the kallikrein-kinin system in inflammatory exudation and pain: lessons from studies on kininogen-deficient rats. *J. Pharmacol. Sci. 93*, 1–20.
[95] Verdot, L.; Lalmanach, G.; Vercruysse, V.; Hartman, S.; Lucius, R.; Hoebeke, J.; Gauthier, F. and Vray, B. (1996). Cystatins up-regulate nitric oxide release from interferon-gamma-activated mouse peritoneal macrophages *J. Biol. Chem. 271*, 28077–28081.
[96] Lowenstein, C. J., Dinerman, J. L. and Snyder, S. H. (1994). Nitric oxide: a physiological messenger. *Ann. Intern. Med. 120*, 227–237.
[97] MacMicking, J.; Xie, Q. W. and Nathan, C. (1997). Nitric oxide and macrophage function. *Annu. Rev. Immunol. 15*, 323–350.
[98] Venturini, G.; Salvati, L.; Muolo, M.; Colasanti, M.; Gradoni, L. and Ascenzi, P. (2000). Nitric oxide inhibits cruzipain, the major papain-like cysteine proteinase from *Trypanosoma cruzi*. *Biochem. Biophys. Res. Commun. 270*, 437–441.
[99] Venturini, G.; Colasanti, M.; Salvati, L.; Gradoni, L. and Ascenzi, P. (2000) Nitric oxide inhibits falcipain, the *Plasmodium falciparum* trophozoite cysteine protease. *Biochem. Biophys. Res. Commun. 267*, 190–193.
[100] Salvati, L.; Mattu, M.; Colasanti, M.; Scalone, A.; Venturini, G.; Gradoni, L. and Ascenzi, P. (2001). NO donors inhibit *Leishmania infantum* cysteine proteinase activity. *Biochim. Biophys. Acta 1545*, 357–366.
[101] Verdot. L.; Lalmanach, G.; Vercruysse, V.; Hoebeke, J.; Gauthier, F. and Vray, B. (1999). Chicken cystatin stimulates nitric oxide release from interferon-gamma-activated mouse peritoneal macrophages via cytokine synthesis. *Eur. J. Biochem. 266*, 1111–1117.
[102] Munoz-Fernandez, M. A.; Fernandez, M. A. and Fresno, M. (1992). Synergism between tumor necrosis factor-α and interferon-γ on macrophage activation for the killing of intracellular *Trypanosoma cruzi* through a nitric oxide-dependent mechanism. *Eur. J. Immunol. 22*, 301–307.

[103] Corradin, S. B.; Fasel, N.; Buchmüller-Rouiller, Y.; Ransijn, A.; Smith, J. and Mauël, J. (1993). Induction of macrophage nitric oxide production by interferon-γ and tumor necrosis factor-α is enhanced by interleukin-10. *Eur. J. Immunol. 23*, 2045–2048.

[104] Chesrown, S. E.; Monnier, J.; Visner, G. and Nick, H. S. (1994). Regulation of inducible nitric oxide synthase mRNA levels by LPS, IFN-γ, TGF-β, and IL-10 in murine macrophage cells lines and rat peritoneal macrophages. *Biochem. Biophys. Res. Comm. 200*, 126–134.

[105] Olivares Fontt, E. and Vray, B. (1995). Relationship between granulocyte macrophage-colony stimulating factor, tumor necrosis factor-α and *Trypanosoma cruzi* infection in murine macrophages. *Parasite Immunol. 17*, 135–141.

[106] Riches, D. W. and Underwood, G. A. (1991). Expression of interferon-beta during the triggering phase of macrophage cytocidal activation. Evidence for an autocrine/paracrine role in the regulation of this state. *J. Biol. Chem. 266*, 24785–24792.

[107] Stuehr, D. J. and Marletta, M. A. (1985). Mammalian nitrate biosynthesis: mouse macrophages produce nitrite and nitrate in response to *Escherichia coli* lipopolysaccharide. *Proc. Natl Acad. Sci. USA 82*, 7738–7742.

[108] Vray, B.; Hartmann, S. and Hoebeke, J. (2002). Immunomodulatory properties of cystatins. *Cell. Mol. Life Sci. 59*, 1503–1512.

[109] Kopitar-Jerala, N. (2006). The role of cystatins in cells of the immune system. *FEBS Lett. 580*, 6295–6301.

[110] Sajid, M. and McKerrow, J. H. (2002). Cysteine proteases of parasitic organisms. *Mol. Biochem. Parasitol. 120*, 1–21.

[111] McKerrow, J. H.; Caffrey, C.; Kelly, B.; Loke, P. and Sajid, M. (2006). Proteases in parasitic diseases. *Annu. Rev. Pathol. 1*, 497–536.

[112] McKerrow, J. H.; Engel, J. C. and Caffrey, C. R. (1999). Cysteine protease inhibitors as chemotherapy for parasitic infections. *Bioorg. Med. Chem. 7*, 639–644.

[113] Turk, B.; Turk, V. and Turk, D. (1997). Structural and functional aspects of papain-like cysteine proteinases and their protein inhibitors. *Biol. Chem. 378*, 141–150.

[114] Sacks, D. and Sher, A. (2002). Evasion of innate immunity by parasitic protozoa. *Nat. Immunol. 3*, 1041–1047.

[115] Ouaissi, A. and Ouaissi, M. (2005). Molecular basis of *Trypanosoma cruzi* and *Leishmania* interaction with their host(s): exploitation of immune and defense mechanisms by the parasite leading to persistence and chronicity, features reminiscent of immune system evasion strategies in cancer diseases. *Arch. Immunol. Ther. Exp. (Warsz) 53*,102–114.

[116] Saha, B.; Das, G.; Vohra, H.; Ganguly, N. K. and Mishra, G. C. (1995). Macrophage-T cell interaction in experimental visceral leishmaniasis: failure to express costimulatory molecules on *Leishmnia*-infected macrophages and its implication in the suppression of cell-mediated immunity. *Eur. J. Immunol. 25*, 2492–2498.

[117] Kemp, M. (1997). Regulator and effector functions of T-cell subsets in human *Leishmania* infections. *APMIS* 68, 1–33.

[118] Wahinya, D. N.; Mbati, P. A.; Jomo, P. M. and Githure, J. I. (1998). Relationship between parasite load and immune responses in early stages of *Leishmania donovani* infection in inbred BALB/c mice. *East Afr. Med. J.* 75, 156–159.

[119] Melby, P. C.; Tryon, V. V.; Chandrasekar, B. and Freeman, G. L. (1998). Cloning of Syrian hamster (*Mesocricetus auratus*) cytokine cDNAs and analysis of cytokine mRNA expression in experimental visceral leishmaniasis. *Infect. Immun. 6*, 2135–2142.

[120] Maekawa, Y.; Himeno, K.; Ishikawa, H.; Hisaeda, H.; Sakai, T.; Dainichi, T.; Asao, T.; Good, R. A. and Katunuma, N. (1998). Switch of CD4+ T cell differentiation from T_H2 to T_H1 by treatment with cathepsin B inhibitor in experimental leishmaniasis. *J. Immunol. 161*, 2120–2127.

[121] Mottram, J. C.; Souza, A. E.; Hutchison, J. E.; Carter, R.; Frame, M. J. and Coombs, G. H. (1996). Evidence from disruption of the lmcpb gene array of *Leishmania mexicana* that cysteine proteinases are virulence factors. *Proc. Nat. Acad. Sci. USA* 93, 6008–6013.

[122] Frame, M. J.; Mottram, J. C. and Coombs, G. H. (2000). Analysis of the roles of cysteine proteinases of *Leishmania mexicana* in the host–parasite interaction. *Parasitology 121*, 367–377.

[123] Denise, H.; McNeil, K.; Brooks, D. R.; Alexander, J.; Coombs, G. H. and Mottram, J. C. (2003). Expression of multiple *CPB* genes encoding cysteine proteases is required for *Leishmania mexicana* virulence *in vivo*. *Infect. Immun. 71*, 3190–3195.

[124] Soares, R. M. A.; Santos, A. L. S.; Bonaldo, M. C.; Andrade, A. F. B.; Alviano, C. S.; Angluster, J. and Goldenberg, S. (2003). *Leishmania (Leishmania) amazonensis*: differential expression of proteinases and cell-surface polypeptides in avirulent and virulent promastigotes. *Exp. Parasitol. 104*, 104–112.

[125] Lima, A. K. C.; Elias, C. G. R.; Souza, J. E. O.; Santos, A. L. S. and Dutra, P. M. L. (2009). Dissimilar peptidase production by avirulent and virulent promastigotes of *Leishmania braziliensis*: inference on the parasite proliferation and interaction with macrophages. *Parasitology 136*, 1179–1191.

[126] Das, L.; Datta, N.; Bandyopathyay, S. and Das, P. K. (2001). Successful therapy of lethal murine visceral leishmaniasis with cystatin involves up-regulation of nitric oxide and a favorable T cell response. *J. Immunol. 166*, 4020–4028.

[127] Mukherjee, S.; Ukil, A. and Das, P.K. (2007). Immunomodulatory peptide from cystatin, a natural cysteine protease inhibitor, against leishmaniasis as a model macrophage disease. *Antimicrob. Agents Chemother. 51*, 1700–1707.

[128] Olivier, M.; Gregory, D. J. and Forget, G. (2005). Subversion mechanisms by which *Leishmania* parasites can escape the host immune response: a signaling point of view. *Clin. Microbiol. Rev. 18*, 293–305.

[129] Martiny, A.; Meyer-Fernandes, J. R.; de Souza, W. and Vannier-Santos, M. A. (1999). Altered tyrosine phosphorylation of ERK1 MAP kinase and other macrophage molecules caused by *Leishmania* amastigotes. *Mol. Biochem. Parasitol. 102*, 1–12.

[130] Ghosh, S.; Bhattacharya, S.; Sirkar, M.; Sa, G. S.; Das, T.; Majumder, D.; Roy, S. and Majumder, S. (2002). *Leishmania donovani* suppresses activated protein 1 and NF-κB activation in host macrophages via ceramide generation: involvement of extracellular signal-regulated kinase. *Infect. Immun. 70*, 6828–6838.

[131] Awasthi, A.; Mathur, R.; Khan, A.; Joshi, B. N.; Jain, N.; Sawant, S.; Boppana, R.; Mitra, D. and Saha, B. (2003). CD40 signaling is impaired in *L. major*-infected macrophages and is rescued by a p38MAPK activator establishing a host-protective memory T cell response. *J. Exp. Med. 197*, 1037–1043.

[132] Kar, S.; Ukil, A.; Sharma, G. and Das, P. K. (2010). MAPK-directed phosphatases preferentially regulate pro- and anti-inflammatory cytokines in experimental visceral leishmaniasis: involvement of distinct protein kinase C isoforms. *J. Leukoc. Biol.* 88, 9–20.

[133] Xie, Q. W.; Whisnant, R. and Nathan, C. (1993). Promoter of the mouse gene encoding calcium-independent nitric oxide synthase confers inducibility by interferon gamma and bacterial lipopolysaccharide. *J. Exp. Med. 177*, 1779–1784.

[134] Bergeron, M. and Olivier, M. (2006). *Trypanosoma cruzi*-mediated IFN-γ-inducible nitric oxide output in macrophages is regulated by iNOS mRNA stability. *J. Immunol. 177*, 6271–6280.

[135] Yang, J.; Lin, Y.; Guo, Z.; Cheng, J.; Huang, J. and Deng, L. (2001). The essential role of MEKK3 in TNF-induced NF-κB activation. *Nat. Immunol. 2*, 620–624.

[136] Kar, S.; Ukil, A. and Das, P. K. (2009). Signaling events leading to the curative effect of cystatin on experimental visceral leishmaniasis: involvement of ERK1/2, NF-kappaB and JAK/STAT pathways. *Eur. J. Immunol.* 39, 741–751.

[137] Ordóñez-Gutiérrez, L.; Martínez, M.; Rubio-Somoza, I.; Diaz, I.; Mendez, S. and Alunda, J. M. (2009). *Leishmania infantum*: antiproliferative effect of recombinant plant cystatins on promastigotes and intracellular amastigotes estimated by direct counting and real-time PCR. *Exp. Parasitol. 123*, 341–346.

[138] Bhoola, K. D.; Figueroa, C. D. and Worthy, K. (1992). Bioregulation of kinins: kallikreins, kininogens, and kininases. *Pharmacol. Rev.* 44, 1–80.

[139] Aliberti, J.; Viola, J. P.; Vieira-de-Abreu, A.; Bozza, P. T.; Sher, A. and Scharfstein, J. (2003). Cutting edge: bradykinin induces IL-12 production by dendritic cells: a danger signal that drives T_H1 polarization. *J. Immunol. 170*, 5349–5353.

[140] Farmer, S. G. and Burch, R. M. (1992). Biochemical and molecular pharmacology of kinin receptors. *Annu. Rev. Pharmacol. Toxicol. 32*, 511–536.

[141] Kozik, A.; Moore, R. B.; Potempa, J.; Imamura, T.; Rapala-Kozik, M. and Travis, J. (1998). A novel mechanism for bradykinin production at inflammatory sites. Diverse effects of a mixture of neutrophil elastase and mast cell tryptase versus tissue and plasma kallikreins on native and oxidized kininogens. *J. Biol. Chem. 273*, 33224–33229.

[142] Svensjö, E.; Batista, P. R.; Brodskyn, C. I.; Silva, R.; Lima, A. P.; Schmitz, V.; Saraiva, E.; Pesquero, J. B.; Mori, M. A.; Müller-Esterl, W. and Scharfstein, J. (2006). Interplay between parasite cysteine proteases and the host kinin system modulates microvascular leakage and macrophage infection by promastigotes of the *Leishmania donovani* complex. *Microbes Infect. 8*, 206–220.

[143] Engel, J. C.; Doyle, P. S.; Hsieh, I. and McKerrow, J. H. (1998). Cysteine protease inhibitors cure an experimental *Trypanosoma cruzi* infection. *J. Exp. Med. 188*, 725–734.

[144] Chagas, C. (2008). A new disease entity in man: a report on etiologic and clinical observations. *Int. J. Epidemiol. 37*, 694–695.

[145] Gürtler, R. E.; Diotaiuti, L. and Kitron, U. (2008). Commentary: Chagas disease: 100 years since discovery and lessons for the future. *Int. J. Epidemiol. 37*, 698–701.

[146] Scharfstein, J.; Schechter, M.; Senna, M.; Peralta, J. M.; Mendonça-Previato, L. and Miles, M. A. (1986). *Trypanosoma cruzi*: characterization and isolation of a 57/51,000

m.w. surface glycoprotein (GP57/51) expressed by epimastigotes and bloodstream trypomastigotes. *J. Immunol. 137*, 1336–1341.

[147] Arnholdt, A. C. and Scharfstein, J. (1991). Immunogenicity of *Trypanosoma cruzi* cysteine proteinase. *Res. Immunol. 142*, 146–151.

[148] Gazzinelli, R. T.; Leme, V. M.; Cancado, J. R.; Gazzinelli, G. and Scharfstein, J. (1990). Identification and partial characterization of *Trypanosoma cruzi* antigens recognized by T cells and immune sera from patients with Chagas' disease. *Infect. Immun. 58*, 1437–1444.

[149] Arnholdt, A. C.; Piuvezam, M. R.; Russo, D. M.; Lima, A. P.; Pedrosa, R. C.; Reed, S. G. and Scharfstein, J. (1993). Analysis and partial epitope mapping of human T cell responses to *Trypanosoma cruzi* cysteinyl proteinase. *J. Immunol. 151*, 3171–3179.

[150] Cazzulo, J. J.; Labriola, C.; Parussini, F.; Duschak, V. G.; Martinez, J. and Franke de Cazzulo, B. M. (1995). Cysteine proteinases in *Trypanosoma cruzi* and other trypanosomatid parasites. *Acta Chim. Slovenica 42*, 409–418.

[151] McGrath, M. E.; Eakin, A. E.; Engel, J. C.; McKerrow, J. H.; Craik, C. S. and Fletterick, R. J. (1995). The crystal structure of cruzain: a therapeutic target for Chagas' disease. *J. Mol. Biol. 247*, 251–259.

[152] Huang, L.; Brinen, L. S. and Ellman, J. A. (2003). Crystal structures of reversible ketonebased inhibitors of the cysteine protease cruzain. *Bioorg. Med. Chem. 11*, 21–29.

[153] Barr, S. C., Warner, K. L.; Kornreic, B. G.; Piscitelli, J.; Wolfe, A.; Benet, L. and McKerrow, J. H. (2005). A cysteine protease inhibitor protects dogs from cardiac damage during infection by *Trypanosoma cruzi*. *Antimicrob. Agents Chemother. 49*, 5160–5161.

[154] Doyle, P. S.; Zhou, Y. M.; Engel, J. C. and McKerrow, J. H. (2007). A cysteine protease inhibitor cures Chagas' disease in an immunodeficient-mouse model of infection. *Antimicrob. Agents Chemother. 51*, 3932–3939.

[155] Mott, B. T.; Ferreira, R. S.; Simeonov, A.; Jadhav, A.; Ang, K. K. H.; Leister, W.; Shen, M.; Silveira, J. T.; Doyle, P. S.; Arkin, M. R.; McKerrow, J. H.; Inglese, J.; Austin, C. P.; Thomas, C. J.; Shoichet, B. K. and Maloney, D. J. (2010) Identification and optimization of inhibitors of trypanosomal cysteine proteases: cruzain, rhodesain, and TbCatB. *J. Med. Chem. 53*, 52–60.

[156] Parodi, A. J.; Labriola, C. and Cazzulo, J. J. (1995). The presence of complex-type oligosaccharides at the C-terminal domain glycosylation site of some molecules of cruzipain. *Mol. Biochem. Parasitol. 69*, 247–255.

[157] Martinez, J.; Henriksson, J.; Ridaker, M.; Pettersson, U. and Cazzulo, J. J. (1998). Polymorphisms of the genes encoding cruzipain, the major cysteine proteinase from *Trypanosoma cruzi*, in the region encoding the C-terminal domain. *FEMS Microbiol. Lett. 159*, 35–39.

[158] Cazzulo, J. J.; Stoka, V. and Turk, V. (1997). Cruzipain, the major cysteine proteinase from the protozoan parasite *Trypanosoma cruzi*. *Biol. Chem. 378*, 1–10.

[159] Monteiro, A. C. S.; Abrahamson, M.; Lima, A. P. C. A.; Vannier-Santos, M. A. and Scharfstein, J. (2001). Identification, characterization and localization of chagasin, a tight-binding cysteine proteases inhibitor in *Trypanosoma cruzi*. *J. Cell Sci. 114*, 3933–3942.

[160] Benson, D. A.; Karsch-Mizrachi, I.; Lipman, D. J.; Ostell, J. and Wheeler, D. L. (2006). GenBank. *Nucleic Acids Res. 34*, 16–20.

[161] Rigden, D. J.; Monteiro, A. C. and Grossi de Sa, M. F. (2001). The protease inhibitor chagasin of *Trypanosoma cruzi* adopts an immunoglobulin-type fold and may have arisen by horizontal gene transfer. *FEBS Lett. 504*, 41–44.
[162] Salmon, D.; do Aido-Machado, R.; Diehl, A.; Leidert, M.; Schmetzer, O.; Lima, A. A. P.; Scharfstein, J.; Oschkinat, H. and Pires, J. R. (2006). Solution structure and backbone dynamics of the *Trypanosoma cruzi* cysteine protease inhibitor chagasin. *J. Mol. Biol. 357*, 1511–1521.
[163] Figueiredo da Silva, A. A.; Carvalho Vieira, L. D.; Krieger, M. A.; Goldenberg, S.; Tonin Zanchin, N. I. and Guimaraes, B. G. (2007) Crystal structure of chagasin, the endogenous cysteine-protease inhibitor from *Trypanosoma cruzi*. *J. Struct. Biol. 157*, 416–423.
[164] Ljunggren, A.; Redzynia, I.; Alvarez-Fernandez, M.; Abrahamson, M.; Mort, J. S.; Krupa, J. C.; Jaskolski, M. and Bujacz, G. (2007). Crystal structure of the parasite protease inhibitor chagasin in complex with a host target cysteine protease. *J. Mol. Biol. 371*, 1511–1521.
[165] Rigden, D. J.; Mosolov, V. V. and Galperin, M. Y. (2002). Sequence conservation in the chagasin family suggests a common trend in cysteine proteinase binding by unrelated protein inhibitors. *Protein Sci. 11*, 1971–1977.
[166] Santos, C. C.; Sant'anna, C.; Terres, A.; Cunha-e-Silva, N. L.; Scharfstein, J. and Lima, A. P. (2005). Chagasin, the endogenous cysteine-protease inhibitor of *Trypanosoma cruzi*, modulates parasite differentiation and invasion of mammalian cells. *J. Cell Sci. 118*, 901–915.
[167] Besteiro, S.; Coombs, G. H. and Mottram, J. C. (2004). A potential role for ICP, a leishmanial inhibitor of cysteine peptidases, in the interaction between host and parasite. *Mol. Microbiol. 54*, 1224–1236.
[168] Duschak, V. G.; Barboza, M. and Couto, A. S. (2003). *Trypanosoma cruzi*: partial characterization of minor cruzipain isoforms non-adsorbed to Concanavalin A-Sepharose. *Exp. Parasitol. 104*, 122–130.
[169] Duschak, V. G.; Barboza, M.; García, G. A.; Lammel, E. M.; Couto, A. S. and Isola, E. L. (2006). Novel cysteine proteinase in *Trypanosoma cruzi* metacyclogenesis. *Parasitology* 132, 345–355.
[170] Barrett, A. J.; Rawlings, N. D.; Davies, M. E.; Machleidt, W.; Salvesen, G. and Turk, V. (1986). Proteinase inhibitors: cysteine proteinase inhibitors of the cystatin superfamily (Barrett, A. J. and Salvesen, G., eds.), pp. 515–569, Elsevier, Amsterdam.
[171] Lima, A. P.; Scharfstein, J.; Storer, A. C. and Ménard, R. (1992). Temperature-dependent substrate inhibition of the cysteine proteinase (GP57/51) from *Trypanosoma cruzi*. *Mol. Biochem. Parasitol. 56*, 335–338.
[172] Moreau, T.; Gutman, N., El Moujahed, A.; Esnard, F. and Gauthier, F. (1988). Cysteine-proteinase-inhibiting function of T kininogen and of its proteolytic fragments. *Eur. J. Biochem. 173*, 185–190.
[173] Stoka, V.; Nycander, M.; Lenarcic, B.; Labriola, C.; Cazzulo, J. J.; Björk, I. and Turk, V. (1995). Inhibition of cruzipain, the major cysteine proteinase of the protozoan parasite, *Trypanosoma cruzi*, by proteinase inhibitors of the cystatin superfamily. *FEBS Lett. 370*, 101–104.
[174] Serveau, C.; Lalmanach, G.; Juliano, M. A.; Scharfstein, J.; Juliano, L. and Gauthier, F. (1996). Investigation of the substrate specificity of cruzipain, the major cysteine

proteinase of *Trypanosoma cruzi*, through the use of cystatin-derived substrates and inhibitors. *Biochem. J. 313*, 951–956.

[175] Lalmanach, G.; Mayer, R.; Serveau, C.; Scharfstein, J. and Gauthier, F. (1996). Biotin-labelled peptidyl diazomethane inhibitors derived from the substrate-like sequence of cystatin: targeting of the active site of cruzipain, the major cysteine proteinase of *Trypanosoma cruzi*. *Biochem. J. 318*, 395–399.

[176] Lalmanach, G.; Serveau, C.; Brillard-Bourdet, M.; Chagas, J. R.; Juliano, L.; Mayer, R. and Gauthier, F. (1995). Conserved cystatin segments as models for designing specific substrates and inhibitors of cysteine proteinases. *J. Protein Chem. 14*, 645–653.

[177] Gauthier, F.; Moreau, T.; Lalmanach, G.; Brillard-Bourdet, M.; Ferrer-Di Martino, M. and Juliano, L. (1993). A new, sensitive fluorogenic substrate for papain based on the sequence of the cystatin inhibitory site. *Arch. Biochem. Biophys. 306*, 304–308.

[178] Serveau, C.; Juliano, L.; Bernard, P.; Moreau, T.; Mayer, R. and Gauthier, F. (1994). New substrates of papain, based on the conserved sequence of natural inhibitors of the cystatin family. *Biochimie 76*, 153–158

[179] Scharfstein, J. (2006). Parasite cysteine proteinase interactions with alpha 2-macroglobulin or kininogens: differential pathways modulating inflammation and innate immunity in infection by pathogenic trypanosomatids. *Immunobiology 211*, 117–125.

[180] Morrot, A.; Strickland, D. K.; de Lourdes Higuchi, L.; Reis, M.; Pedrosa, R. and Scharfstein, J. (1997). Human T cell responses against the major cysteine proteinase (cruzipain) of *Trypanosoma cruzi*: role of the multifunctional alpha 2-macroglobulin receptor in antigen presentation by monocytes. *Int. Immunol. 9*, 825–834.

[181] Scharfstein, J.; Abrahamson, M.; de Souza, C.B.; Barral, A. and Silva, I. V. (1995). Antigenicity of cystatin-binding proteins from parasitic protozoan. Detection by a proteinase inhibitor based capture immunoassay (PINC-ELISA). *J. Immunol. Methods 182*, 63–72.

[182] Wiser, M. F.; Lonsdale-Eccles, J. D.; D'Alessandro, A. and Grab, D. J. (1997). A cryptic protease activity from *Trypanosoma cruzi* revealed by preincubation with kininogen at low temperatures. *Biochem. Biophys. Res. Commun. 240*, 540–544.

[183] Del Nery, E.; Juliano, M.A.; Lima, A.P.; Scharfstein, J. and Juliano, L. (1997). Kininogenase activity by the major cysteinyl proteinase (cruzipain) from *Trypanosoma cruzi*. *J. Biol. Chem. 272*, 25713–25718.

[184] Scharfstein, J.; Schmitz, V.; Morandi, V.; Capella, M. M.; Lima, A. P.; Morrot, A.; Juliano, L. and Muller-Esterl, W. (2000). Host cell invasion by *Trypanosoma cruzi* is potentiated by activation of bradykinin $B_{(2)}$ receptors. *J. Exp. Med. 192*, 1289–1300.

[185] Hasan, A. A.; Zisman, T. and Schmaier, A. H. (1998). Identification of cytokeratin 1 as a binding protein and presentation receptor for kininogens on endothelial cells. *Proc. Natl. Acad. Sci. USA 95*, 3615–3620.

[186] Herwald, H.; Hasan, A. A.; Godovac-Zimmermann, J.; Schmaier, A. H. and Muller-Esterl, W. (1995). Identification of an endothelial cell binding site on kininogen domain D3. *J. Biol. Chem. 270*, 14634–14642.

[187] Lima, A. P.; Almeida, P. C.; Tersariol, I. L.; Schmitz, V.; Schmaier, A. H.; Juliano, L.; Hirata, I. Y.; Muller-Esterl, W.; Chagas, J. R. and Scharfstein, J. (2002). Heparan sulfate modulates kinin release by *Trypanosoma cruzi* through the activity of cruzipain. *J. Biol. Chem. 277*, 5875–5881.

[188] Ferreira, S. H.; Greene, L. J.; Alabaster, V. A.; Bakhle, Y. S. and Vane. J. R. (1970). Activity of various fractions of bradykinin potentiating factor against angiotensin I converting enzyme. *Nature 225*, 379–380.

[189] Todorov, A. G.; Andrade, D.; Pesquero, J. B.; de Carvalho Araujo, R.; Bader, M.; Stewart, J.; Gera, L.; Muller-Esterl, W.; Morandi, V.; Goldenberg, R. C.; Neto, H. C. and Scharfstein, J. (2003). *Trypanosoma cruzi* induces edematogenic responses in mice and invades cardiomyocytes and endothelial cells *in vitro* by activating distinct kinin receptor (B_1/B_2) subtypes. *FASEB J. 17*, 73–75.

[190] Scharfstein, J.; Schmitz, V.; Svensjö, E.; Granato, A. and Monteiro, A. C. (2007). Kininogens coordinate adaptive immunity through the proteolytic release of bradykinin, an endogenous danger signal driving dendritic cell maturation. *Scand. J. Immunol. 66*, 128–136.

[191] Schmitz, V.; Svensjö, E.; Serra, R. R.; Teixeira, M. M. and Scharfstein, J. (2009). Proteolytic generation of kinins in tissues infected by *Trypanosoma cruzi* depends on CXC chemokine secretion by macrophages activated via Toll-like 2 receptors. *J. Leukoc. Biol. 85*, 1005–1014.

[192] Monteiro, A. C.; Schmitz, V.; Morrot, A.; de Arruda, L. B.; Nagajyothi, F.; Granato, A.; Pesquero, J. B.; Müller-Esterl, W.; Tanowitz, H. B. and Scharfstein, J. (2007). Bradykinin B_2 Receptors of dendritic cells, acting as sensors of kinins proteolytically released by *Trypanosoma cruzi*, are critical for the development of protective type-1 responses. *PLoS Pathog. 3*, e185.

[193] Mbawa, Z. R.; Webster, P. and Lonsdale-Eccles, J. D. (1991). Immunolocalisation of a cysteine protease within the lysosomal system of *Trypanosoma congolense*. *Eur. J. Cell Biol. 56*, 243–250.

[194] Russo, D. C. W.; Williams, D. J. L. and Grab, D. J. (1994). Directional movement of variable surface glycoprotein-antibody complexes in *Trypanosoma brucei*. *Parasitol. Res. 80*, 487–492.

[195] Authié, E.; Duvallet, G.; Robertson, C. and Williams, D. J. L. (1993). Antibody response to a 33 kDa cysteine protease of *Trypanosoma congolense*: relationship to 'trypanotolerance' in cattle. *Parasite Immunol. 15*, 465–474.

[196] Rautenberg, P.; Schädler, R.; Reinwald, E. and Risse, H. J. (1982). Study on a proteolytic enzyme from *Trypanosoma congolense*. Purification and some biochemical properties. *Mol. Cell Biochem.* 47, 151–159.

[197] Lonsdale-Eccles, J. D. and Mpimbaza, G. W. (1986). Thiol-dependent proteases of African trypanosomes. Analysis by electrophoresis in sodium dodecyl sulphate/polyacrylamide gels co-polymerized with fibrinogen. *Eur. J. Biochem. 155*, 469–473.

[198] Lonsdale-Eccles, J. D. and Grab, D. J. (1987). Lysosomal and non-lysosomal peptidyl hydrolases of the bloodstream forms of *Trypanosoma brucei brucei*. *Eur. J. Biochem. 169*, 467–475.

[199] Lonsdale-Eccles, J. D.; Mpimbaza, G. W.; Nkhungulu, Z. R.; Olobo, J.; Smith, L.; Tosomba, O. M. and Grab, D. J. (1995). Trypanosomatid cysteine protease activity may be enhanced by a kininogen-like moiety from host serum. *Biochem. J. 305*, 549–556.

[200] Mbawa, Z. R.; Gumm, I. D.; Shaw, E. and Lonsdale-Eccles, J. D. (1992). Characterisation of a cysteine protease from bloodstream forms of *Trypanosoma congolense*. *Eur. J. Biochem. 204*, 371–379.

[201] Authié, E.; Muteti, D. K.; Mbawa, Z. R.; Lonsdale-Eccles, J. D.; Webster, P. and Wells, C. W. (1992). Identification of a 33-kilodalton immunodominant antigen of *Trypanosoma congolense* as a cysteine protease. *Mol. Biochem. Parasitol. 56*, 103–116.

[202] Troeberg, L.; Pike, R. N.; Morty, R. E.; Berry, R. K.; Coetzer, T. H. and Lonsdale-Eccles, J. D. (1996). Proteases from *Trypanosoma brucei brucei*. Purification, characterisation and interactions with host regulatory molecules. *Eur. J. Biochem. 238*, 728–736.

[203] Pike, R. N.. Coetzer, T. H. T. and Dennison, C. (1992). Proteolytically active complexes of cathepsin L and a cysteine proteinase inhibitor; purification and demonstration of their formation *in vitro*. *Arch. Biochem. Biophys. 294*, 623–629.

[204] Camargo, E.P.; Kastelein, P. and Roitman, I. (1990). Trypanosomatid parasites of plants (*Phytomonas*). *Parasitol. Today 6*, 22–25.

[205] Breganó, J.W.; Picão, R.C.; Graça, V.K.; Menolli, R.A.; Jankevicius, S.I.; Filho, P.P. and Jankevicius, J.V. (2003). *Phytomonas serpens*, a tomato parasite, shares antigens with *Trypanosoma cruzi* that are recognized by human sera and induce protective immunity in mice. *FEMS Immunol. Med. Microbiol. 39*, 257–264.

[206] Ashall, F.; Harris, D.; Roberts, H.; Healy, N. and Shaw, E. (1990). Substrate specificity and inhibitor sensitivity of a trypanosomatid alkaline peptidase. *Biochim. Biophys. Acta 1035*, 293–299.

[207] Troeberg, L.; Morty, R. E.; Pike, R. N.; Lonsdale-Eccles, J. D.; Palmer, J. T.; McKerrow, J. H. and Coetzer, T. H. T. (1999). Cysteine proteinase inhibitors kill cultured bloodstream forms of *Trypanosoma brucei brucei*. *Exp. Parasitol. 91*, 349–355.

[208] Fuchs, H. and Cleveland, J. (1998). A structural scaffolding of intermediate filaments in health and disease. *Science 279*, 514–519.

In: Cystatins: Protease Inhibitors …
Editors: John B. Cohen and Linda P. Ryseck
ISBN: 978-1-61209-343-7

Chapter 3

CALPAIN-LIKE PROTEINS IN TRYPANOSOMATIDS: EFFECTS OF CALPAIN INHIBITORS ON THE PARASITES' PHYSIOLOGY AND MOTIVATIONS FOR THEIR POSSIBLE APPLICATION AS CHEMOTHERAPEUTIC AGENTS

André Luis Souza dos Santos, Claudia Masini d'Avila-Levy and Marta Helena Branquinha

Departamento de Microbiologia Geral, Instituto de Microbiologia Prof. Paulo de Góes (IMPPG), Centro de Ciências da Saúde (CCS), Universidade Federal do Rio de Janeiro (UFRJ), Ilha do Fundão, Rio de Janeiro, RJ, Brazil
Laboratório de Biologia Molecular e Doenças Endêmicas, Instituto Oswaldo Cruz, Fundação Oswaldo Cruz, Rio de Janeiro, RJ, Brazil

ABSTRACT

Calpains are neutral calcium-dependent cysteine proteases that have been extensively studied in mammalians and that exist in two major isoforms, m-calpain and μ-calpain, which require millimolar and micromolar concentrations of calcium ions, respectively, for their activation. Calpains are involved in several physiological events in eukaryotic cells. However, significant activation of calpains can be detected under several pathological conditions including cancer, neurological disorders, spinal cord injury, atherosclerosis, diabetes and cataract. In order to control these human disorders, the scientific community and pharmaceutical industries have developed bioactive compounds with capability to inhibit the calpain activity. Calpain-like molecules are also produced by pathogenic microorganisms, especially the protozoan parasites belonging to the Trypanosomatidae family. The commercial calpain inhibitors have been tested in order to block some crucial events in the trypanosomatid cells as well as their interaction with their hosts. These studies were encouraged by the publication of the complete genome sequences of three human pathogenic trypanosomatids, *Trypanosoma brucei, Trypanosoma cruzi* and *Leishmania major*, which allowed several *in silico* analyses that in turn directed the identification of numerous genes with interesting chemotherapeutic characteristics. In this sense, a large family of calpain-related proteins was described: 12 genes were identified in *T. brucei*, 15 in *T. cruzi* and 17 in *L. major*. It is interesting to note that, with few exceptions, most organisms outside the animal kingdom have only a single calpain gene, while in the trypanosomatids there is a surprising expansion of genes, which may reflect parasite plasticity to face distinct environments, such as the mammalian host and the insect vector. Calpain-related molecules produced by trypanosomatids are involved in virulence and relevant physiological processes, such as cytoskeleton rearrangement, proliferation, cellular differentiation and interaction with host structures. Interestingly, homologous of calpain have been detected in non-pathogenic trypanosomatids, suggesting a possible conservation of these important molecules during the evolution of the Trypanosomatidae family. Current therapy against both *Trypanosoma* and *Leishmania* is suboptimal due to toxicity of the available therapeutic agents and the emergence of drug resistance. In this sense, parasite cysteine proteases are regarded as a promising target in the therapeutic treatment of trypanosomiasis and leishmaniasis. This review will survey the available information on trypanosomatid calpain-related proteins and prospects for exploitation of this class of cysteine proteases as a novel drug target.

Keywords: Calpain, calpain inhibitors, trypanosomatids, Leishmania, Trypanosoma, development, virulence, chemotherapy

Calpains and Their Inhibitors

Calpains are intracellular, non-lysosomal, neutral calcium-dependent cysteine proteases (belonging to the clan CA, family C2 – MEROPS database – http://merops.sanger.ac.uk), originally studied in mammals. The calpain family contains a group of several well-conserved, ubiquitously expressed and tissue-specific isoforms [1, 2]. The best characterized ones are the so-called μ-calpain and m-calpain, which refer to the requirement for micromolar (3–50 μM Ca^{2+}) and millimolar (0.2–1 mM Ca^{2+}) concentrations of calcium, respectively, to confer catalytic activity. Both isoforms are ubiquitously expressed, being found as heterodimers consisting of a large catalytic subunit (82 kDa in calpain 1, 80 kDa in calpain 2) and a small regulatory subunit (28 kDa, also known as calpain 4). The isoform μ-calpain contains calpain 1, while the large subunit calpain 2 is found in m-calpain, and they share approximately 60% sequence homology. Four domains can be characterized in both calpains. The short and conserved N-terminal sequence corresponds to domain I. Domain II contains the active site, represented by the catalytic triad characteristic of all cysteine proteases, which is composed of Cys-His-Asn residues, and it is highly conserved among calpain family members. Domain III may associate with phospholipids and calcium, and it is supposed to regulate calpain activity by electrostatic interactions with domain II. Finally, domain IV contains five EF-hand motifs, from which the first four are thought to act as calcium-binding sites, while the fifth C-terminal motif is involved in the dimerization with the small subunit. The latter is composed of two domains, the N-terminal glycine-rich domain V and the C-terminal domain VI: while no function has been attributed yet to domain V, domain VI contains five EF-hand motifs that function in a similar way to the five EF-hand motifs found in domain IV [2, 3].

Most studies performed up to date concentrate on μ- and m-calpain, although at least another 13 calpain isoforms have been identified in mammals, based on the homology to domain II. Among these, there are two subgroups, the "typical" (calpains 3, 8, 9, 11–13) and "atypical" (5–7, 10, 14, 15) calpains: the former possess a similar domain structure to calpains 1 and 2, while domain IV is absent in the latter. It remains to be determined how the atypical calpains have their activity regulated by calcium. In addition, these mammalian isoforms may not be associated to the regulatory small subunit and are usually expressed in a tissue-specific manner [4]. Calpain 6 is the only isoform in which the catalytic triad is not conserved: the cysteine residue is replaced by lysine, and so it is suggested that this isoform is not proteolytically active. In addition, several atypical calpain homologues have been described in invertebrates and lower eukaryotes, such as *Drosophila melanogaster*, *Caenorhabditis elegans* and *Saccharomyces cerevisiae*. These contain a cysteine protease domain that shows more similarities to the mammalian calpains than to other cysteine proteases, although their other domains do not necessarily resemble those of conventional calpain large subunit – instead, they must possess unique domains, possibly responsible for any specific functions they may display [1, 5].

Analysis of the three-dimensional crystal structure of m-calpain in the absence of calcium indicated that domain I interact with domain VI of the regulatory subunit, being involved with the heterodimer assembly or regulation. Domain II is spatially subdivided into two sub-domains, IIa and IIb, which disrupt the active site triad in a calcium-free state. The active site Cys^{105} is located in sub-domain IIa, while His^{262} and Asn^{286} are located in sub-domain IIb. Domain III contains an acidic loop that interacts with sub-domain IIb, so that calcium or phospholipid binding to domain III may disrupt the electrostatic interaction between domain III and sub-domain IIb, thereby enabling the latter to fuse with sub-domain IIa to form a functional catalytic site. Therefore, the activation of calpain may depend upon a conformational switch mechanism [4, 6]. Besides the interaction with domain III, calcium also binds to several sites, including the EF-hand motifs located in domains IV and VI. However, mutational studies indicate that calcium can still promote calpain activity in the absence of binding to EF-hand motifs [7]. In addition, domain II can bind two calcium ions, causing the fusion of both catalytic sub-domains [8].

The calcium levels required for activation of calpains *in vitro* are considered generally above the physiological levels found in the cytosol of living cells (<1.0 μM). Consequently, it has been proposed that distinct mechanisms may reduce the calcium concentration required for calpain activation. In this sense, the association of phospholipids with calpains reduces the calcium levels required for the enzyme activation [9]. Calpain translocation to the plasma membrane via domain III interaction with phospholipids may also induce conformational interactions in calpain structure, increasing its sensitivity to calcium [10].

The precise physiological role of calpains remains to be fully established [2]. Calpain-mediated proteolysis is not involved in the complete hydrolysis of proteins, but rather it is considered as a major pathway of post-translational modification of proteins through the limited cleavage of its substrates [11]. In this sense, a broad spectrum of cellular proteins are considered substrates to these enzymes, including cytoskeletal structural proteins, signaling proteins, membrane receptors and transcription factors. Consequently, calpain activity may influence many aspects of the cell physiology including cell adhesion, migration, proliferation, differentiation, signal transduction and apoptosis (Figure 1) [2, 4, 5]. It has been speculated that, under normal conditions, only a small fraction of calpain is active in order to perform essential housekeeping roles. However, significant activation of calpains can be detected under several pathological conditions, often in association with a loss of calcium homeostasis. In this sense, calpain activity has been implicated in the pathology of several human diseases including cancer, neurological disorders, ischemia, spinal cord injury, atherosclerosis, diabetes, cataract formation and muscular dystrophy (Figure 1) [1, 2, 12–14].

The activity of μ- and m-calpain is tightly regulated *in vivo* by calpastatin (clan II, family I27), the unique endogenous protein inhibitor that is exclusively specific for calpains [15]. As pointed out by Carragher [4], since calpastatin is a highly effective calpain inhibitor, understanding the mechanism of its inhibition may provide insights into the development of effective small molecule inhibitors of calpain. The structure of calpastatin is subdivided into five domains: the N-terminal domain L, which has been correlated to the sub-cellular localization of the protein to biological membranes, followed by four repeated inhibitory domains (I–IV), each one containing three sub-domains: A, B and C. These sub-domains are able to bind to and inhibit the proteolytic activity of calpains: while sub-domain A binds to domain IV in calpains, sub-domain C interacts with domain VI of the regulatory sub-unit. The stable calpain/calpastatin complex enables sub-domain B to associate itself to the catalytic domain II of calpains, thus inhibiting the proteolytic activity [16]. While expression of sub-domains A and C alone have no inhibitory activity, sub-domain B is essential for inhibitory activity, although optimal inhibition of calpain by calpastatin requires all three sub-domains [16, 17]. Several calpastatin isoforms have been already identified, through alternative splicing and distinct initiation sites from a single gene, besides proteolytic processing by caspases [18].

Calpastatin can be phosphorylated *in vivo* by both protein kinase A (PKA) and protein kinase C (PKC), which increases the amount of membrane-associated calpastatin and reduces its inhibitory activity against μ- and m-calpain [19, 20]. In addition, PKA-induced phosphorylation triggers its aggregation close to the cell nucleus, which facilitates calpain activation by compartmentalizing the inhibitor away from the cytosol [21, 22]. The raise in the intracellular free calcium level induces dephosphorylation of calpastatin through the activity of a phosphatase, which promotes the cytosolic distribution of the inhibitor [19].

Calpastatin is usually found at much higher levels in the cytosol than calpains. In addition, the four inhibitory domains of calpastatin can inhibit simultaneously up to four calpain molecules. It has been suggested that the sub-cellular translocation and compartmentalization of either calpains or calpastatin may regulate the enzyme activity [23]. Modulation of the balance between protein levels of calpain relative to calpastatin could also regulate calpain activity, and calpain-mediated degradation of calpastatin has been associated with calpain activation *in vivo* [1, 4]. In many of the pathophysiological processes cited above, increased calpain activity is associated with either calpastatin depletion or the presence of anti-calpastatin antibodies [25, 26]. In a similar way, overexpression of calpastatin may prevent many of these pathologies [27, 28].

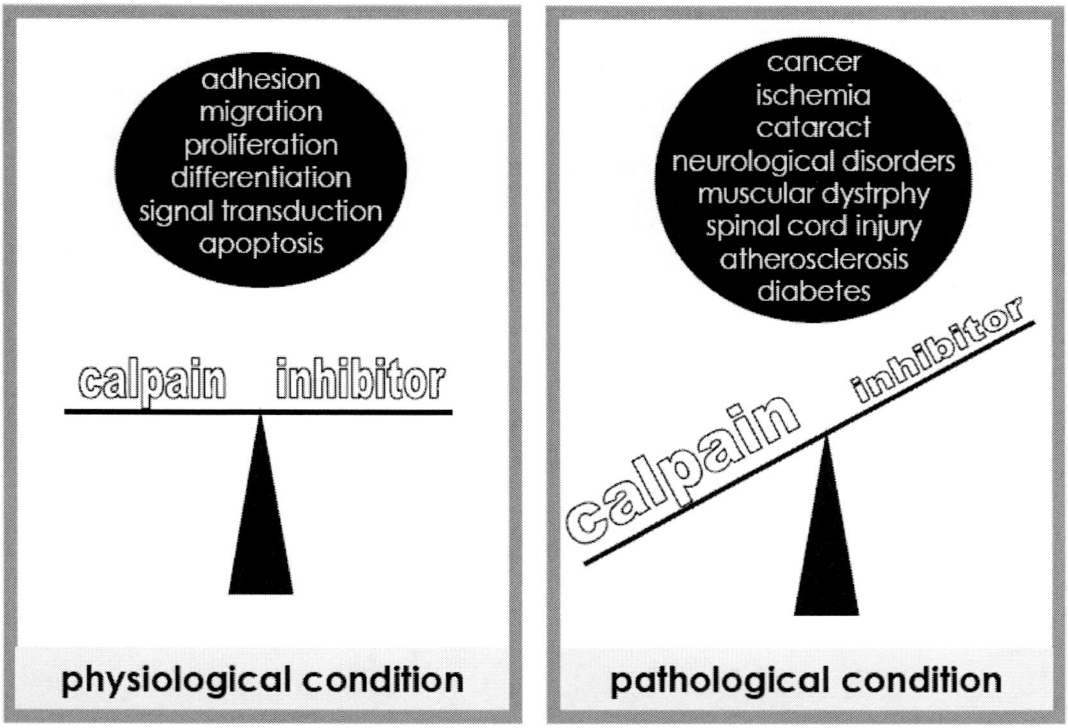

Figure 1. Physiological and pathological events in which calpain molecules have effective participation in humans. The equilibrium between calpain activity and its natural inhibitor promotes crucial events in order to maintain the homeostasis of a eukaryotic cell. On the contrary, when the calpain expression is exacerbated in relation to its inhibitor, several disorders can be triggered.

In keeping with the above information and in view of the multiple functions displayed by calpains, some beginning to be understood but many remaining to be elucidated, the development of specific calpain inhibitors should facilitate the potential to address the thorough characterization of these enzymes and, consequently, the treatment of some important disorders [29]. Initially, the majority of the studies investigating the cellular functions of calpains have employed general protease inhibitors, such as calcium chelator agents like ethylenediamine tetraacetic acid (EDTA) and ethylene glycol tetraacetic acid (EGTA), iodoacetic acid, mersalyl, isocoumarins and diisopropylphosphofluoridate, although many of these lack specificity and potency and suppress the activity of other cysteine proteases, which made them unattractive as biomedical tools for studying calpain-mediated events. Current knowledge about calpain structure has made possible the development of more potent and selective calpain inhibitors [4, 30].

The majority of calpain inhibitors developed to date are peptide analogues, which can be sub-divided in peptidyl epoxides, peptidyl aldehydes and peptidyl α-ketoamides (Figure 2) [4, 30–32]. Peptidyl epoxides form an irreversible thioester bond with the thiol group of the cysteine residue in the active site. These molecules are highly selective for cysteine proteases in general, but do not show specificity against calpains [33]. In this group, the major compound is *trans*-epoxysuccinyl-L-leucylamido-4-guanidino-butane (E-64). Since E-64 is not cell-permeable, modifications in its structure by esterification of its carboxyl group and replacement of the guanidinium group by an alkyl group resulted in the cell-permeable E-64d, which is converted *in vivo* to its active form E-64c by hydrolysis of the ester group [33]. Administration of peptidyl epoxides has been reported to reduce muscular dystrophy, cataract formation and ischemic injury and provided significant neuroprotection following spinal cord injury in distinct animal models of human diseases [reviewed in 4].

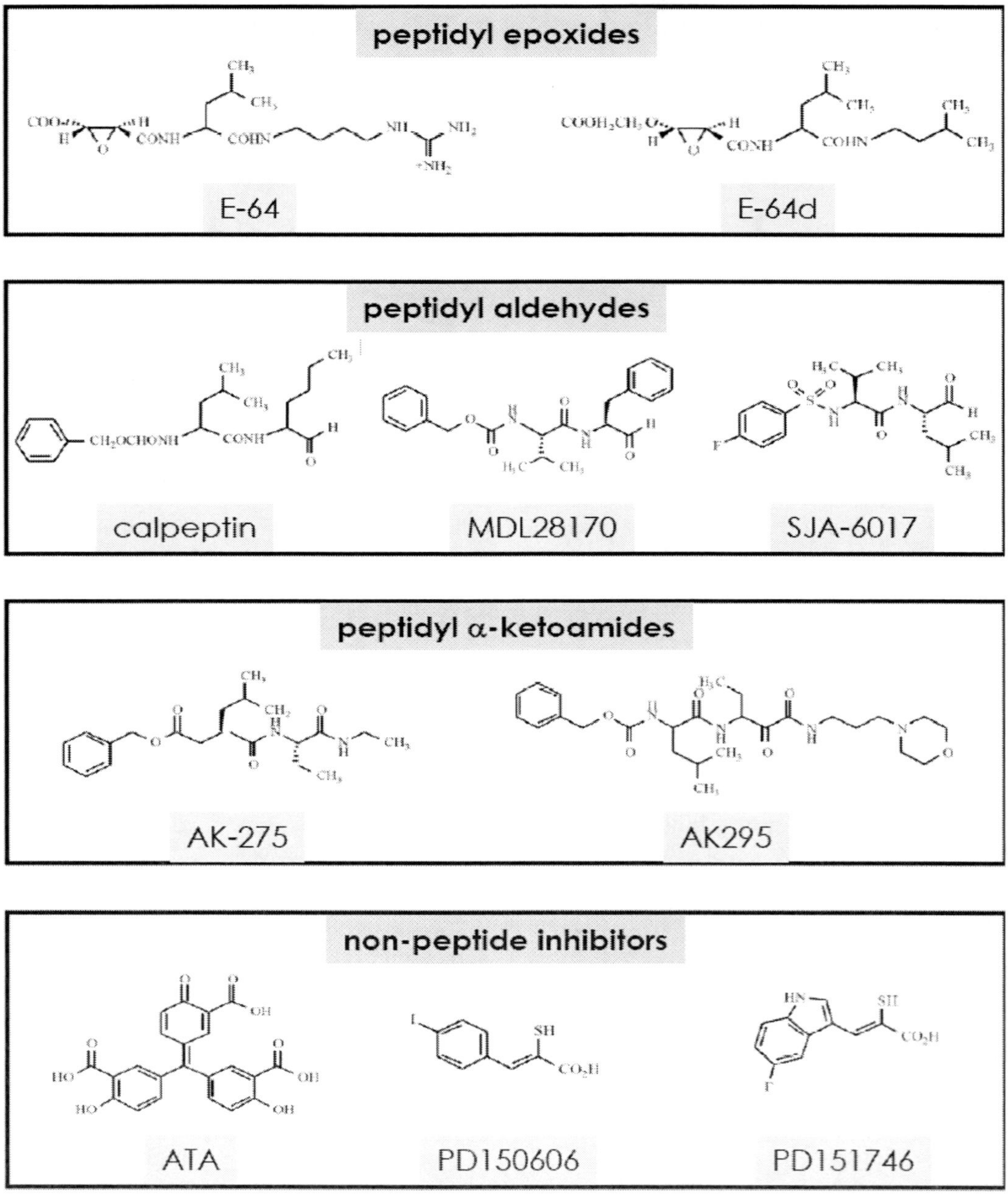

Figure 2. Distinct classes of calpain inhibitors: peptidyl epoxides, peptidyl aldehydes, peptidyl α-ketoamides and non-peptide inhibitors.

Peptidyl aldehydes (Figure 2) are the largest group of calpain inhibitors, being able to react reversibly with the thiol group of the cysteine residue in the active site through the generation of a hemithioacetal intermediate. The first peptidyl aldehyde to be characterized was leupeptinis, a tripeptide derivative with an acetylated α-amino group (N-acetyl-L-leucyl-L-leucyl-L-argininal; Ac-Leu-Leu-Arg-H) [34], which is not cell permeable due to the presence of a positively charged guanidinium group. Modification of its structure enhanced cell permeability, which led to the development of distinct peptidyl aldehydes such as calpeptin (Z-Leu-norleucinal or N-benzyloxycarbonyl-L-leucyl-norleucinal; Z-Leu-Nle-H), which has a benzyloxycarbonyl moiety as the N-terminal lipophilic group [35]. New synthetic and cell permeable dipeptidyl aldehydes were already developed, including MDL28170 (also known as calpain inhibitor III) and SJA-6017 (also known as calpain inhibitor VI) [36, 37], and the tripeptidyl aldehydes calpain inhibitor I and calpain inhibitor II [38]. MDL28170 (N-benzyloxycarbonylvalylphenylalaninal; Z-Val-Phe-CHO) was formed by the incorporation of the benzyloxycarbonyl group into Val-Phe-H, whereas SJA-6017 (N-(4-fluorophenylsulfonyl) -L-valyl-L-leucinal) was synthesized by the linkage of N-(4-fluorophenylsulphonyl) to Val-Leu-H. On the other hand, an acetyl moiety was used as the N-terminal substituent in calpain inhibitor I (Ac-Leul-Leu-Nle-H) and calpain II (Ac-Leu-Leu-Met-H). Both di- and tripeptidyl aldehydes have been tested in different animal models, and among other functions their administration reduced neuronal damage in brain and spinal cord injuries as well as in cerebral ischemia [reviewed in 4]. Many of these compounds, however, still lack specificity and are readily oxidized under physiological conditions because they are aldehydes [30].

Peptidyl α-ketoamides (Figure 2) are also reversible calpain inhibitors, which bind to the cysteine residue in the active site of the enzyme through the presence of an electrophilic, carbonyl-containing group, presenting improved cell permeability. Representative members of this group include AK275 (Z-Leu-aminobutyric acid-CONH-CH_2-CH_3) and AK295 (Z-Leu-aminobutyric acid-CONH-$(CH_2)_3$-morpholine) [39], which have demonstrated protection against neuronal damage in different animal models [4], and a variety of novel derivatives exhibited calpain inhibitory activity, such as calpain inhibitor XII (Z-Leu-Nva-CONH-CH_2-2-pyridyl) [40, 41]. Nevertheless, peptidyl α-ketoamides also lack a great selectivity for calpains relative to other cysteine proteases [30, 42].

A series of non-peptide calpain inhibitors (Figure 2) have also been developed, and may provide greater specificity against calpains [4, 30, 43]. These compounds mainly differ from peptide analogues in their mode of action: non-peptide inhibitors act as reversible, non-competitive inhibitors that are not targeted against the active site of the enzyme, instead they interact with allosteric sites that impair the conformational switch required for calpain activation. This group includes aurintricarboxylic acid [44], quinoline carboximides [31] and α-mercaptoacrylic acids such as PD150606 (3-[4-iodophenyl]-2-mercapto-[Z]-2-propenoic acid) and PD151746 (3-[5-fluoro-3-indolyl]-2-mercapto-[Z]-2-propenoic acid), the latter group showing neuroprotective effects [43]. It is proposed that PD150606 may inhibit calpain activity in a manner similar to calpastatin, since it occupies the same region on calpain domain VI as the sub-domain C of calpastatin [16]. In this sense, it is worth mentioning that a synthetic 27-mer peptide consisting of the central conserved sequence of sub-domain B of domain I of human calpastatin has been demonstrated as a potent and highly selective reversible calpain inhibitor [45, 46].

As reviewed by Carragher [4] and Pietsch and co-workers [47], the future development of both peptide- and non-peptide-based inhibitors of calpain activity, combined with the further elucidation of their binding sites on calpain and mechanism of action, will result in the generation of more potent and specific inhibitors of this class of cysteine protease. This next generation of calpain inhibitors may provide significant therapeutic benefit not only against a variety of human diseases but also against parasitic infections in which calpains may play an important role.

Trypanosomatids and the Significance of Their Proteases

Among protozoa, trypanosomatids constitute a group of uniflagellated parasites characterized by the presence of the kinetoplast, the mitochondrial DNA that is unique in structure, function and mode of replication [48–50]. A fraction of these has evolved to parasitize humans, such as *Trypanosoma brucei,* the etiologic agent of African sleeping sickness; *Trypanosoma cruzi,* the causative agent of Chagas' disease or American trypanosomiasis; and different species of the genus *Leishmania*, responsible for a wide spectrum of clinical manifestations varying from cutaneous and mucocutaneous to visceral leishmaniasis [51–53]. Likewise, some species belonging to the *Phytomonas* genus can induce serious diseases in plants, which indicates the economical importance of these trypanosomatids [54]. Moreover, trypanosomatids parasitizing only insects in their life cycle are very common and are distributed worldwide. This latter group comprises the genera *Blastocrithidia*, *Crithidia*, *Herpetomonas* and *Leptomonas* [49]. The possible zoonotic proclivity of the insect trypanosomatids was first suggested by Laveran and Franchini [55] and exemplified by possible infection of humans, especially immunossupressed individuals [48, 56–60]. Furthermore, insect trypanosomatids possess similar biochemical and/or immunological similarities with pathogenic counterparts including *Trypanosoma* and *Leishmania*, being used as experimental models in several biological, biochemical and molecular studies about the Trypanosomatidae family [48, 49, 61, 62].

Despite the great advances in combating infectious diseases over the past century, these parasites continue to inflict a tremendous social and economic burden on human societies, particularly in the developing world. In this sense, there is a general lack of effective and inexpensive chemotherapeutic agents for treating protozoan diseases. Current therapy is limited to a handful of drugs that suffer from unacceptable toxicity, difficulties of administration and increasing treatment failures, since resistance to these compounds has become a severe problem [63–67]. In view of this scenario, new drugs must be developed. The need for alternatives to treat these infectious diseases has led to the performance of several tests utilizing compounds chosen empirically, or through studies that identify metabolic targets in the parasite, such as proteomic analysis. In addition, the variable protein expression by the different trypanosomatid life cycle stages that alternate between vertebrate and invertebrate hosts as well as by distinct strains needs to be taken into account when new therapeutic targets are proposed.

Proteases of human protozoan pathogens, specifically in trypanosomatids, have attracted the attention of many laboratories because these enzymes are not only involved in "housekeeping" tasks common to many eukaryotes as well as due to their many roles in highly specific functions to the parasites' life cycles, including pathogenesis [68–72]. Consequently, the study of these enzymes has led to the design of novel proteolytic inhibitors against these pathogens, and the simultaneous evaluation of the anti-trypanosomatid activity of molecules originally developed for distinct targets and/or cell models [65, 73, 74]. For instance, K777, a vinyl sulfone protease inhibitor of cruzipain, the major cysteine protease from *Trypanosoma cruzi,* was effective in curing or alleviating the parasitic infection in preclinical proof-of-concept studies and has now entered formal preclinical drug development investigations [65].

The recent publication of the complete genome sequences of *Trypanosoma brucei, Trypanosoma cruzi* and *Leishmania major* led to the discovery of many unexamined cysteine proteases that are predicted to have important biological roles [75]. It is supposed that these cysteine proteases might turn out to be excellent drug targets, and unappreciated relationships can be investigated with tools and techniques developed for nearby cysteine protease homologues [75]. In this sense, a large family of calpain-related proteins has been described in the Trypanosomatidae family. Twelve genes have been identified in *T. brucei*, 15 in *T. cruzi* and 17 in *L. major* [76]. Calpain-related proteins are defined by the 'catalytic' core sequence, however, several of the calpains that differ from the classic calpains lack one or more of the essential catalytic amino acid residues, suggesting functions unrelated to proteolysis. It has been speculated that these proteins are involved in regulatory processes [76, 77], for instance, calpain-6 is involved in microtubules stabilization through a non-catalytic mechanism [78]. It is interesting to note that with few exceptions, most organisms outside the animal kingdom have only a single calpain gene, while in the Trypanosomatidae family, there is a surprising expansion of genes, which may reflect parasite plasticity to face distinct environments, such as the mammalian/plant host and the insect vector [76, 79]. These data reinforce the exploitation of cysteine protease inhibitors, including calpain inhibitors, as possible tools to evaluate the role of these molecules in trypanosomatids.

CALPAIN–RELATED MOLECULES IN *TRYPANOSOMA BRUCEI*

Calpain-related proteins were first described in the Trypanosomatidae family in *T. brucei*. A protein named CAP5.5 or TbCALP4.1 was detected in procyclic forms (tse-tse midgut stage) closely associated to the cytoskeleton, and showed similarity to the catalytic region of classical calpains [80]. CAP5.5 has been shown to be both myristoylated and palmitoylated, suggesting a stable interaction with the cell membrane through interactions with the underlying microtubule cytoskeleton as well. Confirming these data, *in vivo* labeling of *T. brucei* CAP5.5 with radioactive fatty acids confirmed the presence of these fatty acid modifications [80]. Similar to other calpain-like proteins, CAP5.5 lacks a typical domain IV and has replaced the catalytically active amino acids cysteine and histidine with serine and tyrosine [80]. Deviations from the classical Cys-His-Asn catalytic triad pattern are observed not only in kinetoplastids, but also in a number of calpains of higher eukaryotes. For example, in the human calpain CAPN6 the triad is changed to Lys-Tyr-Asn [81].

Through RNA interference (RNAi) experiments, it was shown the loss of CAP5.5 protein from the posterior cell-end, organelle mis-positioning giving rise to aberrant cytokinesis, and disorganization of the sub-pellicular microtubules that define trypanosome cell shape [79]. The protein CAP5.5 is expressed only during the insect stage of the life cycle and the differentiation process [82]. How can such a critical protein be stage-specific? The stage-specificity of CAP5.5 expression can be explained by the presence of a paralogue, CAP5.5V, which is required for cell morphogenesis in bloodstream forms of *T. brucei*; RNAi of bloodstream CAP5.5V renders a similar phenotype described for procyclic CAP5.5 [79].

After the characterization of CAP5.5 in *T. brucei* procyclic forms [80], Ersfeld and co-workers [76] categorized calpain-like proteins in *T. brucei* into five groups, based on their structural features. Members of groups 1 and 2 present four domains and are termed calpain-like proteins, being distinguished by their N-terminal domains: while group 1 contains the domain I^K, which is highly conserved in kinetoplastids, group 2 contains the non-conserved domain I^H. Calpain-like proteins from both groups have, with a single exception in *T. brucei*, one or more alteration of essential residues in the catalytic triad residues. Members of group 3 are termed small kinetoplastid calpain-related proteins (SKCRPs), consisting only of the kinetoplastid-exclusive domain I^K. Six SKRCPs were found in *T. brucei*, nine in *T. cruzi* and ten in *L. major*, with an average length of approximately 200 amino acids. This domain showed no similarities to other known proteins, so there is no indication as to its function. Groups 4 and 5 contain highly divergent calpain-like proteins not studied yet [76, 83]. Considering this classification, the expression pattern in distinct life cycle forms of *T. brucei* and cellular localization of selected members of groups 1-3 was performed. Two transcripts, SKCRP5.1 and CALP8.1/CAP5.5V, were differentially expressed in bloodstream stages, while, transcripts SKCRP1.6, SKCRP7.2 and CALP4.1/CAP5.5 were differentially expressed in procyclic forms. Although RNAi of CAP5.5 or CAP5.5V induces the same phenotype, it is not known if sequence differences confer life cycle-specific functions. Three main cellular localizations were described for these calpain-related molecules in *T. brucei*: the flagellum, the cell body and the periphery of the cell body [83].

CALPAIN–RELATED PROTEINS IN *TRYPANOSOMA CRUZI*

The transcriptome of several *T. cruzi* populations that represents distinct steps in parasite *in vitro* metacyclogenesis revealed a gene encoding a calpain-like protein, which was recently named TcCALPx11. This gene was up-regulated in epimastigote forms under nutritional stress [84], a requirement for differentiation to metacyclic trypomastigotes [85]. This protein is a member of group 1, which is the most conserved group of calpain-like proteins in trypanosomatids [76], and its mRNA is 2.5 times more abundant in epimastigotes under nutritional stress than in epimastigotes growing in complex culture medium. As expected due to the molecular results, the 80 kDa protein was shown to be epimastigote-specific, with its expression enhanced about two times after only one hour of nutritional stress, corroborating its role in the metacyclogenesis process. However, this pattern was also observed after any kind of stress (e.g. nutritional, temperature or acidic pH), indicating that this protein might actually have a general role in the *T. cruzi* stress response [84]. The over-production of this protein in transfected cells did not alter the morphology, the growth rates or the differentiation rates.

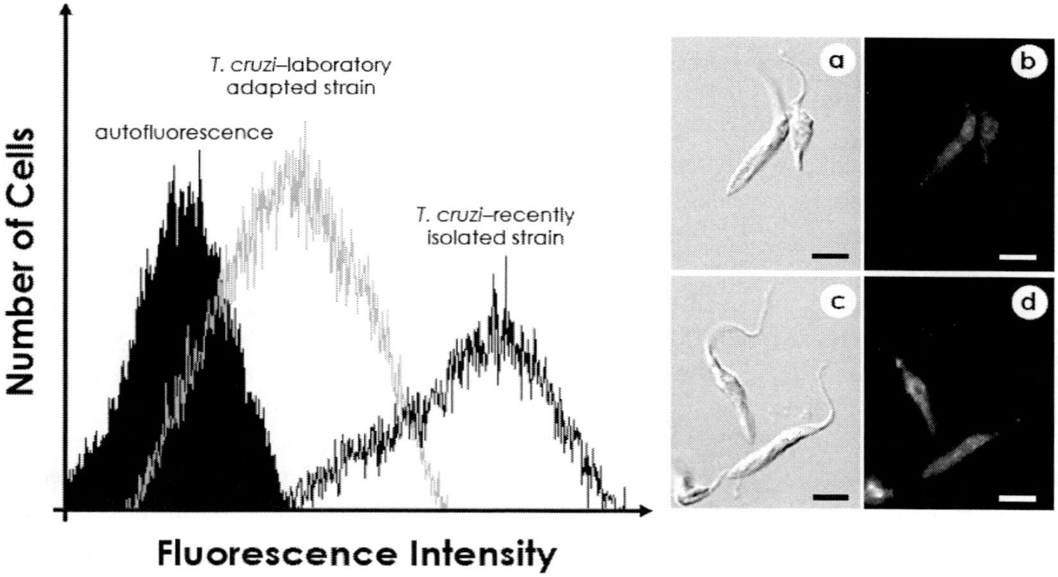

Figure 3. Detection of binding of anti-calpain antibodies to *Trypanosoma cruzi* (Dm28c strain) epimastigote forms by flow cytometric analysis. Epimastigotes regularly kept in brain heart infusion culture medium (*T. cruzi*-laboratory adapted strain) and epimastigote cells obtained from the differentiation of trypomastigotes after a blood passage in mouse (*T. cruzi*-recently isolated strain) were fixed with paraformaldehyde and incubated in the absence (autofluorescence) or in the presence of anti-*Dm*-calpain antibody followed by incubation with secondary antibody conjugated with fluoresceine. For simplicity, only the autofluorescence of recently isolated cells is shown, since the adapted strain presented similar values (data not shown). When treated only with the secondary-fluoresceinated antibody, both strains generated similar curves to that observed in the autofluorescence of cells (not shown). Note that laboratory-adapted strain had significant diminished expression of calpain-like molecules when compared to parasites obtained after passage in mouse. Fluorescence microscopy corroborates the cytometry assay and shows the distribution of calpain-like molecules in both adapted (b) and recently isolated strain (d). The corresponding images under differential interferential contrast observation were reproduced in panels a and c. The bars represent 1 μM.

The bioinformatics analysis gave no indication of putative acylation motifs in TcCALPx11, suggesting that it is not membrane-associated, although it partitioned in the insoluble fraction after detergent extraction, suggesting an association with membranes. Finally, no proteolytic activity was detected either in the soluble recombinant fusion protein or the native protein immunoprecipitated from *T. cruzi* extracts [84].

Sangenito and co-workers [86] reported that *T. cruzi* calpain-like proteins presented a strong cross-reactivity with anti-*Drosophila melanogaster* calpain antibody (anti-*Dm*-calpain) [87] and anti-cytoskeleton-associated protein from *T. brucei* antibody [80], and the labeling was found mainly in intracellular compartments all over the cellular body but not in the flagellum. In addition, the anti-*Dm*-calpain antibody recognized an 80 kDa protein in *T. cruzi* epimastigote lysate. Based on sequence analyses, that study identified four *T. cruzi* calpain-like proteins of around 80 kDa that share the same conserved domain (cd00044) with the fragment of the protein CAA55297.1 that was employed to generate the anti-*Dm*-calpain antibody [86]. Furthermore, it is already known that anti-*Dm*-calpain antibody does not recognize mammalian μ-calpain and m-calpain [87]. Corroborating this finding, *T. cruzi* cells did no present any cross-reactivity with anti-human brain calpain antibodies, suggesting significant immunological and/or biochemical differences between parasite and human calpain-related molecules. Interestingly, the expression of calpain-like proteins was decreased in *T. cruzi* cells kept for long periods in axenic cultures in comparison to a strain recently isolated from mice, which may suggest that there is a relationship between the expression of calpain-like molecules and the parasite virulence (Figure 3) [86]. As is well known, the loss of *T. cruzi* virulence is associated to modifications of biological properties of this parasite that might lead to changes in the expression of some relevant proteins [88].

A large-scale approach to assess possible proteins involved in *T. cruzi* resistance to benznidazole revealed that a calpain-like protein (XP_806365) is among the proteins involved in this process [89]. The indication that calpain-like proteins are involved in relevant events for the trypanosomatids by discovery-driven science such as transcriptome and proteome highlights its importance.

It is well established that *T. cruzi* strains display a high level of biological divergence [90–92]. In this context, populations of *T. cruzi* can be clustered in two major phylogenetic lineages: *T. cruzi* I (TCI) and *T. cruzi* II (TCII). TCI has been mainly associated with the sylvatic transmission cycle, being also observed in domestic cycles [93], and TCII parasites are proposed to be associated with the domestic transmission cycle [94]. Furthermore, a third group denominated zymodeme III (or Z3) was also described [90, 95, 96]. However, the position of Z3 is still under review, and some isolates of *T. cruzi* present a hybrid profile, since they show genotypic characteristics of both TCI and TCII [95]. For these reasons, researchers have developing many criteria for grouping the *T. cruzi* strains, including protease production [97–99]. Corroborating the *T. cruzi* heterogeneity data, different levels of calpain-like proteins expression were also detected in distinct *T. cruzi* phylogenetic lineages: higher levels (2-fold) were found for Y strain (lineage TCII) and 10-times lower in INPA4167 strain (Z3 zymodeme) in comparison to the Dm28c strain (lineage TCI) [86].

CALPAIN–RELATED PROTEINS IN *LEISHMANIA*

The first unequivocal report of the presence of calpain-related proteins in the Trypanosomatidae family was in *T. brucei* [80]. Although *in silico* analysis raises concerns about the real possibility of a calpain with proteolytic activity in the Trypanosomatidae family, it was described in 1993 a neutral cysteine protease dependent on calcium in *L. donovani* promastigote. Not only was its activity detected, but also, a possible specific inhibitor, named caldonostatin [100, 101]. Caldonopain was found to be localized in cytosol along with its specific endogenous inhibitor caldonostatin. The ratio of caldonopain-caldonostatin unit was higher in the infected macrophage compared to the parasitic protozoa and BALB/c macrophage alone [100]. The authors postulated that the amount of both calcium and its protein inhibitor may have a direct impact on the caldonopain-induced biological process to regulate cellular action of this pathogen.

Evidences of the presence of calpain-related molecules in *Leishmania* was also reported by d'Avila-Levy and co-workers [102] who showed that anti-*Dm*-calpain strongly recognized a polypeptide migrating at approximately 80 kDa in *L. amazonensis* parasite lysate. Comparable to *T. cruzi*, no common epitopes were found between mammalian calpains and *L. amazonensis* polypeptides. This calpain-like molecule was detected majority throughout the cytoplasmic and flagellum of *L. amazonensis* promastigote cells, as demonstrated by fluorescence microscopy analysis using the anti-*Dm*-calpain antibody [102]. Curiously, some parasite regions have presented punctuated labeling, with accumulated fluorescence intensity, suggesting the presence of calpan-rich portions in *L. amazonensis* [102]. It has been demonstrated that leishmanolysin and lipophosphoglycan from *Leishmania* are arranged in functional micro-domains at the plasma membrane [103].

Interestingly, broad spectrum techniques revealed that calpains may be involved in several critical steps for the parasite life cycle and interaction with the mammalian host. For instance, when highly sensitive gene expression microarray technology was employed to identify genes that are differentially expressed in *L. donovani* isolated from post kala-azar dermal leishmaniasis (PKDL) patients in comparison with those from visceral leishmaniasis, a 2-fold higher expression of five proteins in PKDL parasites was reported, including a short calpain-like protein with significant homology to a *T. brucei* calpain [104]. Moreover, a comparative proteomics screen also revealed that a calpain-related protein SKCRP14.1 is down-regulated in antimonial-resistant strains of *L. donovani* and modulates the susceptibility to antimonials and miltefosine by interfering with drug-induced programmed cell death pathways: when overexpressed, this calpain-like protein significantly increased the sensitivity of the resistant strain to antimonials, being able to promote cell death, but the opposite effect was seen in miltefosine-treated cells, in which this calpain-like protein protected against miltefosine-induced death. It was suggested that SKCRP14.1 is likely to be a regulator of programmed cell death and since an altered expression of the same protein can have such different outcomes on drug-induced death pathways in *Leishmania* spp. [105].

Finally, an evaluation of differential gene expression in *L. major* procyclics and metacyclics using DNA microarray analysis revealed that one calpain-related protein named LmCALP20.2 was up-regulated in the procyclic promastigote insect stage, while the LmCALP20.1, coded by the adjacent gene, was up-regulated in the subsequent metacyclic insect stage [106]. Moreover, the flagellar calpain-like protein LmCALP20.10/SMP-1 was detectable only in promastigote stages of *Leishmania* containing a well-developed flagellum, and not in amastigotes, which contain only a highly truncated flagellum [107]. Therefore, as reported in *T. brucei*, it seems that calpain-related protein expression is tightly regulated during the life cycle of these parasites.

Calpain–Related Proteins in Monoxenic Trypanosomatids

Monoxenic trypanosomatids still do not have their genomes sequenced. This limits the available information and tools for the identification and characterization of calpain-related proteins. In this sense, classical biochemical approaches can provide critical data. In *Crithidia deanei*, an endosymbiont-harboring insect trypanosomatid, a released protease was purified and partially characterized. The enzyme presented interesting properties, for instance, neutral pH and inhibition by both classical cysteine protease inhibitors, such as E-64, and also metal chelators, such as EGTA. In addition, dialysis against EGTA totally restrained the enzymatic activity, and $CaCl_2$ was capable of fully restoring the proteolytic activity, while other divalent ions restored it only partially [108]. The enzyme is a homotrimer of 80 kDa protein, which cross react with the anti-*Dm*-calpain antibody. Taken together, these data strongly suggest that this enzyme may be a calpain-related protein, however, since its microsequencing was not done, it remained an open question.

Cysteine-like proteases of protozoa have been implicated in a variety of biological events, and the expression of these enzymes is modulated in response to distinct stimuli, including environmental changes and differentiation. An 80 kDa calpain-like protein that cross-reacted with the anti-*Dm*-calpain antibody was detected in both the cellular body and flagellum of *Herpetomonas samuelpessoai* promastigote cells, and its presence was enhanced after cellular differentiation induced *in vitro* by dimethylsulfoxide [109]. Interestingly, an additional ~30 kDa calpain-related polypeptide was exclusively observed in differentiated paramastigote cells. Collectively, these results suggest a stage-specific expression of calpain-like molecules in *H. samuelpessoai*. The *Herpetomonas* genus represents an interesting model to study cellular differentiation, since it displays three developmental stages during its life cycle, promastigote, paramastigote and opisthomastigote [49], which can be easily induced *in vitro* [110]. Therefore, monoxenic trypanosomatids represents an interesting model, which may help to elucidate the specific functions performed by this intriguing class of proteases.

EFFECT OF CALPAIN INHIBITORS ON TRYPANOSOMATIDS

Several classes of inhibitors, including peptidyl epoxide, aldehyde, and ketoamide inhibitors, targeting the active site have proven effective against the calpains and are in the process of evaluation in animal models of human diseases [4]. However, a major limitation to the clinical use of such inhibitors is their lack of specificity among cysteine proteases and other proteolytic enzymes. The development of a new class of calpain inhibitors that interact with domains outside of the catalytic site of calpain may provide greater specificity and therapeutic potential [4]. Indeed, in the trypanosomatids, there are several lines of evidence indicating that calpains may participate in relevant cellular processes in a proteolytic-independent manner [79, 80].

Trypanosoma

Our research group started to study the effect of the calpain inhibitor MDL28170 on *T. cruzi*. When the MDL28170 was added to the replicative epimastigote forms, a higher 48 h-IC_{50} was observed for three distinct *T. cruzi* phylogenetic lineages assayed: 31.7 μM for Dm28c (TCI), 34.3 μM for Y strain (TCII) and 37.4 μM for INPA4167 strain (Z3) [86]. The anti-trypanosomal activity of this inhibitor was shown to be reversible. Optical microscopy observations of *T. cruzi* Dm28c cells treated with the calpain inhibitor in the 20–70 μM range revealed, irrespective to the drug concentration, an increase in the cell volume, with the flagellates becoming round and some of them presenting no detectable flagellum, but no cell lysis was observed, corroborating the trypanostatic effect on the epimastigote form [86]. MDL28170 was also capable of significantly reducing the viability of bloodstream trypomastigotes, presenting an IC_{50}/24 h value of 20.4 μM. Also, parasites pre-treated with the inhibitor, at sub-inhibitory drug concentrations, prior to macrophage infection presented a clear dose-dependent inhibition profile of this cellular interaction, where the inhibition increased from 20 to 50% (in relation to control) as MDL28170 concentration rose from 6.25 to 50 μM. In addition, macrophages experimentally infected with *T. cruzi* trypomastigotes that were pos-treated with the calpain inhibitor presented a significant reduction in the percentage of intracellular amastigotes, resulting in a diminished infection [111]. A clear time- and dose-dependent effect of the MDL28170 on the macrophage infection rate was observed, while the highest drug concentration (25 μM) was capable of reducing almost the entire infection after 72 h [111]. Taken together, these findings robustly confirmed that the calpain inhibitor MDL28170 acted against *T. cruzi* clinically relevant forms, trypomastigotes and amastigotes, without displaying any relevant cytotoxic effect on mammalian host cells [111].

The MDL28170 compound also inhibited *in vitro T. cruzi* metacyclogenesis, promoted several ultrastructural alterations, such as disorganization of the reservosomes (these organelles have the particular ability to concentrate proteins and lipids obtained from medium together with the main proteolytic enzymes originated from the secretory pathway, being at the same time a storage organelle and the main site of protein degradation [112], intimately implicated in epitmastigote to trypomastigote differentiation [113]), Golgi and plasma membrane disruption, as well as impaired epimastigites adhesion to the mid gut of the insect vector *Rhodnius prolixus* [114]. The latter effect was also observed when parasites were treated with anti-calpain antibodies [114]. However, these effects cannot be directly associated to calpain activity inhibition. Calpain activity has never been demonstrated in *Leishmania* spp. or *T. cruzi*. Nevertheless, it could be assumed that calpains could participate in the aforementioned events through a non-catalytic mechanism, as demonstrated in other cellular systems [78]. In spite of it, inhibition of other parasite cysteine proteases by MDL28170 cannot be excluded, particularly because this enzymatic class is abundant in total parasite extracts [115, 116].

Curiously, *T. cruzi* epimastigote cells treated with MDL28170 presented a reduced expression of calpain-like proteins, concomitant with an increased expression of cruzipain, the major cysteine protease produced by this parasite [86]. These biochemical changes may demonstrate a correlation of the expression levels of cysteine proteases in *T. cruzi*, and it is tempting to speculate that, as previously suggested by Yong and co-workers [117], the overexpression of a certain cysteine protease may be necessary in order to titrate out the toxic levels of the inhibitor compound that is active against a different protease from the same class. These results pointed out the necessity of use a combined therapy in order to block the expression of both calpain-like molecules and cruzipain.

Leishmania

The calpain inhibitor MDL28170 also arrested the growth of *L. amazonensis* in a dose-dependent manner. The IC_{50} after 48 h was found to be 23.3 μM. The anti-leishmanial activity was irreversible, since protozoa pretreated for 72 h with the calpain inhibitor at 30 μM did not resume growth when sub-cultured in fresh medium. In addition, optical microscopy observations showed a massive deterioration of promastigote cells, in which the treatment with the calpain inhibitor initially promoted an increase in the cell volume and then lysis of the parasite cell (Figure 4) [102].

Cell death is now well defined in higher eukaryote cells. In fact, many studies have subdivided programmed cell death into the three categories of apoptosis (type I), autophagy (type II) and necrosis (type III) based on criteria such as morphological alterations, initiating death signal, or the implication of a family of aspartate-directed cysteine proteases, the caspases [118–120]. Apoptosis triggered by two anti-leishmanial drugs, pentostam and amphotericin B, in both amastigote and promastigote forms of *L. donovani* was partially inhibited by caspase inhibitors (Z-VAD-FMK, ZDEVD-FMK, Boc-D-FMK), but not inhibitors of other cysteine proteases such as cathepsin (Mu-Phe-HPh-FMK) or calpain (calpeptin) [121]. In a similar way, potassium antimonyl tartrate, a Sb(III)-containing drug, induced the oligonucleosomal DNA fragmentation, a classical hallmark of apoptosis, in amastigotes of *L. infantum* under a calpain-independent pathway [122]. These finding sustained the hypothesis that calpain inhibitors can be used against *Leishmania* without interferes with other metabolic dysfunction induced by classical anti-leishmania compounds, which opens a new window in order to study the possible synergistic action of distinct anti-trypanosomatids agents. In this same line of thinking, calpain inhibitor I was not effective in block the Ca^{2+}-dependent cell death pathway elicited by reactive oxygen species in *T. brucei* [123].

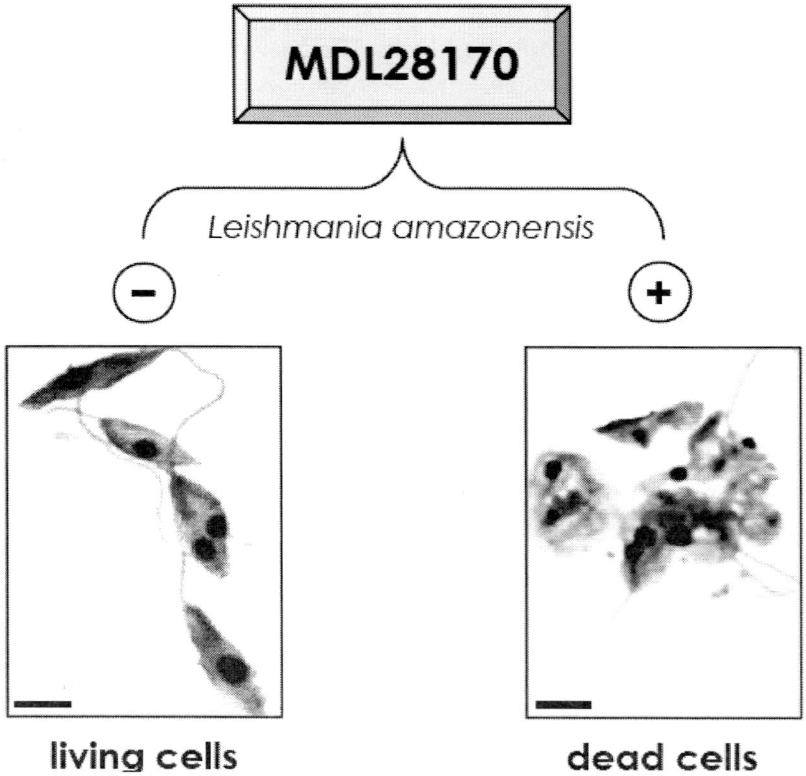

Figure 4. Microscopic observations of the morphology of promastigote forms of *L. amazonensis* incubated for 48 h in the absence (–) or in the presence (+) of 30 μM MDL28170. Note the complete lysis of the parasite cells after incubation with calpain inhibitor. The bars represent 1 μM.

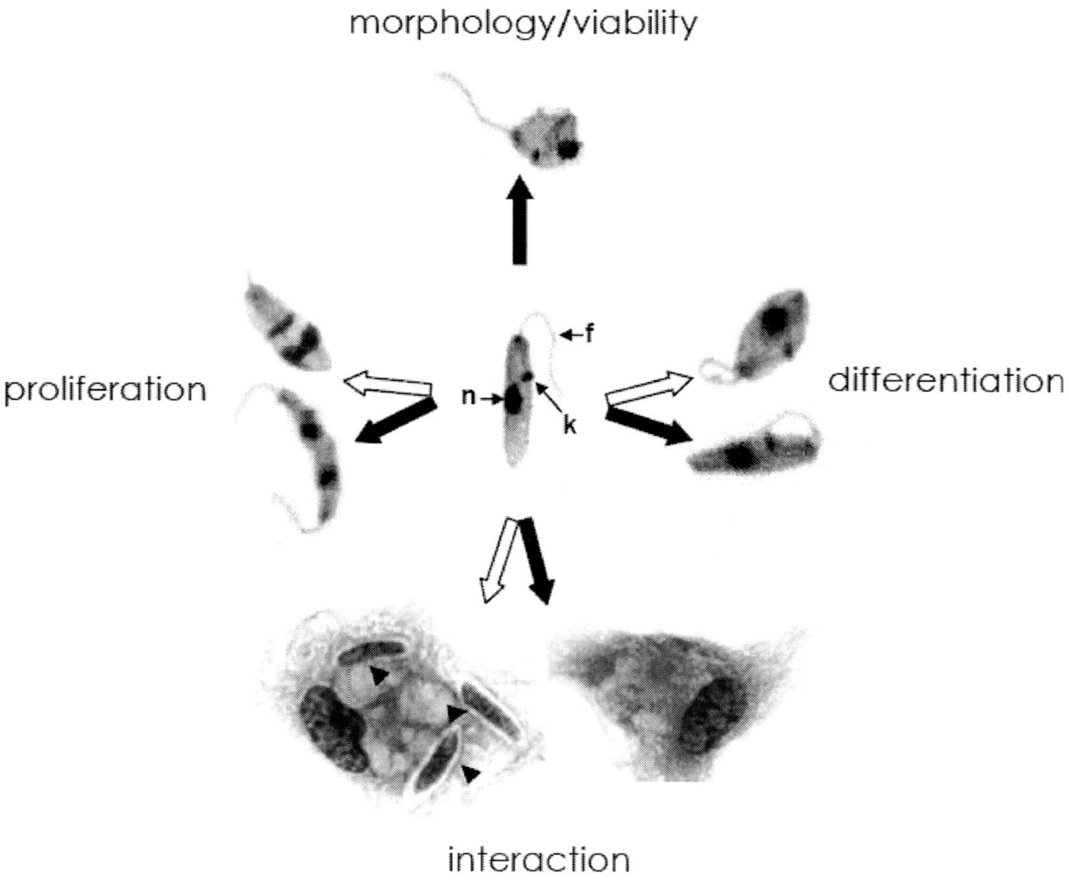

Figure 5. Proposition of different roles performed by calpain molecules produced by trypanosomatids. Calpain-like proteins participate directly and/or indirectly in numerous vital biological processes of parasite cells as well as during the different steps of interaction with host structures (white arrows). These processes can be blocked in different extensions by the action of calpain inhibitors (black arrows). In this sense, calpain inhibitors alter the parasite morphology/viability (note the altered position of the nucleus in comparison to the non-treated parasite), proliferation (note the distorted plane of cell division) and differentiation (note the changed position of the kinetoplast in relation to the nucleus and the shift of the cell body architecture). Note: n, nucleus; k, kinetoplast; f, flagellum. Arrowheads show the parasites inside the host cell.

Nitric oxide (NO) has been demonstrated to be the principal effector molecule mediating intracellular killing of *Leishmania*, both *in vitro* and *in vivo* [124–126]. Specific detection methods revealed a rapid and extensive cell death with morphological features of apoptosis in axenic amastigotes exposed to NO donors, in intracellular amastigotes inside the activated mouse macrophages and also in activated macrophages of regressive lesions in a leishmaniasis-resistant mouse model [127]. From the biochemical point of view, NO-mediated *Leishmania* amastigotes apoptosis did not seem to be controlled by caspase activity as indicated by the lack of effect of cell permeable inhibitors of caspases and cysteine proteases, in contrast to specific proteasome inhibitors, such as lactacystin or calpain inhibitor I [128]. Similarly, calpain inhibitor I was shown to interfere with apoptotic DNA fragmentation in *L. donovani* promastigote death induced by miltefosine [129]. Together, these data suggest that different biochemical pathways could be exploited by *Leishmania* parasites to die by apoptosis. Complicating this scenario, it is difficult to compare all these works since the type of programmed cell death is hard to define in *Leishmania* and there are many differences between the inducers used, the *Leishmania* species studied, the parasitic stage considered (i.e. promastigote or amastigote), the state of maturation of the parasitic stage (i.e. dividing or non-dividing) and its resistance to clinical drugs [127].

CONCLUSION

This review provided a framework indicating that the study of calpains and/or calpain-like proteins inhibition may be an attractive anti-trypanosomatid approach irrespective of whether these proteins are proteolytically active or not. As the up-regulation of several members of the calpain family is involved in a diverse range of biological processes and human diseases, this family of proteases has an important therapeutic potential, and a huge effort has been made in the field of research to develop a means of identifying selective calpain inhibitors [29]. In the immediate term, further studies in trypanosomatid calpains may employ existing drugs developed for human calpains inhibition, since extreme biochemical selectivity may not be necessary for anti-protozoan drugs because of the inherent biologic selectivity in the function and location of protozoan proteases. In addition, the inhibitor concentration necessary to chemically knockout a parasitic enzyme is likely much lower than that predicted for homologous host enzymes [130]. In the long term, knowledge of structural and functional relationships and substrate specificity of these proteins in trypanosomatids should make them candidates for computational-assisted drug design. Therefore, although current data using MDL28170 do not allow to directly infer that calpain-like proteins are in fact involved in the cellular processes described above, some indirect evidence supports the notion that this inhibitor at least interferes in the expression of calpain-like proteins. These results add new *in vitro* insights into the exploitation of calpain inhibitors in treating trypanosomatid infections and add this family of proteases to the list of potential targets for development of more potent and specific inhibitors against the incurable diseases caused by parasites belonging to the *Leishmania* and *Trypanosoma* genera, since numerous biological vital processes in these parasites were drastically altered/inhibited including proliferation, differentiation, and interaction (adhesion, internalization and intracellular susceptibility) with host cells (Figure 5).

ACKNOWLEDGMENTS

This study was supported by MCT/CNPq (Conselho Nacional de Desenvolvimento Científico e Tecnológico), FAPERJ (Fundação Carlos Chagas Filho de Amparo à Pesquisa do Estado do Rio de Janeiro), CAPES (Coordenação de Aperfeiçoamento de Pessoal de Nível Superior) and Fundação Oswaldo Cruz (FIOCRUZ).

REFERENCES

[1] Sorimachi, H.; Ishiura, S. and Suzuki, K. (1997). Structure and physiological function of calpains. *Biochem. J. 328*, 721–732.

[2] Goll, D. E.; Thompson, V. F.; Li, H.; Wei, W. and Cong, J. (2003). The calpain system. *Physiol. Rev. 83*, 731–801.

[3] Hosfield, C. M.; Elce, J. S.; Davies, P. L. and Zia, Z. (1999). Crystal structure of calpain reveals the structural basis for Ca^{2+}-dependent protease activity and a novel mode of enzyme activation. *EMBO J. 18*, 6880–6889.

[4] Carragher, N. O. (2006). Calpain inhibitors: a therapeutic strategy targeting multiple disease states. *Curr. Pharmac. Des. 12*, 615–638.

[5] Croall, D. E. and Ersfeld, K. (2007). The calpains: modular designs and functional diversity. *Gen. Biol. 8*, 1–11.

[6] Strobl, S.; Fernandez-Catalan, C.; Braun, M.; Huber, R.; Masumoto, H.; Nakagawa, K.; Irie, A.; Sorimachi, H.; Bourenkow, G.; Bartunik, H.; Suzuki, K. and Bode, W. (2000). The crystal structure of calcium-free human m-calpain suggests an electrostatic switch mechanism for activation by calcium. *Proc. Natl. Acad. Sci. USA 97*, 588–592.

[7] Dutt, P.; Arthur, J. S.; Grochulski, P.; Cygler, M. and Elce, J. S. (2000). Roles of individual EF-hands in the activation of m-calpain by calcium. *Biochem. J. 348*, 37–43.

[8] Hata, S.; Sorimachi, H.; Nakagawa, K.; Maeda, T.; Abe, K. and Suzuki, K. (2001). Domain II of m-calpain is a Ca^{2+}-dependent cysteine protease. *FEBS Lett. 501*, 111–114.

[9] Arthur, J. S. and Crawford, C. (1996). Investigation of the interaction of m-calpain with phospholipids: calpain-phospholipid interactions. *Biochim. Biophys. Acta 1293*, 201–206.

[10] Tompa, P.; Emori, Y.; Sorimachi, H.; Suzuki, K. and Friedrich, P. (2001). Domain II of calpain is a Ca^{2+}-regulated phospholipid-binding domain. *Biochem. Biophys. Res. Commun. 280*, 1333–1339.

[11] Friedrich, P. and Bozóky, Z. (2005). Digestive versus regulatory proteases: on calpain action *in vivo*. *Biol. Chem. 386*, 609–612.

[12] Huang, Y. and Wang, K. K. (2001). The calpain family and human disease. *Trends Mol. Med. 7*, 355–362.

[13] Zatz, M. and Starling, A. (2005). Calpains and disease. *N. Engl. J. Med. 352*, 2413–2423.

[14] Bertipaglia, I. and Carafoli, E. (2007). Calpains and human disease. *Subcell. Biochem. 45*, 29-53.

[15] Crawford, C.; Brown, N. R. and Willis, A. C. (1993). Studies of the active site of m-calpain and the interaction with calpastatin. *Biochem. J. 296*, 135–142.

[16] Todd, B.; Moore, D.; Deivanayagan, C. C.; Lin, G. D.; Chattopadhyay, D.; Maki, M.; Wang, K. K. and Narayana, S. V. (2003). A structural model for the inhibition of calpain by calpastatin: crystal structures of the native domain VI of calpain and its complexes with calpastatin peptide and a small molecule inhibitor. *J. Mol. Biol. 328*, 131–146.

[17] Maki, M.; Takano, E.; Mori, H.; Kannagi, R.; Murachi, T. and Hatanaka, M. (1987). Repetitive region of calpastatin is a functional unit of the proteinase inhibitor. *Biochem. Biophys. Res. Commun. 143*, 300–308.

[18] Takano, J.; Watanabe, M.; Hitomi, K. and Maki, M. (2000). Four types of calpastatin isoforms with distinct amino-terminal sequences are specified by alternative first exons and differentially expressed in mouse tissues. *J. Biochem. 128*, 83–92.

[19] Adachi, Y.; Ishida-Takahashi, A.; Takahashi, C.; Takano, E.; Murachi, T. and Hatanaka, M. (1991). Phosphorylation and subcellular distribution of calpastatin in human hematopoietic system cells. *J. Biol. Chem. 266*, 3968–3972.

[20] Averna, M.; De Tullio, R.; Salamino, F.; Melloni, F. and Pontremoli, S. (1999). Phosphorylation of rat brain calpastatins by protein kinase C. *FEBS Lett. 450*, 13–16.

[21] Tullio, R. D.; Passalacqua, M.; Averna, M.; Salamino, F.; Melloni, E. and Pontremoli, S. (1999). Changes in intracellular localization of calpastatin during calpain activation. *Biochem. J. 343*, 467–472.

[22] Averna, M.; Tullio, R. D.; Passalacqua, M.; Salamino, F.; Pontremoli, S. and Melloni, E. (2001). Changes in intracellular calpastatin localization are mediated by reversible phosphorylation. *Biochem. J. 354*, 25–30.

[23] Lane, R. D.; Allan, D. M. and Mellgren, R. L. (1992). A comparison of the intracellular distribution of mu-calpain, m-calpain and calpastatin in proliferating human A431 cells. *Exp. Cell Res. 203*, 5–16.

[24] Sorimachi, Y.; Harada, K.; Saido, T. C.; Ono, T.; Kawashima, S. and Yoshida, K. (1997). Downregulation of calpastatin in rat heart after brief ischemia and reperfusion. *J. Biochem. 122*, 743–748.

[25] Mimori, T.; Suganuma, K.; Tanami, Y.; Nojima, T.; Matsumura, M.; Fuji, T.; Yoshizawa, T.; Suzuki, K. and Akizuki, M. (1995). Autoantibodies to calpastatin (an endogenous inhibitor for calcium-dependent neutral protease, calpain) in systemic rheumatic diseases. *Proc. Natl. Acad. Sci. USA 92*, 7267–7271.

[26] Newcomb, J. K.; Pike, B. R.; Zhao, X.; Banik, N. L. and Hayes, R. L. (1999). Altered calpastatin protein levels following traumatic brain injury in rat. *J. Neurotrauma 16*, 1–11.

[27] Bardsley, R. G.; Allcock, S. M.; Dawson, J. M.; Dumelow, N. W.; Higgins, J. A.; Lasslett, Y. V.; Lockley, A. K.; Parr, T. and Buttery, P. J. (1992). Effect of beta-agonists on expression of calpain and calpastatin activity in skeletal muscle. *Biochemie 74*, 267–273.

[28] Tidball, J. G. and Spencer, M. J. (2002). Expression of calpastatin transgene slows muscle wasting and obviates changes in myosin isoform expression during murine muscle disuse. *J. Physiol. 545*, 819–828.

[29] Saez, M. E.; Ramirez-Lorca, R.; Moron, F. J. and Ruiz, A. (2006). The therapeutic potential of the calpain family: new aspects. *Drug Discov. Today 11*, 917–923.

[30] Donkor, I. O. (2000). A survey of calpain inhibitors. *Curr. Med. Chem. 7*, 1171–1188.

[31] Wang, K. K. and Yuen, P. W. (1997). Development and therapeutic potential of calpain inhibitors. *Adv. Pharmacol. 37*, 117–152.

[32] Leung-Toung, R.; Zhao, Y.; Li, W.; Tam, T. F.; Karimiam, K. and Spino, M. (2006). Thiol proteases: inhibitors and potential therapeutic targets. *Curr. Med. Chem. 13*, 547–581.

[33] Barrett, A. J.; Kembhavi, A. A.; Brown, M. A.; Kirschke, H.; Knight, C. G.; Tamai, M. and Hanada, K. (1982). L-*trans*-epoxysuccinyl-leucylamido(4-guanidino) butane (E-64) and its analogues as inhibitors of cysteine proteinases including cathepsins B, H and L. *Biochem. J. 201*, 189–198.

[34] Sasaki, T.; Kikuchi, T.; Yumoto, N.; Yoshimura, N. and Murachi, T. (1984). Comparative specificity and kinetic studies on porcine calpain I and calpain II with naturally occurring peptides and synthetic fluorogenic substrates. *J. Biol. Chem. 259*, 12489–12494.

[35] Yano, Y.; Shiba, E.; Kambayashi, J.; Sakon, M.; Kawasaki, T.; Fujitani, K.; Kang, J. and Mori, T. (1993). The effects of calpeptin (a calpain specific inhibitor) on agonist induced microparticle formation from the platelet plasma membrane. *Thromb Res. 71*, 385–396.

[36] Mehdi, S.; Angelastro, M. R. and Wiseman, J. S. (1988). Inhibition of the proteolysis of rat erythrocyte membrane proteins by a synthetic inhibitor of calpain. *Biochem. Biophys. Res. Commun. 157*, 1117–1123.

[37] Inoue, J.; Nakamura, M.; Cui, Y. S.; Sakai, Y.; Sakai, O.; Hill. J. R.; Wang, K. K. and Yuen, P. W. (2003). Structure-activity relationshio study and drug profile of N-(4-fluorophenylsulfonyl)-L-valyl-L-leucinal (SJA6017) as a potent calpain inhibitor. *J. Med. Chem. 46*, 868–871.

[38] Sasaki, T.; Kishi, M.; Saito, M.; Tanaka, T.; Higuchi, N.; Kominami, F.; Katunuma, N. and Murachi, T. (1990). Inhibitory effect of di- and tripeptidyl aldehydes on calpains and cathepsins. *J. Enzyme Inhib. 3*: 195–201.

[39] Harbeson, S. L.; Abelleira, S. M.; Akiyama, A.; Barrett, R.; Carroll, R, M.; Straub, J. A.; Tkacz, J. N.; Wu, C.and Musso, G. F. (1994). Stereospecific synthesis of peptidyl alpha-keto amides as inhibitors of calpain. *J. Med. Chem. 37*, 2918–2929.

[40] Trumbeckaite. S.; Neuhof, C.; Zierz, S. and Gellerich. F. N. (2003). Calpain inhibitor (BSF 409425) diminishes ischemia/reperfusion-induced damage of rabbit heart mitochondria. *Biochem. Pharmacol. 65*, 911–916.

[41] Lubisch, W.; Beckenbach, F.; Bopp, S.; Hoffmann, H. P.; Kartal, A.; Kastel, C.; Lindner, T.; Metz-Garrecht, M.; Reeb, J.; Regner, F.; Vierling, M. and Möller, A. (2003). Benzoylalanine-derived ketoamides carrying vinylbenzyl amino residues: discovery of potent water-soluble calpain inhibitors with oral bioavailability. *J. Med. Chem. 46*, 2404–2412.

[42] Neffe, A. T. and Abell, A. D. (2005). Developments in the design and synthesis of calpain inhibitors. *Curr. Opin. Drug Discov. Dev. 8*, 684–700.

[43] Wang, K. K.; Nath, R.; Posner, A.; Raser, K. J.; Buroker-Kilgore, M.; Hajimohammadreza, I.; Probert, A. W. Jr.; Marcoux, F. W.; Ye, Q.; Takano, E.; Hatanaka, M.; Maki, M.; Caner, H.; Collins, J. L.; Fergus, A.; Lee, K. S.; Lunney, E. A.; Hays, S. J. and Yuen, P. (1996). An alpha-mercaptoacrylic acid derivative is a selective nonpeptide cell-permeable calpain inhibitor and is neuroprotective. *Proc. Natl. Acad. Sci. USA 93*, 6687–6692.

[44] Posner, A.; Raser, K. J.; Hajimohammadreza, I.; Yuen, P. W. and Wang, K. K. (1995). Aurintricarboxylic acid is an inhibitor of mu- and m-calpain. *Biochem. Mol. Biol. Int. 36*, 291–299.

[45] Gil-Parrado, S.; Assfalg-Machleidt, I.; Fiorino, F.; Deluca, D.; Pfeiler, D.; Schaschke, N.; Moroder, L. and Machleidt, W. (2003). Calpastatin exon 1B-derived peptide, a

selective inhibitor of calpain: enhancing cell permeability by conjugation with penetratin. *Biol. Chem. 384*, 395–402.
[46] Jiao, W.; McDonald, Q.; Coxon, J. M. and Parker, E. J. (2010). Molecular modeling studies of peptide inhibitors highlight the importance of conformational prearrangement for inhibition of calpain. *Biochemistry 49*, 5533–5539.
[47] Pietsch, M.; Chua, K. C. and Abell, A. D. (2010). Calpains: attractive targets for the development of synthetic inhibitors. *Curr. Top. Med. Chem. 10*, 270–293.
[48] McGhee, R. B. and Cosgrove, W. B. (1980). Biology and physiology of the lower Trypanosomatidae. *Microbiol. Rev. 44*, 140–173.
[49] Vickerman, K. (1994). The evolutionary expansion of the trypanosomatid flagellates. *Int. J. Parasitol. 24*, 1317–1331.
[50] De Souza, W.; Attias, M. and Rodrigues, J. C. F. (2009). Particularities of mitochondrial structure in parasitic protists (Apicomplexa and Kinetoplastida). *Int. J. Biochem. Cell Biol. 41*, 2069–2080.
[51] Handman, E. (1999). Cell biology of *Leishmania*. *Adv. Parasitol.* 44, 1–39.
[52] De Souza, W. (2002). Basic cell biology of *Trypanosoma cruzi*. *Curr. Pharmaceut. Des.* 8, 269–285.
[53] Matthews, K. R. (2005). The developmental cell biology of *Trypanosoma brucei*. *J. Cell Sci. 118,* 283–290.
[54] Camargo, E. P. (1999). *Phytomonas* and other trypanosomatid parasites of plants and fruit. *Adv. Parasitol. 42*, 29–112.
[55] Laveran, A. and Franchini, G. (1913). Infection expérimentales de mammifères par des flagellés du tube digestif de *Ctenocephalus canis* et a'*Anopheles maculipennis*. *C. R. Acad. Sci. 157*, 744–747.
[56] Dedet, J. P.; Roche, B.; Pratlong, F.; Cales-Quist, D.; Jouannelle, J.; Benichou, J. C. and Huerre, M. (1995). Diffuse cutaneous infection caused by a presumed monoxenous trypanosomatid in a patient infected with HIV. *Trans. R. Soc. Trop. Med. Hyg. 89*, 644–646.
[57] Jiménez, M. I.; López-Vélez, R.; Molina, R.; Cañavate, C. and Alvar, J. (1996). HIV co-infection with a currently non-pathogenic flagellate. *Lancet 347*, 264–265.
[58] Pacheco, R. S.; Marzochi, M. C. A.; Pires, M. Q.; Brito, C. M. M.; Madeira, M. F. and Barbosa-Santos, E. G. O. (1998). Parasite genotypically related to a monoxenous trypanosomatid of dog's flea causing opportunistic infection in an HIV positive patient. *Mem. Inst. Oswaldo Cruz 93*, 531–537.
[59] Chicharro, C. and Alvar, J. (2003). Lower trypanosomatids in HIV/AIDS patients. *Ann. Trop. Med. Parasitol. 97*, 75–78.
[60] Morio, F.; Reynes, J.; Dollet, M.; Pratlong, F.; Dedet, J. P. and Ravel, C. (2008). Isolation of a protozoan parasite genetically related to the insect trypanosomatid *Herpetomonas samuelpessoai* from a human immunodeficiency virus-positive patient. *J. Clin. Microbiol. 46*, 3845–3847.
[61] Santos, A. L. S.; Branquinha, M. H. and d'Avila-Levy, C. M. (2006). The ubiquitous gp63-like metalloprotease from lower trypanosomatids: in the search for a function. *An. Acad. Bras. Ciênc. 78*, 687–714.
[62] Santos, A. L. S.; d'Avila-Levy, C. M.; Elias, C. G. R.; Vermelho, A. B. and Branquinha, M. H. (2007). *Phytomonas serpens* immunological similarities with the human trypanosomatid pathogens. *Microbes Infect. 9*, 915–921.

[63] Fairlamb, A. H. (2003). Chemotherapy of human African trypanosomiasis: current and future prospects. *Trends Parasitol. 19*, 488–494.

[64] Cavalli, A. and Bolognesi, M. L. (2009). Neglected tropical diseases: multi-target-directed ligands in the search for novel lead candidates against *Trypanosoma* and *Leishmania*. *J. Med. Chem. 52*, 7339–7359.

[65] McKerrow, J. H.; Doyle, P. S.; Engel, J. C.; Podust, L. M.; Robertson, S. A.; Ferreira, R.; Saxton, T.; Arkin, M.; Kerr, I. D.; Brinen, L. S. and Craik, C. S. (2009). Two approaches to discovering and developing new drugs for Chagas disease. *Mem. Inst. Oswaldo Cruz 104*, 263–269.

[66] Wilkinson, S. R. and Kelly, J. M. (2009). Trypanocidal drugs: mechanisms, resistance and new targets. *Expert Rev. Mol. Chem. 11*, 1–24.

[67] Urbina, J. A. (2010). Specific chemotherapy of Chagas disease: Relevance, current limitations and new approaches. *Acta Trop. 115*, 55–68.

[68] McKerrow, J. H.; Sun, E.; Rosenthal, P. J. and Bouvier, J. (1993). The proteases and pathogenicity of parasitic protozoa. *Annu. Rev. Microbiol. 47*, 821–853.

[69] Cazzulo, J. J. (2002). Proteinases of *Trypanosoma cruzi:* patential targets for the chemotherapy of Chagas disease. *Curr. Top. Med. Chem. 2*, 1261–1271.

[70] Klemba, M. and Goldberg, D. E. (2002). Biological roles of proteases in parasitic protozoa. *Annu. Rev. Biochem. 71*, 275–305.

[71] Lalmanach, G.; Boulangé, A.; Serveau, C.; Lecaille, F.; Scharfstein, J.; Gauthier, F. and Authié, E. (2002). Congopain from *Trypanosoma congolense*: drug target and vaccine candidate. *Biol. Chem. 383*, 739–749.

[72] Yao, C. (2010). Major surface protease of trypanosomatids: one size fits all? *Infect. Immun. 78*, 22-31.

[73] Vermelho, A. B.; Giovanni-de-Simone, S.; d'Avila-Levy, C. M.; Santos, A. L. S.; Melo, A. C. N.; Silva Jr, F. P.; Bon, E. P. S. and Branquinha, M. H. (2007). Trypanosomatidae peptidases: a target for drugs development. *Curr. Enz. Inhib. 3*, 19–48.

[74] Vermelho, A. B.; Branquinha, M. H.; d'Ávila-Levy, C. M.; Santos, A. L. S.; Paraguai de Souza, E. and Nogueira de Melo, A. C. (2010). Biological roles of peptidases in trypanosomatids. *Open Parasitol. J. 4*, 5–23.

[75] Atkinson, H. J.; Babbitt, P. C. and Sajid, M. (2009). The global cysteine peptidase landscape in parasites. *Trends Parasitol. 25*, 573–581.

[76] Ersfeld, K.; Barraclough, H. and Gull, K. (2005). Evolutionary relationships and protein domain architecture in an expanded calpain superfamily in kinetoplastid parasites. *J. Mol. Evol. 61*, 742–757.

[77] Pils, B. and Schultz, J. (2004). Inactive enzyme-homologues find new function in regulatory processes. *J. Mol. Biol. 340*. 399–404.

[78] Tonami, K.; Kurihara, Y.; Aburatani, H.; Uchijima, Y.; Asano, T. and Kurihara, H. (2007). Calpain 6 is involved in microtubule stabilization and cytoskeletal organization. *Mol. Cell Biol. 27*, 2548–2561.

[79] Olego-Fernandez, S.; Vaughan, S.; Shaw, M. K.; Gull, K. and Ginger, M. L. (2009). Cell morphogenesis of *Trypanosoma brucei* requires the paralogous, differentially expressed calpain-related proteins CAP5.5 and CAP5.5V. *Protist 160*, 576–590.

[80] Hertz-Fowler, C.; Ersfeld, K. and Gull, K. (2001). CAP5.5, a life-cycle-regulated, cytoskeleton-associated protein is a member of a novel family of calpain-related proteins in *Trypanosoma brucei. Mol. Biochem. Parasitol. 116*, 25–34.
[81] Dear, N.; Matena, K.; Vingron, M. and Boehm, T. (1997). A new subfamily of vertebrate calpains lacking a calmodulin-like domain: implications for calpain regulation and evolution. *Genomics 45*, 175–184.
[82] Matthews, K. R. and Gull, K. (1994). Evidence for an interplay between cell cycle progression and the initiation of differentiation between life cycle forms of African trypanosomes. *J. Cell Biol. 125*, 1147–1156.
[83] Liu, W.; Apagyi, K.; McLeavy, L. and Ersfeld, K. (2010). Expression and cellular localisation of calpain-like proteins in *Trypanosoma brucei. Mol. Biochem. Parasitol. 169*, 20–26.
[84] Giese, V.; Dallagiovanna, B.; Marchini, F. K.; Pavoni, D. P.; Krieger, M. A. and Goldenberg, S. (2008). *Trypanosoma cruzi*: a stage-specific calpain-like protein is induced after various kinds of stress. *Mem. Inst. Oswaldo Cruz 103*, 598–601.
[85] Bonaldo, M. C.; Souto-Padron, T.; de Souza, W. and Goldenberg, S. (1988). Cell-substrate adhesion *during Trypanosoma cruzi* differentiation. *J. Cell Biol. 106*, 1349–1358.
[86] Sangenito, L. S.; Ennes-Vidal, V.; Marinho, F.A.; da Motta, F. F.; Santos, A. L. S.; d'Avila-Levy, C.M. and Branquinha, M. H. (2009). Arrested growth of *Trypanosoma cruzi* by the calpain inhibitor MDL28170 and detection of calpain homologues in epimastigote forms. *Parasitology 136*, 433–441.
[87] Emori, Y. and Saigo, K. (1994) Calpain localization changes in coordination with actin-related cytoskeletal changes during early embryonic development of *Drosophila. J. Biol. Chem. 269*, 25137–25142.
[88] Contreras, V. T.; Lima, A. R. and Zorrilla, G. (1998). *Trypanosoma cruzi:* maintenance in culture modify gene and antigenic expression of metacyclic trypomastigotes. *Mem. Inst. Oswaldo Cruz 93*, 753–760.
[89] Andrade, H. M.; Murta, S. M. F.; Chapeaurouge, A.; Perales, J.; Nirdé, P. and Romanha, A. (2008). Proteomic analysis of *Trypanosoma cruzi* resistance to benznidazole. *J. Proteome Res. 7*, 2357–2367.
[90] Miles, M.A.; Souza, A.; Povoa, M.; Shaw, J. J.; Lainson, R. and Toye, P. J. (1978). Isozymic heterogeneity of *Trypanosoma cruzi* in the first autochthonous patients with Chagas' disease in Amazonian Brazil. *Nature 272*, 819–821.
[91] Souto, R. P.; Fernandes, O.; Macedo, A. M.; Campbell, D. A. and Zingales, B. (1996). DNA markers define two major phylogenetic lineages of *Trypanosoma cruzi. Mol. Biochem. Parasitol. 83*, 141–152.
[92] Kikuchi, S. A.; Sodré, C. K.; Kalume, D. E.; Elias, C. G. R.; Santos, A. L. S.; Soeiro, M. N.; Meuser, M.; Chapeaurouge, A.; Perales, J. and Fernandes, O. (2010). Proteomic analysis of two *Trypanosoma cruzi* zymodeme 3 strains. *Exp. Parasitol. 126*, 540–551.
[93] Brisse, S.; Dujardin, J. C. and Tibayrenc, M. (2000). Identification of six *Trypanosoma cruzi* lineages by sequence-characterised amplified region markers. *Mol. Biochem. Parasitol. 111*, 95–105.
[94] Macedo, A. M.; Machado, C. R.; Oliveira, R. P. and Pena, S. D. (2004). *Trypanosoma cruzi*: genetic structure of populations and relevance of genetic variability to the pathogenesis of Chagas disease. *Mem. Inst. Oswaldo Cruz 99*, 1–12.

[95] Fernandes, O.; Santos, S.S.; Cupolillo, E.; Mendonça, B.; Derre, R.; Junqueira, A. C.; Santos, L. C.; Sturm, N. R.; Naiff, R. D.; Barret, T. V.; Campbell, D. A. and Coura, J. R. (2001). A mini-exon multiplex polymerase chain reaction to distinguish the major groups of *Trypanosoma cruzi* and *T. rangeli* in the Brazilian Amazon. *Trans. R. Soc. Trop. Med. Hyg. 95*, 97–99.

[96] Brandão, A. and Fernandes, O. (2006). *Trypanosoma cruzi*: mutation in 3′-untranslated region of calmodulin gene are specific for lineages *T. cruzi* I, *T. cruzi* II and zymodeme III isolates. *Exp. Parasitol. 112*, 247–252.

[97] Fampa, P.; Lisboa, C. V.; Jansen, A. M.; Santos, A. L. S. and Ramirez, M. I. (2008). Protease expression analysis in recently field-isolated strains of *Trypanosoma cruzi*: a heterogeneous profile of cysteine protease activities between TC I and TC II major phylogenetic groups. *Parasitology 135*, 1093–1100.

[98] Fampa, P.; Santos, A. L. S. and Ramirez, M. I. (2010). *Trypanosoma cruzi*: ubiquity expression of surface cruzipain molecules in TCI and TCII field isolates. *Parasitol. Res. 107*, 443–447.

[99] Gomes, S. A. O.; Misael, D.; Silva, B. A.;, Feder, D.; Silva, C. S.; Gonçalves, T. C. M.; Santos, A. L. S. and Santos-Mallet, J. R. (2009). Major cysteine protease (cruzipain) in Z3 sylvatic isolates of *Trypanosoma cruzi* from Rio de Janeiro-Brazil. *Parasitol. Res. 105*, 743–749.

[100] Bhattacharya, J.; Dey, R. and Datta, S. C. (1993). Calcium dependent thiol protease caldonopain and its specific endogenous inhibitor in *Leishmania donovani*. *Mol. Cell Biochem. 126*, 9–16.

[101] Dey, R.; Bhattacharya, J. and Datta, S. C. (2006). Calcium-dependent proteolytic activity of a cysteine protease caldonopain is detected during *Leishmania* infection. *Mol. Cell Biochem. 281*, 27–33.

[102] d'Avila-Levy, C. M.; Marinho, F. A.; Santos, L. O.; Martins, J. L. M.; Santos, A. L. S. and Branquinha, M. H. (2006). Antileishmanial activity of MDL28170, a potent calpain inhibitor. *Int. J. Antimicrob. Agents 28*, 138–142.

[103] Denny, P. W.; Field, M. C. and Smith, D. F. (2001). GPI-anchored proteins and glycoconjugates segregate into lipid rafts in kinetoplastida. *FEBS Lett. 491*, 148–153.

[104] Salotra, P.; Duncan, R. C.; Singh, R.; Raju, B. V. S.; Sreenivas, G. and Nakhasi, H. L. (2006). Upregulation of surface proteins in *Leishmania donovani* isolated from patients of post kala-azar dermal leishmaniasis. *Microbes Infect. 8*, 637–644.

[105] Vergnes, B.; Gourbal, B.; Gorard, I.; Sundar, S.; Drummelsmith, J. and Ouellette, M. (2007). A proteomics screen implicates HSP83 and a small kinetoplastid calpain-related protein in drug resistance in *Leishmania donovani* clinical field isolates by modulating drug-induced programmed cell death. *Mol. Cell. Proteomics 6*, 88–101.

[106] Saxena, A.; Worthey, E. A.; Yan, S.; Leland, A.; Stuart, K. D. and Myler, P. J. (2003). Evaluation of differential gene expression in *Leishmania major* Friedlin procyclics and metacyclics using DNA microarray analysis. *Mol. Biochem. Parasitol. 129*, 103–114.

[107] Tull, D.; Vince, J. E.; Callaghan, J. M.; Naderer, T.; Spurck, T.; McFadden, G. I.; Currie, G.; Ferguson, K.; Bacic, A. and McConville, M. J. (2004) SMP-1, a member of a new family of small myristoylated proteins in kinetoplastid parasites, is targeted to the flagellum membrane in *Leishmania*. *Mol. Biol. Cell 15*, 4775–4786.

[108] d'Avila-Levy, C. M.; Souza, R. F.; Gomes, R. C.; Vermelho, A. B. and Branquinha, M. H. (2003). A novel extracellular cysteine proteinase from *Crithidia deanei*. *Arch. Biochem. Biophys. 420*, 1–8.

[109] Pereira, F. M.; Elias, C. G. R.; d'Avila-Levy, C. M.; Branquinha, M. H. and Santos, A. L. S. (2009). Cysteine peptidases in *Herpetomonas samuelpessoai* are modulated by temperature and dimethylsulfoxide-triggered differentiation. *Parasitology 136*, 45–54.

[110] Castellanos, G. B.; Angluster, J. and de Souza, W. (1981). Induction of differentiation in *Herpetomonas samuelpessoai* by dimethylsulphoxide. *Acta Trop. 38*, 29–37.

[111] Ennes-Vidal, V.; Menna-Barreto, R. F.; Santos, A. L.; Branquinha, M. H. and d'Avila-Levy, C. M. (2010). Effects of the calpain inhibitor MDL28170 on the clinically relevant forms of *Trypanosoma cruzi in vitro*. *J. Antimicrob. Chemother. 65*, 1395–1398.

[112] Sant'Anna, C.; Nakayasu, E. S.; Pereira, M. G.; Lourenço, D.; de Souza, W.; Almeida, I. C. and Cunha-e-Silva, N. L. (2009). Subcellular proteomics of *Trypanosoma cruzi* reservosomes. *Proteomics 9*, 1782–1794.

[113] Soares, M. J. (1999). The reservosome of *Trypanosoma cruzi* epimastigotes: an organelle of the endocytic pathway with a role on metacyclogenesis. *Mem. Inst. Oswaldo Cruz 94 Suppl 1*, 139–141.

[114] Ennes-Vidal, V.; Menna-Barreto, R. F.; Santos, A. L. S.; Branquinha, M. H. and d'Avila-Levy, C. M. (2010). MDL28170, a potent calpain inhibitor, affects *Trypanosoma cruzi* metacyclogenesis, ultrastructure and attachment to the luminal midgut surface of *Rhodnius prolixus*. Submitted manuscript.

[115] Branquinha, M. H.; Vermelho, A. B.; Goldenberg, S.; Bonaldo, M. C. (1996). Ubiquity of cysteine- and metalloproteinase activities in a wide range of trypanosomatids. *J. Eukaryot. Microbiol. 43*, 131–135.

[116] Santos, A. L. S.; Abreu, C. M.; Alviano, C. S. and Soares, R. M. A. (2005). Use of proteolytic enzymes as an additional tool for trypanosomatid identification. *Parasitology 130*, 79–88.

[117] Yong, V.; Schmitz, V.; Vannier-Santos, M. A.; Lima, A. P. C. A.; Lalmanach, G.; Juliano, L.; Gauthier, F. and Scharfstein, J. (2000). Altered expression of cruzipain and a cathepsin B-like target in a *Trypanosoma cruzi* cell line displaying resistance to synthetic inhibitors of cysteine-proteinases. *Mol. Biochem. Parasitol. 109*, 47–59.

[118] Green, D. R. (2000). Apoptotic pathways: paper wraps stone blunts scissors. *Cell 102*, 1–4.

[119] Green, D. R.; Knight, R. A.; Melino, G.; Finazzi-Agro, A. and Orrenius, S. (2004). Ten years of publication in cell death. *Cell Death Differ. 11*, 2–3.

[120] Bras, M.; Queenan, B. and Susin, S. A. (2005). Programmed cell death via mitochondria: different modes of dying. *Biochemistry (Moscow) 70*, 231–239.

[121] Lee, N.; Bertholet, S.; Debrabant, A.; Muller, J.; Duncan, R. and Nakhasi, H. L. (2002). Programmed cell death in the unicellular protozoan parasite *Leishmania*. *Cell Death Differ. 9*, 53–64.

[122] Sereno, D.; Holzmuller, P.; Mangot, I.; Cuny, G.; Ouaissi, A. and Lemesre, J. L. (2001). Antimonial-mediated DNA fragmentation in *Leishmania infantum* amastigotes. *Antimicrob. Agents Chemother. 45*, 2064–2069.

[123] Ridgley, E. L.; Xiong, Z. H. and Ruben, L. (1999). Reactive oxygen species activate a Ca^{2+}-dependent cell death pathway in the unicellular organism *Trypanosoma brucei brucei*. *Biochem. J. 340*, 33–40.

[124] Green, S. J.; Crawford, R. M.; Hockmeyer, J. T.; Meltzer, M. S. and Nacy, C. A. (1990). *Leishmania major* amastigotes initiate the L-arginine-dependent killing mechanism in IFN-gamma-stimulated macrophages by induction of tumor necrosis factor-alpha. *J. Immunol. 145*, 4290–4297.

[125] Panaro, M. A.; Acquafredda, A.; Lisi, S.; Lofrumento, D. D.; Trotta, T.; Satalino, R.; Saccia, M.; Mitolo, V. and Brandonisio, O. (1999). Inducible nitric oxide synthase and nitric oxide production in *Leishmania infantum*-infected human macrophages stimulated with interferon-gamma and bacterial lipopolysaccharide. *Int. J. Clin. Lab. Res. 29*, 122–127.

[126] Sisto, M.; Brandonisio, O.; Panaro, M. A.; Acquafredda, A.; Leogrande, D.; Fasanella, A.; Trotta, T.; Fumarola, L. and Mitolo, V. (2001). Inducible nitric oxide synthase expression in *Leishmania*-infected dog macrophages. *Comp. Immunol. Microbiol. Infect. Dis. 24*, 247–254.

[127] Holzmuller, P.; Bras-Gonçalves, R. and Lemesre, J. L. (2006). Phenotypical characteristics, biochemical pathways, molecular targets and putative role of nitric oxide-mediated programmed cell death in *Leishmania*. *Parasitology 132 Suppl*, S19–S32.

[128] Holzmuller, P.; Sereno, D.; Cavaleyra, M.; Mangot, I.; Daulouede, S.; Vincendeau, P. and Lemesre, J. L. (2002). Nitric oxide-mediated proteasome-dependent oligonucleosomal DNA fragmentation in *Leishmania amazonensis* amastigotes. *Infect. Immun. 70*, 3727–3735.

[129] Paris, C.; Loiseau, P. M.; Bories, C. and Breard, J. (2004). Miltefosine induces apoptosis-like death in *Leishmania donovani* promastigotes. *Antimicrob. Agents Chemother. 48*, 852–859.

[130] McKerrow, J. H.; Rosenthal, P. J.; Swenerton, R. and Doyle, P. 2008. Development of protease inhibitors for protozoan infections. *Curr. Opin. Infect. Dis. 21*, 668–672.

In: Cystatins: Protease Inhibitors …
Editors: John B. Cohen and Linda P. Ryseck
ISBN: 978-1-61209-343-7

Chapter 4

CYSTATIN C IN HIV PATIENTS: MORE THAN JUST A GFR MARKER

Amandine Gagneux-Brunon, Christophe Mariat and Pierre Delanaye

[1]Service de Néphrologie, CHU de Saint-Etienne, Saint-Etienne, France
[2]Service de Néphrologie, CHU Sart Tilman, Liège, Belgique

With the development of highly active antiretroviral therapy (HAART), chronic kidney disease has become a relevant cause of morbidity in individuals infected by HIV. In this context, cystatin C is emerging as an interesting biomarker both for the evaluation of glomerular filtration rate (GFR) and the detection of drug-induced kidney injury.

In this chapter, we will first focus on serum cystatin C as a GFR marker in HIV infected patients. We will compare the respective advantages and limitations of serum creatinine and cystatin C in the context of HIV infection and will review the very first clinical studies on the use of cystatin C in this specific setting. Secondly, we will discuss the potential interest of urine cystatin C as a biomarker to detect renal tubular injuries associated with nucleotide reverse transcriptase inhibitors therapies. We will conclude by examining the questions that need to be answered in order to clarify the real added value of cystatin C for the management of HIV infected patients.

The United Nations Programs on HIV/AIDS (UNAIDS) estimated that 33 Millions of people were living with an HIV-Infection in 2007 [1]. The development of Highly Active Antiretroviral Therapies (HAART) has resulted in an improved survival among HIV seropositive individuals. With advancing age and HAART-related metabolic effects (hypertension [2], diabetes mellitus [3], dyslipidemia [4]), Chronic Kidney Disease (CKD) has become one of the major comorbidities in HIV-seropositive individuals [5]. Prevalence of CKD (defined by an eGFR below 60 mL/min/1.73m^2) in HIV infected is variable from 3 to 24 %. [6, 7] Risk Factors for CKD in HIV-infected patients are female sex, black race [8], Acquired Immuno-Deficiency Syndrome (AIDS), lower CD4 nadir, older age, HCV infection, hypertension, diabetes mellitus [6,9], and injection drug use [10]. Exposure to HAART is associated with an increased risk for CKD. Most frequently incriminated antiretroviral drugs are tenofovir, didanosine, atazanavir, lopinavir and indinavir. [11,12]

CKD is associated with an increased risk of both mortality [13,14] and cardiovascular events (CVE) [15] in HIV-seropositive individuals. The odds ratio for CVE is 1.2 for every 10 mL/min per $1.73m^2$ decrease in eGFR. Proteinuria is a risk factor of AIDS (Hazard ratio 1.31) and death, an increase in serum creatinine is associated with an increase risk of AIDS defining illness. [16] The etiologies of CKD in HIV-infected individuals are multiple and presented in table 1. [17]

Table 1. Causes of Chronic Kidney Disease in HIV-Infected patients

Kidney Disease	Description & causes
HIV-associated nephropathies	Characteristically a disease of African descent Focal or global collapsed glomeruli Mesangial hyperplasia, and mesangial matrix deposition Tubular interstitial inflammation
HIV immune complex disease	IgA nephritis Postinfectious GN Membranous nephritis (HBV, HCV, syphilis) Membranoproliferative GN (HBV, HCV and or cryoglobuminemia) Mesangial proliferative GN (HCV) Fibrillary GN Lupus like nephritis
Interstitial nephritis	
Minimal change glomerulonephritis	
Diabetic nephropathy	
Hypertensive nephropathy	
HAART-associated kidney disease [85]	
Tubulopathies	Fanconi's syndrome: tenofovir, didanosine, ritonavir Renal tubular acidosis: lamivudine, stavudine Nephrogenic diabetes insipidus: tenofovir, didanosine, indinavir
Nephrolithiasis	indinavir, atazanavir, nelfinavir, amprenavir, saquinavir, efavirenz
Chronic interstitial nephritis	Indinavir, Tenofovir

Because of the high prevalence of CKD in HIV-infected patients, screening, prevention and management of CKD have become a medical challenge. In 2005, the Infectious Diseases Society of America (IDSA) published guidelines for CKD screening in HIV-infected patients. At initial diagnosis of HIV-infection, kidney function must be assessed by measuring proteinuria and by estimating GFR. The use of the Modification Diet in Renal Disease (MDRD) study equation is recommended. [18] In case of CKD risk factors, kidney function should be assessed every 6 months. [19].

MDRD, Cockcroft-Gault and Chronic Kidney Disease Epidemiology Collaboration (CKD-EPI) equations are based on serum creatinine. Many factors influence serum creatinine concentration independently of any GFR change [20,21]. In HIV-infected patients, serum creatinine is affected by AIDS wasting syndrome, hepatic disease, alteration in protein intake and trimethoprim use. GFR measurement by a gold standard method (inuline clearance, iohexol plasma clearance, ^{99}mTc-DTPA and [51] Cr-EDTA clearances) is cumbersome and costly in all HIV-infected patients. Novel biomarkers are needed to facilitate CKD diagnosis in HIV-infected patients.

Cystatin C is produced by all nucleated cells at a constant rate, freely filtered in the glomerulus, reabsorbed and fully reabsorbed and metabolized by the proximal tubular cells [22, 23]. In normal conditions, cystatin C is not detectable in urine. Serum cystatin C could be an alternative endogenous biomarker of kidney function especially in population with "abnormal" muscular mass. Serum cystatin C might be more accurate compared to serum creatinine to predict kidney function in HIV-infected patients. As several factors (C-reactive protein, current cigarette smoking, white blood cell count, diabetes, serum albumin) affect cystatin C levels, this biomarker needs to be evaluated in HIV-infected patients. [24,25]

Another challenge in HIV-infection is the early diagnosis of HAART-induced nephropathies. Urinary cystatin C is a marker of tubular dysfunction independent of GFR. [26,27] Tenofovir is one of the most incriminated antiretrovirals agents inducing tubulopathies. Protease inhibitors as indinavir and atazanavir cause crystalluria, which may lead to intratubular crystal formation and renal injury. [28] Urinary Cystatin C was described as a marker of tenofovir induced-nephropathy. [29]

Serum Creatinine: An Imperfect Marker of Kidney Function in HIV-Infected Individuals

Serum creatinine level is related to GFR. Creatinine results from creatine, 98 % of creatine is muscular. Serum creatinine level is thus dependent of muscular mass. [30]

In HIV-infected patients, creatinine generation is expected to be decreased. First, HIV-infection induces modifications of the body composition. HIV-infected men have a lower Fat Free Mass compared to HIV-negative men. [31,32] HIV infection causes wasting syndromes with depletion in lean and fat tissue. HIV wasting syndrome is associated with an increase in mortality. [33] Higher viral load and lower CD4+ lymphocytes count are associated with decreased lean mass in case of HIV-infection. [34] HIV-infected subjects are more often malnourished than the general population. [35] Moreover, the impact of HAART on body composition is now well known. The HAART-related mitochondrial toxicity induces lipodystrophy during treatment [36], its impact on serum creatinine level is not well established.

Secondly, creatinine production in patients with liver disease is decreased. [37] Viral hepatitis are frequent among HIV-infected subjects. About 30% of HIV infected patients in developed Countries are co-infected with HCV. In case of HCV-HIV co-Infection, progression to liver fibrosis is much faster than in HIV seronegative individuals. [38] Prevalence of HBV infection in US HIV-infected individuals approaches 35%. [39] In HIV-infected individuals, HCV co-infection was associated with a decrease in serum creatinine level. [40] Furthermore, HCV co-infection is associated with an increased risk of CKD. [41]

Serum creatinine level was measured in two major studies cohorts and compared with matched controls. In HIV-infected subjects on HAART in Nutrition for Healthy Living Cohort (NFHL), serum creatinine level was significantly lower than in controls from National Health and Nutrition Examination Survey (NHANES) cohort, however there was no difference in lean mass assessed by Bioelectrical Impedance. [40] Compared with NHANES subjects, NFHL subjects were older and had a greater prevalence in liver disease, more often hypertension and diabetes mellitus. The African American percentage of subjects was superior in NFHL cohort than in NHANES cohort. In the Fat Redistribution and Metabolic Change in HIV-infection (FRAM) study, in comparison with HIV-negative individuals from Coronary Artery Risk Development in Young Adults (CARDIA) study, there was no difference in serum creatinine level, but body mass index was lower in FRAM cohort, there was no difference for lean body mass assessed by body magnetic resonance imaging. [42] The African American percentage of individuals was equivalent in the two cohorts.

The use of drugs inhibiting creatinine secretion like trimethoprim affects serum creatinine level in HIV-infected subjects independently of a decrease in GFR. [43] Serum creatinine elevation after trimethoprim introduction is around 15 to 30 %. [44,45] Trimethoprim/Sulfamethoxazol (TMP/SMX) is used in HIV-positive patients in primary and secondary prophylaxis of the infections due to *Toxoplasma gondii* and *Pneumocystis jiroveci.* [46] Primary prophylaxis is recommended if CD4 positive cell count is below 200/mm^3. Patients receiving TMP/SMX should be excluded of the studies evaluating the predictive performance of serum creatinine to estimate kidney function in HIV-positive patients.

Table 2. Principal factors influencing serum creatinine level in HIV-infected subjects

	Effects on creatinine	HIV
Lean mass	Creatinine production	AIDS wasting-syndrome, mitochondrial toxicity of HAART
Dietary intake	Creatinine production	Strong prevalence of Malnutrition in HIV-infected subjects
Liver disease	Creatine production	Prevalence of HCV infection in HIV-infected subjects around 30%
African American ethnicity	Reduction of the tubular excretion of creatinine	
Trimethoprim	Reduction of the tubular excretion of creatinine	*Pneumocystis jiroveci* infections prophylaxis

Serum Cystatin C as a Promising Alternative Biomarker to Assess Kidney Function in HIV-Positive Subjects

Multiple factors affect serum levels of cystatin C. These factors were examined in two major studies (table 3). The magnitude of the associations with height and weight is greater for serum creatinine than for serum cystatin C levels. Therefore, serum cystatin c level is well associated with body mass index, indicating than cystatin C is associated with fat mass. Factors influencing cystatin C may influence the predictive value of serum cystatin C in HIV-positive patients.

First, markers of inflammation are chronically elevated in HIV-positive adults. High-sensitivity (hs) CRP was 55% higher in HIV-positive patients from Strategies for Management of Anti-Retroviral Therapy study (SMART) than in controls from Multi-Ethnic Study of Atherosclerosis (MESA) and Coronary Artery Development in Young Adults (CARDIA). [47] In this last study, Cystatin C was 27.2 % higher in the SMART-study than in the MESA-study; the analysis did not allow to conclude if higher cystatin C level was independent from higher CRP level and smoking status. The association between hs-CRP and cystatin C level is inconsistent. In the Jaroszewicz *et al.* [48] study, there was no significant correlation between CRP and cystatin C level, but CRP level was significantly greater in HIV-positive subjects than in controls. Only 78 patients were included in this study. Secondly, HIV-positive individuals were more often smokers than HIV non-positive individuals. Forty percent, 22.2 % and 15 % of participants in the SMART (HIV-infected participants) CARDIA and MESA studies were respectively current smokers. [47] The increase in cystatin C level may reflect a loss in kidney function or be dependent of smoking status and CRP level. Serum cystatin C level was also associated with CRP level and smoking status in the FRAM study and NFHL cohort. [40, 42]

Table 3. factors influencing serum cystatin C level

	Study population	GFR evaluation method	Mean GFR	Factors associated with cystatin C (after adjustment with GFR or 24-h creatinine clearance)	Effects on cystatin C level
Knight *et al.* 2004 [24]	8058 subjects from Gröningen Netherlands 95 % Caucasian Age from 28 to 75 years 4% of patients with diabetes	24-h urine creatinine clearance	102 ± 27 mL/min	Age Male gender Weight , Height Current cigarette smoking C-Reactive Protein	Increase Increase Increase Increase
Stevens *et al.* 2009 [25]	MDRD study (N=1085) AASK (N=1205) CSG (N=266) NephroTest (N=438) All patients with CKD stage 2-5 53.5% blacks 13.9 % patients with diabetes	Iothalamate Urinary clearance EDTA Urinary clearance	48 mL/min/1.73m^2 (15-95)	Age Female gender Height, weight, BMI Urine protein Diabetes CRP White blood cell count Serum albumin	Decreasse Decrease Increase Increase Increase Increase Increase Decrease

Other studies suggest that HIV infection has an impact on Cystatin C level. Cystatin C level was correlated with HIV-RNA in blood plasma and associated with HAART duration. This association was independent of age, Body Mass Index (BMI), CD4+ lymphocytes count and serum CRP concentration. [48] The interruption of HAART in SMART study was associated with an increase in serum cystatin C concentrations. Mocroft *et al.* could not conclude if this increase in cystatin C concentration was associated with a worsening kidney function. [49] Higher baseline HIV viral load and an increase in HIV viral load during the follow-up period was associated with a decline in GFR estimated with equations based on cystatin C in the FRAM study cohort. [50] Yet, Kalayjian *et al.* showed that estimated GFR with MDRD was improved after viral suppression by HAART. [51] HIV replication should affect kidney function. After HAART introduction, cystatin C was decreased, which might reflect an improving kidney function. The increase in Cystatin C level associated with HIV replication should be dependant of a temporary decrease in GFR. All of these studies have a limitation; GFR was not measured by a gold standard method.

In HIV-positive patients, higher cystatin C level was associated with higher Alanine aminotransferase concentration, with HCV and HBV co-infection and with history of injection drug use. [40,42] These associations could be dependant of GFR. HCV and HBV co-infection are risk factors for developing CKD in HIV-positive patients. Serum Cystatin C could be a more accurate biomarker of kidney function than serum creatinine in these situations. In cirrhotic patients, serum Cystatin C had a higher sensitivity than creatinine for detecting abnormal kidney function: GFR was measured by inulin clearance, cystatin C based equations showed a higher precision than Cockroft-Gault and MDRD equations. [52] However, all estimators based on cystatin C or creatinine overestimated measured GFR. [53]

The influence of CD4+ cell count on cystatin C level is still unclear. In NFHL cohort, a lower CD4+ cell count was associated with higher serum cystatin C level [40]. This association was also found in the FRAM study, and also demonstrated by Mauss *et al.* [54]. Jaroszewicz *et al.* did not find this association. [48] This association was not found in HIV-positive children. [55] CD4 + lymphopenia reflects the severity of HIV infection. The association between lower CD4+ cell count and higher serum cystatin could be due to the impact of HIV on kidney function.

Cystatin C is produced by most of the nucleated cells. HIV is able to enter in Monocytes, Macrophages, dendritic cells, Langherans' cells and lymphocytes. By the analysis of HIV-1 infected secretomes, Ciborowski *et al.* found that the infected macrophage secreted less cystatin C than a non-infected one. [56] There is no study to evaluate the impact of this observation on serum cystatin C level in HIV-positive subjects.

CYSTATIN C: A PREDICTIVE MARKER OF MORTALITY IN HIV-INFECTED SUBJECTS

In the general population, serum cystatin C is a predictor of cardiovascular events and of mortality particularly in elderly. [57, 58] Higher serum cystatin C level is associated with an increase risk of cardiovascular disease, this association is stronger for serum cystatin C than for reduced GFR estimated by MDRD. [59, 60] GFR was not measured in these studies, cystatin C could be an early biomarker of GFR decrease. The association between cystatin C and mortality could be dependent of GFR.

The SMART trial was a study designed to compare two strategies of anti-retroviral treatment. In the drug conservation arm, treatment was guided by immunological response and HAART was stopped when CD4+ lymphocyte count was above $350/mm^3$. In the second arm (viral suppression), HAART was maintained. [61] In the SMART trial, hs-CRP, IL-6 and D-dimers were associated with all cause mortality [62] and with risk of opportunistic diseases

[63]. In the SMART study, serum cystatin C was higher in the drug conservation arm [49], and the risk of cardiovascular disease was also higher in the drug conservation arm. [64] HIV-infected subjects with high cardiovascular risk assessed by Framingham score had higher serum cystatin C level. [65] eGFR based on cystatin C was associated with all cause mortality in HIV-infected subjects, although eGFR by MDRD formula was not associated with mortality in this population. 5 year-mortality rate of patients with an eGFRcys< 60mL/min/1.73m^2 was 22,9%. HIV-positive subjects with an eGFR based on cystatin C <60ml/min/1.73m^2 had an approximate 10% higher absolute risk of dying in the 5-year follow-up in the FRAM study. [66]

Serum Creatinine and Serum Cystatin C: Which is the Best Biomarker of Kidney Function in HIV-Positive Individuals?

Major cohort studies measuring cystatin C and creatinine are presented in table 4. In most of them, serum creatinine level was not different or lower in the study groups than in the control groups. Serum Cystatin C was often higher in the study group than in the controls groups. Study and controls group are often different in the prevalence of HCV infection, diabetes, proteinuria and hypertension, in smoking status, African American origins, in CRP level.All of these studies have in common an important limitation: the absence of GFR measurement by a gold standard method.

Currently, only 4 studies about cystatin C in HIV-infected subjects, measured GFR by a gold standard method. They are presented in table 5.

Barraclough *et al.* showed that the predictive performance of cystatin C to estimate GFR was inferior to all creatinine based methods: MDRD, CG, 24-hour creatinine clearance. MDRD and 24-hour creatinine clearance performed with the best level of precision and accuracy. All the patients were receiving HAART. A minority was malnourished and a majority had abnormal body composition. Most subjects had normal measured GFR. The first limitation is the small sample size: only 27 patients were included. Cystatin C was measured by a quantitative sandwich enzyme immunoassay (ELISA) which is now not recommended for measuring cystatin C. [67] Cystatin C must be determined by particle-enhanced immunonephelometric assay, the most evaluated assay.

For Beringer *et al.,* equations combining cystatin C and creatinine provided the least biased and most precise estimates in comparison with CG and equation with cystatin C only in HIV-positive subjects [68]. However, there was no statistically significant difference among the four evaluated equations. All of these equations underestimated measured GFR by Iothalamate clearance. Only 22 patients were recruited, most of them received HAART. They had relatively normal muscle mass assessed by Dual Energy X-ray Absoprtiometry (DEXA-Scan). The degree of underestimation of GFR with cystatin C was greater in patients with detectable HIV viral load than in patients with HIV viral load inferior to 400 copies/mL (-28.8% *vs* -14.3%). This difference was not statistically significant. [68]

In *Bonjoch et al.* study, cystatin C showed the strongest correlation (r=-0.76) with measured GFR by an isotopic method followed by CG, CKD-EPI, MDRD and 24h urine-creatinine clearance. GFR was measured only in 15 patients, all of them received HAART, and 80% of them had a suppressed viral load for at mean 90 months. [69] As cystatin C showed a good correlation with isotopic GFR, a subsequent analysis included 106 patients to validate other estimators. Serum Cystatin C was higher in patients exhibiting a detectable HIV viral load. Furthermore, in this study, an increase in cystatin C level was associated with a lower CD4+ cells count. In the 15 patients for whom isotopic GFR was measured, the only predictor for a decrease in the isotopic GFR was hypertension. Unexpectedly, HCV infection, HIV-infection severity were not associated with a decrease in GFR. This observation is probably due to the small sample size. Among these 15 patients, there were no patients with CD4+ cell counts inferior to $200/mm^3$, and only five patients with an HCV or HBV infection.

Table 4. Major cohort studies of cystatin C in HIV-infected subjects

	Cohort of HIV-infected subjects	Control Cohort	Differences between HIV-positive cohort and controls	Results	Risk factors for higher serum cystatin C level in HIV-participants
Odden et al. 2007 [42]	The FRAM study n=519 Between the ages of 33 and 45 years	CARDIA study n=290	Controls were older, more women in control cohort, BMI higher in control cohort smoking status, hypertension, dyslipidemia, proteinuria, HCV infection	Serum Cystatin C higher in the FRAM study cohort The prevalence of cystatin C higher than 1 mg/L was 31 % in the HIV-infected participants and only 4% in controls No difference in serum creatinine level between HIV-infected partcipants and controls	Lower HDL-c level, higher uric acid level, proteinuria, hypertension, higher CRP Current smoking CD4 lymphopenia Co-infection HCV, current heroin use, Longer duration of efavirenz and indinavir use
Mauss et al. 2008 [54]	treatment naïve, caucasian patients (n=261)	Healthy volunteers (n=193)	More male in the study group Older age, higher BMI in control group CRP level, HCV co-infection, smoking status were not reported in the patients' characteristics	Mean eGFRcreat by MDRD higher in the study group Mean eGFR cys[a] lower in the study group Prevalence of CKD (stage 2 and more) in the study group with MDRD= 23%, with eGFR cys= 41%	Positive correlation between cystatin C level and HIV viral load
Jones et al. 2008 [40]	The NFHL cohort n=250	The NHANES n=2628	More african american subjects in NFHL, Greater prevalence of hypertension, diabetes, liver disease in NFHL, higher CRP level in NFHL, lower albumin level in NFHL	Serum creatinine level lower in NFHL, Serum cystatin C level higher in NFHL cohort prevalence of eGFRcys[b]<60mL/min/1.73m^{2} =15.2 % prevalence of eGFRcreat <60 mL/min/1.73m^2 using MDRD=2.4%	HCV infection, liver disease, lower CD4+ lymphocyte count, HIV viral load current injection drug use lower serum albumin level
Neuhaus et al. 2010 [47]	SMART (n=494)	MESA study (n=5386)	Older age, more women in control cohort, more black in SMART, lower BMI in SMART study, prevalence of dyslipidemia, current smoking, diabetes, hypertension greater in SMART	Cystatin C was 27.2 % higher in SMART study participants	

[a] eGFRcys(mL/min)=74,835/(cystatin C(mg/L)$^{1.333}$)
[b] eGFRcys(mL/min/1.73m^2)=76,7 (cystatin C$^{-1.18}$)

Table 5. Comparison of the 4 studies about the predictive performance of cystatin C in estimation of Glomerular Filtration Rate in HIV-infected subject

	n patients	GFR measurement	eGFRCysC equation	eGFRcreat	Results
Barraclough et al. 2009 [67]	27	Clearance of DTPA	eGFR=86.7/cysC-4.2	MDRD, CG	eGFRCys less accurate than eGFRcreat
Beringer et al. 2010 [68]	22	Clearance of Iothalamate	$eGFR=127.7*CysC^{-1.17}*age^{-0.13}*(0.91$ if female)*(1.06 if African American) eGFRcysC,creat= $177.6*Creat^{-0.65}*CysC^{-0.57}*age^{-0.20}$ *0.82 if female and/or 1.11 if African-American	MDRD, CG	No statiscally significant differences
Bonjoch et al. 2010 [86]	15	Isotopic Clearance	CysC alone, No equation	MDRD, CG, CKD-EPI	CysC well correlated with isotopic GFR
van Deventer et al. 2010 [70]	20	^{51}Cr-EDTA clearance	$eGFR=10^{2.35}*10^{(CysC*-0.33)}*10^{(-0.003*age)}$	MDRD, CKD-EPI	eGFRCysC more accurate than eGFcreat

In a South African cohort, van Deventer *et al.* sought to develop new prediction equations for estimating GFR, and to determine whether Cystatin C based equations offer an advantage over MDRD, CKD-EPI (Chronic Kidney Disease Epidemiology Collaboration) equation and developed eGFRcr equations. One hundred patients were included, 50 in the development dataset and 50 in the test dataset. Twenty patients were HIV-positive, 11 of them in development dataset and 9 of them in the test dataset. The development dataset permitted to formulate a new equation based on serum cystatin C and another one based on serum creatinine. In the overall study population, eGFRcys equation was more precise than eGFR based on creatinine, particularly in patients with GFR>60 mL/min/1.73m^2. The accuracy within 30% of serum cystatin C equation was 84% *vs* 74% for MDRD. In this cohort, GFR was measured in 100 patients by a plasma clearance of ^{51}Cr EDTA. In the 20 patients HIV-positive, 15 had a Body Mass index <20kg/m^2. The equation based on cystatin C is the most accurate in HIV-positive patients (70% within 30% *vs* 65 % for MDRD). HIV associated variables as viral load and CD4+ cell count were not reported. [70]

All of these studies have one major limitation. The sample size is small. More studies are needed to evaluate predictive performance of cystatin C in HIV-positive patients using gold-standard methods to measure GFR and a standardized technique of serum cystatin C measurement.

Urinary Cystatin C as a Marker of HAART-Induced Nephropathies

Cystatin C may be an alternative biomarker of GFR in HIV-infected patients. The other challenge is the diagnosis of HAART-induced nephropathies. Uchida and Gotoh were the first to suggest that Urinary Cystatin C/ Urinary Creatinine (UCysC/UCr) could be a biomarker of tubular dysfunction. [71] Increased UCysC concentrations allow an accurate detection of tubular dysfunction. [72] UCysC/UCr was described as an accurate biomarker to identify patients requiring renal replacement therapy in case of Acute Tubular Necrosis, and was a good predictor of mortality in Intensive Care Unit (ICU). [73,74] UCysC/UCr reflects tubular dysfunction independently of GFR. [73]

HAART provides renal dysfunction. Tenofovir disoproxil fumarate (TDF) is a nucleotide reverse transcriptase inhibitor, secreted at the level of the proximal tubule. TDF can induce a toxic mitochondriopathy in the proximal tubule. [75, 76] Exposure to TDF is an independent risk factor for CKD development in HIV-positive patients. [11, 77,9] Tubular impairment induced by TDF use is not always reversible after TDF interruption. [78,79] TDF tubular toxicity causes a "Fanconi Syndrome" with a renal leak of phosphates, small proteins, glucose, uric acid and bicarbonates. [80] There is a need for novel biomarkers to faster detect TDF induced tubulopathies. UCysC should be a biomarker to monitor TDF renal toxicity. *Jaafar et al.* assessed the predictive value of UCysC/Ucr for the diagnosis a Fanconi syndrome induced by TDF. [29] UCysC was measured in thirty-seven patients referred for a suspicion of TDF or adefovir induced nephropathy. Only eleven of them met the diagnostic criteria of Fanconi syndrome. All patients in the Fanconi positive group had detectable UCysC. With a threshold for UCysC/UCr of 14µg/mmol, the positive predictive value is 76.9%, and the negative predictive value of 95.8% in Fanconi Syndrome diagnosis. UCysC/UCr is an efficient biomarker for TDF tubulopathy. Studies with larger samples and comparing the predictive value of urinary cystatin C with other biomarkers of kidney injury as ß2-microglobulin, NGAL, KIM-1 might be of interest.

Conclusions

Serum creatinine is far from being a good biomarker of kidney function in HIV-infected subjects principally because of a decrease in creatinine generation. Better assessment of kidney function in HIV-positive subjects is now necessary to prevent cardiovascular disease and deaths. IDSA will publish new recommendations in 2012. CKD in HIV-positive patients must be treated as in the general population. Several studies showed good outcomes of kidney transplantation in HIV-infected subjects, in spite of a greater incidence of acute rejection in HIV-infected subjects than in controls. [81, 82, 83] Moreover, HAART use in patients with CKD is often inadequate, underexposure or inadequate adjustment of HAART result in a 22.5-35.5% excess mortality. [84] Cystatin C might be a promising biomarker to assess kidney function in HIV-positive patients. Yet, the influence of CRP, HIV viral load, smoking status on serum Cystatin C level is a potential limitation to use this biomarker to estimate GFR in HIV-positive patients. Recent studies showed contradictory results. No study has included enough patients to show a significant difference between serum cystatin C and serum creatinine to estimate GFR.

Furthemore, HAART contribute to CKD in HIV-positive patients. The use of urinary cystatin C as an early diagnosis tool in tenofovir-induced tubulopathies is promising. Tenofovir is not the only antiretroviral providing kidney impairment.

Additional studies are needed to determine the predictive value of cystatin C in the estimation of GFR, the sensitivity and the specificity of urinary cystatin C in the diagnosis of HAART induced nephropathies.

REFERENCES

[1] Quinn TC. HIV epidemiology and the effects of antiviral therapy on long-term consequences. *AIDS*. 2008;22 Suppl 3:S7-12.

[2] Seaberg EC, Muñoz A, Lu M, et al. Association between highly active antiretroviral therapy and hypertension in a large cohort of men followed from 1984 to 2003. *AIDS*. 2005;19(9):953-960.

[3] Koster JC, Remedi MS, Qiu H, Nichols CG, Hruz PW. HIV protease inhibitors acutely impair glucose-stimulated insulin release. *Diabetes*. 2003;52(7):1695-1700.

[4] Riddler SA, Li X, Chu H, et al. Longitudinal changes in serum lipids among HIV-infected men on highly active antiretroviral therapy. *HIV Med*. 2007;8(5):280-287.

[5] Schwartz EJ, Szczech LA, Ross MJ, et al. Highly active antiretroviral therapy and the epidemic of HIV+ end-stage renal disease. *J. Am. Soc. Nephrol*. 2005;16(8):2412-2420.

[6] Fernando SK, Finkelstein FO, Moore BA, Weissman S. Prevalence of chronic kidney disease in an urban HIV infected population. *Am. J. Med. Sci*. 2008;335(2):89-94.

[7] Crum-Cianflone N, Ganesan A, Teneza-Mora N, et al. Prevalence and factors associated with renal dysfunction among HIV-infected patients. *AIDS Patient Care STDS*. 2010;24(6):353-360.

[8] Lucas GM, Lau B, Atta MG, et al. Chronic kidney disease incidence, and progression to end-stage renal disease, in HIV-infected individuals: a tale of two races. *J. Infect. Dis*. 2008;197(11):1548-1557.

[9] Overton ET, Nurutdinova D, Freeman J, Seyfried W, Mondy KE. Factors associated with renal dysfunction within an urban HIV-infected cohort in the era of highly active antiretroviral therapy. *HIV Med*. 2009;10(6):343-350.

[10] Vupputuri S, Batuman V, Muntner P, et al. The risk for mild kidney function decline associated with illicit drug use among hypertensive men. *Am. J. Kidney Dis*. 2004;43(4):629-635.

[11] Mocroft A, Kirk O, Reiss P, et al. Estimated glomerular filtration rate, chronic kidney disease and antiretroviral drug use in HIV-positive patients. *AIDS*. 2010;24(11):1667-1678.

[12] Tordato F, Cozzi Lepri A, Cicconi P, et al. Evaluation of glomerular filtration rate in HIV-1-infected patients before and after combined antiretroviral therapy exposure. *HIV Med*. 2011;12(1):4-13.

[13] Estrella MM, Parekh RS, Abraham A, et al. The impact of kidney function at highly active antiretroviral therapy initiation on mortality in HIV-infected women. *J. Acquir. Immune Defic. Syndr*. 2010;55(2):217-220.

[14] Mayor AM, Dworkin M, Quesada L, Ríos-Olivares E, Hunter-Mellado RF. The morbidity and mortality associated with kidney disease in an HIV-infected cohort in Puerto Rico. *Ethn Dis*. 2010;20(1 Suppl 1):S1-163-7.

[15] George E, Lucas GM, Nadkarni GN, et al. Kidney function and the risk of cardiovascular events in HIV-1-infected patients. *AIDS*. 2010;24(3):387-394.

[16] Szczech LA, Hoover DR, Feldman JG, et al. Association between renal disease and outcomes among HIV-infected women receiving or not receiving antiretroviral therapy. *Clin. Infect. Dis*. 2004;39(8):1199-1206.

[17] de Silva TI, Post FA, Griffin MD, Dockrell DH. HIV-1 infection and the kidney: an evolving challenge in HIV medicine. *Mayo Clin. Proc*. 2007;82(9):1103-1116.

[18] Gupta SK, Eustace JA, Winston JA, et al. Guidelines for the management of chronic kidney disease in HIV-infected patients: recommendations of the HIV Medicine Association of the Infectious Diseases Society of America. *Clin. Infect. Dis*. 2005;40(11):1559-1585.

[19] Estrella MM, Fine DM. Screening for chronic kidney disease in HIV-infected patients. *Adv Chronic Kidney Dis*. 2010;17(1):26-35.

[20] Stevens LA, Coresh J, Greene T, Levey AS. Assessing kidney function--measured and estimated glomerular filtration rate. *N. Engl. J. Med*. 2006;354(23):2473-2483.

[21] Perrone RD, Madias NE, Levey AS. Serum creatinine as an index of renal function: new insights into old concepts. *Clin. Chem*. 1992;38(10):1933-1953.

[22] Dharnidharka VR, Kwon C, Stevens G. Serum cystatin C is superior to serum creatinine as a marker of kidney function: a meta-analysis. *Am. J. Kidney Dis*. 2002;40(2):221-226.

[23] Séronie-Vivien S, Delanaye P, Piéroni L, et al. Cystatin C: current position and future prospects. *Clin. Chem. Lab. Med*. 2008;46(12):1664-1686.

[24] Knight EL, Verhave JC, Spiegelman D, et al. Factors influencing serum cystatin C levels other than renal function and the impact on renal function measurement. *Kidney Int*. 2004;65(4):1416-1421.

[25] Stevens LA, Schmid CH, Greene T, et al. Factors other than glomerular filtration rate affect serum cystatin C levels. *Kidney Int*. 2009;75(6):652-660.

[26] Conti M, Moutereau S, Zater M, et al. Urinary cystatin C as a specific marker of tubular dysfunction. *Clin. Chem. Lab. Med*. 2006;44(3):288-291.

[27] Herget-Rosenthal S, van Wijk JAE, Bröcker-Preuss M, Bökenkamp A. Increased urinary cystatin C reflects structural and functional renal tubular impairment independent of glomerular filtration rate. *Clin. Biochem*. 2007;40(13-14):946-951.

[28] Röling J, Schmid H, Fischereder M, Draenert R, Goebel FD. HIV-associated renal diseases and highly active antiretroviral therapy-induced nephropathy. *Clin. Infect. Dis*. 2006;42(10):1488-1495.

[29] Jaafar A, Séronie-Vivien S, Malard L, et al. Urinary cystatin C can improve the renal safety follow-up of tenofovir-treated patients. *AIDS*. 2009;23(2):257-259.

[30] Delanaye P, Cavalier E, Maillard N, et al. [Creatinine: past and present]. *Ann. Biol. Clin. (Paris)*. 2010;68(5):531-543.

[31] Visnegarwala F, Shlay JC, Barry V, et al. Effects of HIV infection on body composition changes among men of different racial/ethnic origins. *HIV Clin Trials*. 2007;8(3):145-154.

[32] Delpierre C, Bonnet E, Marion-Latard F, et al. Impact of HIV infection on total body composition in treatment-naive men evaluated by dual-energy X-ray absorptiometry comparison of 90 untreated HIV-infected men to 241 controls. *J Clin Densitom*. 2007;10(4):376-380.

[33] Grinspoon S, Mulligan K. Weight loss and wasting in patients infected with human immunodeficiency virus. *Clin. Infect. Dis*. 2003;36(Suppl 2):S69-78.

[34] McDermott AY, Terrin N, Wanke C, et al. CD4+ cell count, viral load, and highly active antiretroviral therapy use are independent predictors of body composition

alterations in HIV-infected adults: a longitudinal study. *Clin. Infect. Dis.* 2005;41(11):1662-1670.

[35] Nahlen BL, Chu SY, Nwanyanwu OC, et al. HIV wasting syndrome in the United States. *AIDS.* 1993;7(2):183-188.

[36] Brinkman K, Smeitink JA, Romijn JA, Reiss P. Mitochondrial toxicity induced by nucleoside-analogue reverse-transcriptase inhibitors is a key factor in the pathogenesis of antiretroviral-therapy-related lipodystrophy. *Lancet.* 1999;354(9184):1112-1115.

[37] Cocchetto DM, Tschanz C, Bjornsson TD. Decreased rate of creatinine production in patients with hepatic disease: implications for estimation of creatinine clearance. *Ther Drug Monit.* 1983;5(2):161-168.

[38] Singal A, Anand BS. Management of hepatitis C virus infection in HIV/HCV co-infected patients: clinical review. *World J. Gastroenterol.* 2009;15(30):3713-3724.

[39] Chun HM, Fieberg AM, Hullsiek KH, et al. Epidemiology of Hepatitis B virus infection in a US cohort of HIV-infected individuals during the past 20 years. *Clin. Infect. Dis.* 2010;50(3):426-436.

[40] Jones CY, Jones CA, Wilson IB, et al. Cystatin C and creatinine in an HIV cohort: the nutrition for healthy living study. *Am. J. Kidney Dis.* 2008;51(6):914-924.

[41] Fischer MJ, Wyatt CM, Gordon K, et al. Hepatitis C and the risk of kidney disease and mortality in veterans with HIV. *J. Acquir. Immune Defic. Syndr.* 2010;53(2):222-226.

[42] Odden MC, Scherzer R, Bacchetti P, et al. Cystatin C level as a marker of kidney function in human immunodeficiency virus infection: the FRAM study. *Arch. Intern. Med.* 2007;167(20):2213-2219.

[43] Berglund F, Killander J, Pompeius R. Effect of trimethoprim-sulfamethoxazole on the renal excretion of creatinine in man. *J. Urol.* 1975;114(6):802-808.

[44] Myre SA, McCann J, First MR, Cluxton RJ. Effect of trimethoprim on serum creatinine in healthy and chronic renal failure volunteers. *Ther Drug Monit.* 1987;9(2):161-165.

[45] Maki DG, Fox BC, Kuntz J, Sollinger HW, Belzer FO. A prospective, randomized, double-blind study of trimethoprim-sulfamethoxazole for prophylaxis of infection in renal transplantation. Side effects of trimethoprim-sulfamethoxazole, interaction with cyclosporine. *J. Lab. Clin. Med.* 1992;119(1):11-24.

[46] Mofenson LM, Brady MT, Danner SP, et al. Guidelines for the Prevention and Treatment of Opportunistic Infections among HIV-exposed and HIV-infected children: recommendations from CDC, the National Institutes of Health, the HIV Medicine Association of the Infectious Diseases Society of America, the Pediatric Infectious Diseases Society, and the American Academy of Pediatrics. *MMWR Recomm Rep.* 2009;58(RR-11):1-166.

[47] Neuhaus J, Jacobs DR, Baker JV, et al. Markers of inflammation, coagulation, and renal function are elevated in adults with HIV infection. *J. Infect. Dis.* 2010;201(12):1788-1795.

[48] Jaroszewicz J, Wiercinska-Drapalo A, Lapinski TW, et al. Does HAART improve renal function? An association between serum cystatin C concentration, HIV viral load and HAART duration. *Antivir. Ther. (Lond.).* 2006;11(5):641-645.

[49] Mocroft A, Wyatt C, Szczech L, et al. Interruption of antiretroviral therapy is associated with increased plasma cystatin C. *AIDS.* 2009;23(1):71-82.

[50] Longenecker CT, Scherzer R, Bacchetti P, et al. HIV viremia and changes in kidney function. *AIDS.* 2009;23(9):1089-1096.

[51] Kalayjian RC, Franceschini N, Gupta SK, et al. Suppression of HIV-1 replication by antiretroviral therapy improves renal function in persons with low CD4 cell counts and chronic kidney disease. *AIDS*. 2008;22(4):481-487.
[52] Woitas RP, Stoffel-Wagner B, Flommersfeld S, et al. Correlation of Serum Concentrations of Cystatin C and Creatinine to Inulin Clearance in Liver Cirrhosis. *Clin Chem*. 2000;46(5):712-715.
[53] Pöge U, Gerhardt T, Stoffel-Wagner B, et al. Calculation of glomerular filtration rate based on Cystatin C in cirrhotic patients. *Nephrology Dialysis Transplantation*. 2006;21(3):660 -664.
[54] Mauss S, Berger F, Kuschak D, et al. Cystatin C as a marker of renal function is affected by HIV replication leading to an underestimation of kidney function in HIV patients. *Antivir. Ther. (Lond.)*. 2008;13(8):1091-1095.
[55] Esezobor CI, Iroha E, Oladipo O, et al. Kidney function of HIV-infected children in Lagos, Nigeria: using Filler's serum cystatin C-based formula. *J Int AIDS Soc*. 2010;13:17.
[56] Ciborowski P, Kadiu I, Rozek W, et al. Investigating the human immunodeficiency virus type 1-infected monocyte-derived macrophage secretome. *Virology*. 2007;363(1):198-209.
[57] Shlipak MG, Katz R, Sarnak MJ, et al. Cystatin C and prognosis for cardiovascular and kidney outcomes in elderly persons without chronic kidney disease. *Ann. Intern. Med*. 2006;145(4):237-246.
[58] Shlipak MG, Sarnak MJ, Katz R, et al. Cystatin C and the risk of death and cardiovascular events among elderly persons. *N. Engl. J. Med*. 2005;352(20):2049-2060.
[59] Ix JH, Shlipak MG, Chertow GM, Whooley MA. Association of cystatin C with mortality, cardiovascular events, and incident heart failure among persons with coronary heart disease: data from the Heart and Soul Study. *Circulation*. 2007;115(2):173-179.
[60] Hoke M, Amighi J, Mlekusch W, et al. Cystatin C and the risk for cardiovascular events in patients with asymptomatic carotid atherosclerosis. *Stroke*. 2010;41(4):674-679.
[61] El-Sadr WM, Lundgren JD, Neaton JD, et al. CD4+ count-guided interruption of antiretroviral treatment. *N. Engl. J. Med*. 2006;355(22):2283-2296.
[62] Kuller LH, Tracy R, Belloso W, et al. Inflammatory and coagulation biomarkers and mortality in patients with HIV infection. *PLoS Med*. 2008;5(10):e203.
[63] Rodger AJ, Fox Z, Lundgren JD, et al. Activation and coagulation biomarkers are independent predictors of the development of opportunistic disease in patients with HIV infection. *J. Infect. Dis*. 2009;200(6):973-983.
[64] Phillips AN, Carr A, Neuhaus J, et al. Interruption of antiretroviral therapy and risk of cardiovascular disease in persons with HIV-1 infection: exploratory analyses from the SMART trial. *Antivir. Ther. (Lond.)*. 2008;13(2):177-187.
[65] Falasca K, Ucciferri C, Mancino P, et al. Cystatin C, adipokines and cardiovascular risk in HIV infected patients. *Curr. HIV Res*. 2010;8(5):405-410.
[66] Choi A, Scherzer R, Bacchetti P, et al. Cystatin C, albuminuria, and 5-year all-cause mortality in HIV-infected persons. *Am. J. Kidney Dis*. 2010;56(5):872-882.

[67] Barraclough K, Er L, Ng F, et al. A comparison of the predictive performance of different methods of kidney function estimation in a well-characterized HIV-infected population. *Nephron Clin Pract*. 2009;111(1):c39-48.

[68] Beringer PM, Owens H, Nguyen A, et al. Estimation of glomerular filtration rate by using serum cystatin C and serum creatinine concentrations in patients with human immunodeficiency virus. *Pharmacotherapy*. 2010;30(10):1004-1010.

[69] Bonjoch A, Bayés B, Riba J, et al. Validation of estimated renal function measurements compared with the isotopic glomerular filtration rate in an HIV-infected cohort. *Antiviral Res*. 2010;88(3):347-354.

[70] van Deventer HE, Paiker JE, Katz IJ, George JA. A comparison of cystatin C- and creatinine-based prediction equations for the estimation of glomerular filtration rate in black South Africans. *Nephrol Dial Transplant*. 2010. doi: 10.1093/ndt/gfq621

[71] Uchida K, Gotoh A. Measurement of cystatin-C and creatinine in urine. *Clin. Chim. Acta*. 2002;323(1-2):121-128.

[72] Conti M, Moutereau S, Zater M, et al. Urinary cystatin C as a specific marker of tubular dysfunction. *Clin. Chem. Lab. Med*. 2006;44(3):288-291.

[73] Herget-Rosenthal S, Poppen D, Hüsing J, et al. Prognostic value of tubular proteinuria and enzymuria in nonoliguric acute tubular necrosis. *Clin. Chem*. 2004;50(3):552-558.

[74] Nejat M, Pickering JW, Walker RJ, et al. Urinary cystatin C is diagnostic of acute kidney injury and sepsis, and predicts mortality in the intensive care unit. *Crit Care*. 2010;14(3):R85.

[75] Zimmermann AE, Pizzoferrato T, Bedford J, et al. Tenofovir-associated acute and chronic kidney disease: a case of multiple drug interactions. *Clin. Infect. Dis*. 2006;42(2):283-290.

[76] Kohler JJ, Hosseini SH, Hoying-Brandt A, et al. Tenofovir renal toxicity targets mitochondria of renal proximal tubules. *Lab. Invest*. 2009;89(5):513-519.

[77] Mauss S, Berger F, Schmutz G. Antiretroviral therapy with tenofovir is associated with mild renal dysfunction. *AIDS*. 2005;19(1):93-95.

[78] Kinai E, Hanabusa H. Progressive renal tubular dysfunction associated with long-term use of tenofovir DF. *AIDS Res. Hum. Retroviruses*. 2009;25(4):387-394.

[79] Horberg M, Tang B, Towner W, et al. Impact of tenofovir on renal function in HIV-infected, antiretroviral-naive patients. *J. Acquir. Immune Defic. Syndr*. 2010;53(1):62-69.

[80] Brim NM, Cu-Uvin S, Hu SL, O'Bell JW. Bone disease and pathologic fractures in a patient with tenofovir-induced Fanconi syndrome. *AIDS Read*. 2007;17(6):322-328, C3.

[81] Mazuecos A, Fernandez A, Andres A, et al. HIV infection and renal transplantation. *Nephrol Dial Transplant*. 2010 doi:10.1093/ndt/gfq592

[82] Landin L, Rodriguez-Perez JC, Garcia-Bello MA, et al. Kidney transplants in HIV-positive recipients under HAART. A comprehensive review and meta-analysis of 12 series. *Nephrol. Dial. Transplant*. 2010;25(9):3106-3115.

[83] Stock PG, Barin B, Murphy B, et al. Outcomes of Kidney Transplantation in HIV-Infected Recipients. *N Engl J Med*. 2010;363(21):2004-2014.

[84] Choi AI, Rodriguez RA, Bacchetti P, et al. Low rates of antiretroviral therapy among HIV-infected patients with chronic kidney disease. *Clin. Infect. Dis*. 2007;45(12):1633-1639.

[85] Perazella MA. Tenofovir-induced kidney disease: an acquired renal tubular mitochondriopathy. *Kidney Int*. 2010;78(11):1060-1063.
[86] Bonjoch A, Bayés B, Riba J, et al. Validation of estimated renal function measurements compared with the isotopic glomerular filtration rate in an HIV-infected cohort. *Antiviral Res*. 2010. Available at: *http://www.ncbi.nlm.nih.gov/pubmed/20887753* [Accédé Octobre 17, 2010].

In: Cystatins: Protease Inhibitors … ISBN: 978-1-61209-343-7
Editors: John B. Cohen and Linda P. Ryseck

Chapter 5

CYSTATIN C IN ACUTE CORONARY SYNDROMES: THE INVESTIGATION SHOULD GO ON

S. Ferraro, G. Marano, B. Suardi, P. La Musta, E. M. Biganzoli, P. Boracchi and A. S. Bongo
SCDO, Cardiologia 2, Ospedale Maggiore, Novara, Italy
Istituto Statistica Medica e Biometria, Università degli Studi, Milano Italy

ABSTRACT

Background: Cystatin C(CC) could contribute adding value to traditional cardiovascular risk factors in the prediction of adverse events in Acute Coronary Syndromes (ACS). Aim of present chapter is to assess the evidence on the prognostic value of CC in ACS patients, by reviewing current literature.

Methods and Results: by Pub Med, Embase, Ovid, 29 papers were identified, and 9 longitudinal observational studies in which serum CC was investigated as prognostic marker in ACS, were selected. The reference populations allowed to classify studies in: Group A) 4 studies with 50-60% of ACS patients and a sample size of 450-1030 patients; Group B) 3 studies on non ST Elevation Acute Coronary Syndromes (NSTEACS) patients and a sample size of 380-1120 patients; Group C) 1 study with 160 ST Elevation Myocardial infarction (STEMI) and NSTEACS patients; Group D) 1 study with 71 STEMI. Outcome: The end-point was "time to": 1) cardiovascular death, non fatal myocardial infarction (MI), or stroke, with a median follow up of 3 years; 2) major adverse cardiovascular events (MACE), with a variable median follow up of 3 years, 1 year, and 6 months; 3) death or recurrence of MI, with a median follow up of 3 years, 6 months. Prognostic role: According to results in: - Group A CC levels >1.3 mg/L were a significant risk factor for fatal and non fatal cardiovascular events (HR estimates ranging from 1.72 to 2.27); -Group B, for CC levels >1.25 mg/L, the risk of death was about 12 times greater than that of patients with lower CC levels; in a second study, for CC levels >1.01 mg/L a significant prognostic value on death was found (HR=4.07,(CI:2.16-7.66)). According to the recurrence of AMI evaluated in two studies, only one assessed a significant prognostic role (HR=1.95(CI:1.05-3.63)). For another study , according to a composite endpoint of fatal and non fatal cardiac events, with a CC >0.93 mg/L, a HR of 1.57(CI:1.04-2.49) was shown. For Groups C and D similar results on the prognostic value of CC on MACE were obtained: HR=9.43(CI:4.0-21.8) for CC levels >1.05 mg/L, HR =2.17(CI:1.07-6.98) for CC levels >0.96 mg/L.

Discussion: Despite the low number and the poor level of evidences, there is a general agreement on the prognostic value of CC in ACS, encouraging further studies on homogeneous cases series of STEMI and NSTEMI. In this patients the further exploitation of the marker as therapeutic target could improve their management for secondary prevention purposes.

INTRODUCTION

1.0. Cystatin C: The Clinical Evidence of a New Cardiovascular Risk Factor

1.1. Cystatin C: Evidences for a Putative Marker of Atherosclerosis

Cystatin C (CC) is an endogenous inhibitor of a peculiar family of cystein proteases, cathepsins, that are ubiquitously secreted by nucleated cells, at a constant rate [1]. The dynamic equilibrium between these proteases and their inhibitor modulates the catabolism of all proteins, thereby this is critical in some physiological pathways as antigen presentation, hormone processing and degradation of extracellular matrix (ECM)[2,3]. Some studies reported for a wide spectrum of diseases, that the change from a physiological to a pathological state, could be linked to the unbalance between CC and cathepsins [4]. This last could cause an increase of bone resorption, neutrophil chemotaxis, cell proliferation, or a reduction of the resistance to bacterial and viral infections, or it could affect in general, tissue remodeling and vessel homeostasis [5-12]. Low cellular levels of CC could cause a spread cathepsin activity, that mainly triggers ECM degradation and inflammation, leading processes for the progression of different diseases such as cancer, renal failure, osteoporosis, atherosclerosis and finally cardiovascular disease (CVD) [5,6,8,9,12-14].

The complex inflammatory network, gauged on the balance between CC and cathepsin is unraveled in figure 1, with a peculiar reference to its role in all phases of atherosclerosis evolution [13,15-17]. The interest on CC as marker of atherosclerosis has gained, according to initial evidences on its involvement in the remodeling and degradation of ECM of arterial wall [13]. In fact CC inhibits cathepsins with elastolytic cystein protease activity, responsible for the catabolism of elastin and collagen, that represent the main constituents of extracellular matrix ECM of vascular wall. Thus CC directly down-regulates growth and destabilization of atherosclerotic plaques [18]. According to this role, CC concentration is reported as severely decreased in human atherosclerotic lesions, respect to normal vessel wall, jointly to a spread increase of some cathepsins (S, L, K ,B, H) [16,17,19]. High cathepsin levels could trigger the progression of atherosclerotic process, by enhancing the ECM degradation and by increasing the plaque size [16,19]. Moreover also the contents of macrophanges and T cells in the plaque seems to be increased [16]. As reported in figure 1, the cellular secretion of CC levels, in the bloodstream, is modulated by a complex network of pro-inflammatory cytokines. The increase of CC serum level is induced by some cytokines (i.e. TNFβ1) to counterbalance the evolution of inflammation, since it affects the phagocytosis and chemotaxis of leukocytes [7, 20].

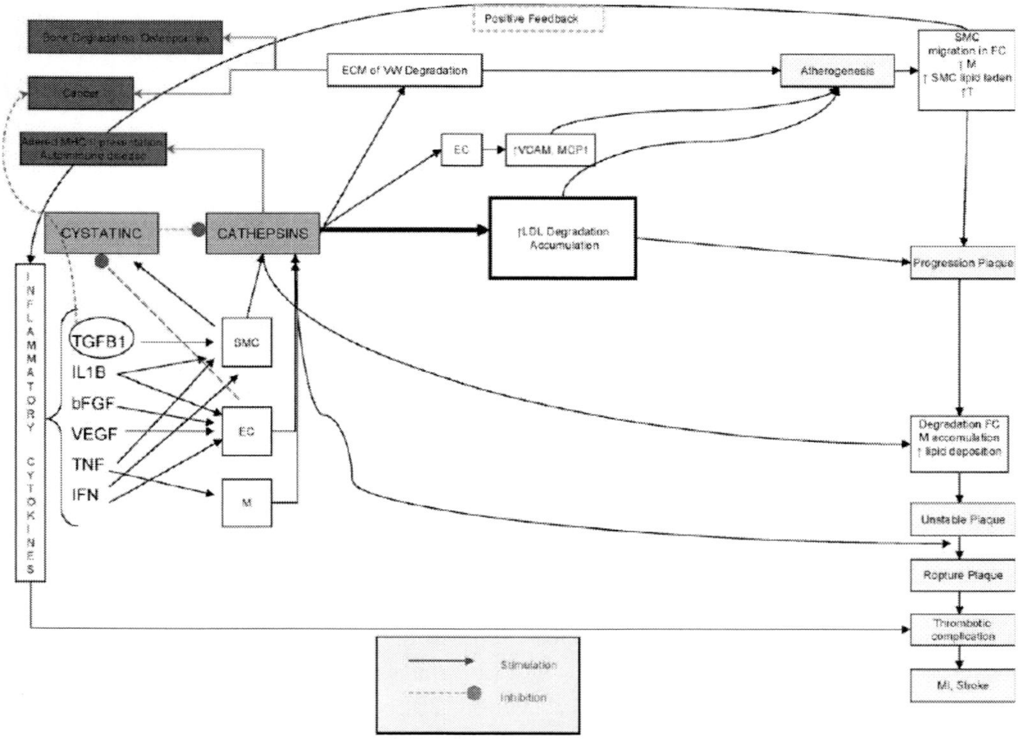

Acronims: M = Macrophages; SMC = Smooth muscle cells; EC = Endothelial cells; VW = Vascular Wall; VCAM-1= Vascular Cellular Adhesium Molecule-1; MCP1= Macrophage Chemoattranct Protein 1; FC = Fibrosis Cap; MHC II = Major Histocompatibility Complex Class II.

Figure 1. In this figure is reported the complex network of pathological mechanisms triggered by the unbalance between Cystatin C and Cathepsins, underlying atherosclerosis and several other diseases with a potential inflammatory hub (cancer, autoimmune disease, osteoporosis, renal impairment). As reported, inflammatory cytokines should cover a critical role to perturb this equilibrium, according to the fact that they could: 1) stimulate Ms, SMCs, ECs to release high levels of cathepsins (i.e. TGFβ1,VEGF, bFGF, TNFα , IFNγ); 2) inhibit ECs (i.e. TNFβ, bFGF) or stimulate SMCs (i.e. TGFβ1) to release CC. The spread increase of Cathepsins promotes the initiation of a wide spectrum of pathological mechanisms such as migration of Ms, ECM degradation, endothelial activation (expression of specific adhesion molecules VCAM1 and MCP1on the surface of vascular ECs), lipid accumulation in VW, causing the atherosclerosis progression. Moreover these pathological mechanisms further enhance the production of proinflammatory cytokines, by a positive feed-back. The interaction of these mechanisms with the environment rich in cathepsins and proinflammatory cytokines is basic to trigger thrombotic complications and the further onset of cardiovascular events (myocardial infarction, stroke).

The translation of CC from research to the clinical context of cardiovascular disease is mainly supported by its physiopathological involvement in atherosclerosis. As a matter of fact, in contrast to the other inflammatory biomarkers clinically available (IL6, C Reactive Protein(CRP)), CC is the one with reliable evidences on a primary role in the plaque destabilization and in the progression of the cardiovascular disease leading to the onset of an acute event. According to the fact that the atherosclerotique progression is associated to an ongoing rowdy inflammatory process that triggers CC release, high serum CC levels, might predict the risk for atherosclerosis. Actually, in the clinical setting, the plausible link between

CC levels and inflammation, relies on the correlation between serum marker levels and CRP levels [21,22]. Nevertheless, the introduction of CC serum measurement in this context, should be supported by the evidence that CC could contribute an added value to the other inflammatory markers in the bulk of the marker tools for atherosclerosis.

Actually there is evidence that the progression of the atherosclerotic plaques, from a stable to a vulnerable state represents one of the main critical processes, promoting thrombotic complications and consequently the onset of an Acute Coronary Syndrome (ACS) [23]. The availability of a serum marker as CC, that besides detecting inflammation triggers the progression of atherosclerotic plaque, gains a great potential to improve the risk assessment for ACS. This is relevant accounting that many stable coronary plaques are angiographically undetectable, thus no clinical methodology currently could predict the site and risk of future infarction.

1.2. Cystatin C: A Potential Marker of Subclinical Renal Impairment

Before the characterization of CC as inflammatory marker, an extensive body of literature reported that high CC serum levels were associated to renal impairment. Several evidences currently agree that CC is a better marker of glomerular filtration rate (GFR) than serum creatinine [24-26]. Creatinine serum concentrations are mainly affected by tubular secretion, age, sex, muscle mass, physical activity and diet [27]. Thus the equations of Cockcroft-Gault and the Modification of Diet in Renal Disease (MDRD) are used to correct creatinine levels and to improve the GFR detection [28].

Although CC is influenced by some of the previous reported factors the clinical use of equations integrating CC instead of creatinine levels could enhance the GFR estimation [21,30,31-33]. This could be due to the fact that in contrast to creatinine, CC is freely filtered by the glomerulus, without secretion or further reabsorption to the blood flow and its levels directly reflect glomerulus function [26]. Moreover CC based equations were reported to detect such a small decrease in GFR as to allow the assessment of renal dysfunction in a subclinical stage [26].

According to the previous reported pathological mechanisms changing CC plasma levels, it is rather difficult to identify the one and reliable cause of marker increase.

Theoretically, in a clinical setting as cardiovascular disease (CVD), characterized by high prevalence of patients with known renal impairment, CC increase could not be associated to the degree of inflammatory process more than to renal dysfunction. However it should be remarked that the association with one of these conditions could carry a relevant physiopathological bias. As a matter of fact the condition of renal impairment is widely reported to be joined to a renal microvascular inflammation [21,38-42] and currently no data are available on CC levels, in condition of normal kidney function, since GFR was not estimated by the gold standard method that is inulin clearance [29,34-37]. Therefore, CC should be considered a systemic marker of an ongoing inflammatory process that is the basic pathological mechanism promoting several diseases. But in the context of cardiovascular diseases its role in the local progression of the disease, reinforces the need for its detection.

1.2. Cystatin C Might Be a Pre- Clinical Marker of CVD

The first studies on the predictive value of CC measurement showed that high marker levels, in apparent healthy subjects, were associated to high risk to develop heart failure, hypertension, diabetes, and metabolic syndrome [43-48]. The evidences on the capability of CC to predict such diseases, in a preclinical stage, could support the assumption to employ the marker measurement in the context of primary prevention, in these different settings [29]. Initially, the most plausible pathological link, between CC elevation and the development of these diseases, seemed renal impairment, according to the fact that, the marker could assess kidney dysfunction in a preclinical state [49-52]. As a matter of fact, a consistent body of literature suggested that, CC measurement could capture the gradient of kidney function, among subjects not matching with the conventional definition of renal disease, with an estimated GFR >60ml/min [29]. An accurate individuation of the potential pathologic mechanisms, promoting CC elevation and the development of such diseases as hypertension and diabetes, is basic to provide the target for the preventive therapeutic intervention [53]. Moreover, since hypertension, diabetes, metabolic syndrome, renal impairment represent conventional risk factors for CVD, several studies conducted across different case series, investigated on the prognostic value of CC levels versus fatal and non fatal cardiovascular events. All these studies on: 1) apparent healthy persons, 2) community-based cohorts of patients without prevalent CVD, 3) patients with a defined cardiac or vascular disease (heart failure, peripheral artery disease) agreed on a positive association between high marker levels and high rate of occurrence of adverse cardiovascular events [47,54-63]. Currently, accounting for preliminary data from different case series, the employment of the marker measurement in a setting of primary prevention of CVD could be hypothesized. This approach could aid the subclinical assessment of diseases representing relevant risk factors for CVD and causing early CC increase. However, CVD covers a too wide spectrum of diseases, characterized by a too heterogeneous pathological background, that for example, heavily differs between ACS and chronic heart failure. The different pathological mechanisms, promoting the rapid evolution of an acute myocardial event or causing the chronic state of the disease, mainly influence the marker release and concentrations [64]. As reported for other cardiovascular and inflammatory markers (i.e. hsTnI, BNP, CRP), with known prognostic role for CVD patients, their release generally differs across the various subsets of patients. In ACS their release follows a rapid kinetics gauged on the time elapsed from the onset of acute symptoms [65]. In contrast to ACS, the increase of these markers, in a chronic condition of CVD, is generally associated to the exacerbation of the symptoms. Therefore, the marker changes in these two cardiovascular frameworks, mainly differ according to the type of release and the time of marker detection [65]. Thus, despite the straightforward evidence for the prognostic value of these markers in the general setting of CVD, this one should heavily differ between the subsets of acute and chronic patients. Indeed, also the prognostic value of CC in a general setting of CVD should be a too rough evaluation. The investigation on patients, carrying homogeneous clinical condition of CVD should be considered to give a reliable clinical evidence of the predictive value of CC versus adverse cardiovascular events, in the perspective to improve the secondary drug prevention measures.

2.0. From CVD to ACS: The Potential of Cystatin C to Improve Secondary Prevention

2.1. Cystatin C Levels in ACS Could Match with Clinical Expectations

Across the too wide setting of CVD, there is a peculiar subset, ACS patients, in which the measurement of CC might be relevant for the management of the disease. There are at least three basic reasons that mainly support the evaluation of CC in ACS : 1) the primary involvement of CC, in the pathological mechanisms promoting the acute event [13,16-18,35-37,55-58]; 2) the need to improve the risk assessment in ACS patients, encouraging the search for novel biomarkers to increase standard risk algorithms [66-70] ;3) the need for new therapeutic strategies to improve secondary prevention [71].

2.2. Cystatin C Could Give New Insights in ACS Pathogenesis

The pathogenesis of Acute Coronary Syndromes covers a huge spectrum of pathological substrates, reflecting the heterogeneous clinical features of patients [72]. According to this, from different clinical studies, some markers of necrosis (i.e. hsTnI), inflammation (i.e. CRP) and hemodynamic stress (i.e. BNP) emerged to aid the prediction of cardiovascular events, even in a sub-clinical phase [73]. However, other studies concluded that these ones could contribute only to a relative little incremental information to traditional risk factors, in the prediction of further adverse events [74,75]. This finding does not encourage their introduction in the clinical setting, in face of a cost benefit perspective. Nevertheless too many factors could affect the evaluation of marker prognostic value versus adverse cardiovascular events such as the features of the reference population considered. Concerning the selection of a case series, the heterogeneity of the pathological mechanisms underlying the onset and the evolution of an acute event, within the ACS setting should be accounted for. As a matter of fact the plaque disruption, the thrombosis, the coronary artery occlusion, the myocardial necrosis, and the consequent inflammatory reaction, heavily differ for features, persistence, degree, and presence in ST Elevation Myocardial Infarction (STEMI), Non ST Elevation Myocardial Infarction (NSTEMI) and unstable angina (UA) [76]. Therefore, the investigation on CC should be further addressed to ACS single subsets, in case series selected for AMI type or UA. Currently, there is a consistent body of literature which highlights that biomarkers introduction, such as BNP, CRP, hsTnI has improved the understanding of complex pathogenesis, highlighting the potential targets of treatment for each ACS subsets [66-70,76]. Actually, the several above reported biomarkers could weigh the different components of the acute event, such as inflammation (CRP), myocardial damage (hsTnI), risk for thrombosis (Fibrinogen), hemodynamic stress (BNP), renal impairment (creatinine), and thus they could trigger a target clinical management [77]. According to evidences from basic research CC levels could contribute to new and specific insights on the evolution of atherosclerosis, and compared with previous markers, it could be a more reliable prognostic tool for adverse cardiovascular events. Therefore, the simultaneous evaluation of multiple biomarkers with predictive value, assessing the degree of inflammation, of cardiomyocite necrosis, of vascular damage, of hemodynamic stress and renal impairment, could give a more complete picture on the risk for the patient than conventional risk factors.

2.3. Cystatin C Could Rationalize the Multimarker Approach

The multimarker strategy, built up by measuring simultaneously a set of biomarkers with different pathological involvement in ACS, as the one previously reported, could suggest the chance to individualize pharmacological treatment on a patient multimarker profile [78]. Actually, the joint evaluation of these biomarkers could allow a more accurate assessment of the profile of risk for each patient and this could be used to support the decision on the therapeutic strategy. Particularly in ACS setting there is a need to re-define, by sensitive markers, the risk for those patients that currently are ascribed to carry a low and intermediate risk of recurrence of cardiovascular events, according to the traditional risk factors [79]. Thus, according to the clinical evidences on the pre-clinical value of CC, it could be speculated that a more complete picture on the evolution of acute event, could be achieved by adding this marker to the ones generally measured for risk assessment [80,81]. According to the fact that CC could be both a marker of inflammation and renal impairment, its integration in a multimarker profile together with BNP and hsTnI could prevent the redundant measurement of CRP and creatinine.

3.0. Aim of the Study

According to:

1. the compelling biological role of CC in renal impairment, inflammatory and atherosclerotic pathways, known as traditional risk factors for CVD development,
2. the general agreement on marker predictive value versus cardiovascular events across different case series of patients,

the aim of the present commentary is to assess the evidence on the prognostic value of CC measurements in ACS patients, by reviewing current available clinical literature. In particular, for these patients the data provided by this work should be basic to improve their therapeutic management, after hospitalization. As a matter of fact this one is tailored on risk classification that mainly resort to traditional cardiovascular risk factors. The risk classification actually seems too rough and unreliable for the set up of an effective therapeutic management in a secondary prevention setting. In fact, despite the success of statins, about 70% of adverse events cannot be prevented by the available drug therapy. Thus in ACS framework the evidence on CC value in risk prediction, could finally improve the management and the survival of ACS patients.

Methods

The Medline search was conducted for reviews and original articles in Pub Med (between 1966 and January 2009), Embase (between 1993 and January 2009), Ovid (between 1966 and January 2009) with the Mesh terms: "Cystatin C", "Acute Coronary Syndromes". By the three databases 29 papers were identified: 28 original studies on CC and one commentary on the prognostic value of CC joint to other biomarkers in cardiovascular disease. Genetic studies or studies in which the measure of the biomarker was associated to other peculiar markers or aspects of ACS, or in which it was reported as surrogate endpoint were excluded. Finally, only the original studies in which serum or plasma CC measurement was investigated as prognostic marker in ACS were selected, according to the adequacy of statistical methods for survival analysis. All selected studies used: 1) Kaplan–Meier method for the estimation of event free survival during follow up, 2) Log- rank test to compare differences between survival curves according to different CC levels, 3) Proportional-hazard model to perform multivariable analyses. Furthermore the levels of evidence was defined according to GRADE classification [82].

5.0. Results

Paper Selection. The first step of the selection brought to 11 longitudinal observational studies papers [83-93] in which the aim of the study was to evaluate the prognostic value of CC levels for the occurrence of fatal and non fatal cardiovascular events, in case series of patients with ACS or with a general Coronary Heart Disease (CHD), including ACS patients. A further selection, according to the appropriateness of the statistical methods allowed to exclude 2 studies, since they simply evaluated the association between marker levels and presence of MACE [89,92]. As a matter of fact these studies compared the means of CC levels between MACE and not MACE groups, in ACS cohort, by using independent sample t-test. Finally, nine longitudinal observational studies were selected [83-88,90,91,93].

Analytical detection and sampling. Except for 1 study [85], the assays for CC detection used were standardized clinical laboratory platforms, for 8 studies the Dade Behring assay (Deerfield,IL,USA) was employed. The sample matrix was serum in 5 studies and plasma in 4, both matrices are of reference for the assay. The collection of samples was made at admission or within 6 hs from presentation at hospital ward. The samples were generally stored at -80°C.

Reference population. Four studies were performed on European [84,88,90,91], 2 on USA [83,87] case series. The setting considered, to perform the study and to draw the first sample for marker detection, was generally CCU for all studies. Just 1 study was conducted in the hospital rehabilitation ward [93]. According to the characteristics of reference population the studies were classified in 4 groups. *Group A:* in 4 studies the case series recruited CHD patients, with a rate of ACS of about 50-60%. In this group studies achieved a sample size ranging from 450 to 1030 patients [83,87,91,93]. *Group B:* in 3 studies the sample was composed by non ST-Elevated Acute coronary Syndromes (NSTEACS) patients. The sample size ranged from 380 to 1120 patients [84,88,90]. *Group C:* 1 study with 160 patients consisting on a matched case series of ST Elevated Myocardial Infarction (STEMI) and NSTEACS patients [85]. *Group D*: 1 study with 71 STEMI patients [86]. All studies were longitudinal observational, 4 were sub-studies respectively from prospective cohort studies [83,87,91] or clinical trials [90]. For all these studies, the enrolment does not rely on sample size estimation, according to an expected incidence of cardiovascular events in the case series. For studies with largest sample size, the case generally series derived: 1) from prospective cohorts recruiting patients from different cooperating clinics [83,84,87,91,93] or 2) from patients enrolled in clinical trials, whose serum samples were stored for sub-studies on biomarkers [90]. On the other hand the studies with lowest sample size were based on case series of ACS patients admitted to a single Department of Cardiology in 1 year of clinical practice [85,86,88]. In both cases the enrolment was continuous. The selection of the case series, when performed, was based on too loose exclusion criteria. Only, in some studies were excluded patients carrying active infections, neoplastic diseases, clinically evident renal dysfunction [84,85], since these factors could be ascribed as confounding conditions affecting marker levels and consequent interpretation.

Outcome. According to the definition of the end-point the studies were ascribed to different groups. *Group 1*: time to cardiovascular death, non fatal myocardial infarction (MI), or stroke for 2 studies [83,87]. *Group 2:* time to major adverse cardiovascular events (MACE) for 4 studies [84-86,93]. *Group 3:* time to death or recurrence of MI for 3 studies [88,90,91].

Follow up. The length of follow up was variable according to the different end-points. In *Group 1* [83,87] the median follow up was about 3 years; in *Group 2* it was variable: in one study of 3 years [93], in two studies of 1 year [84,85], in one study of 6 months [86]. *Group 3* accounted for a follow up of 3 years for 1 study [90], 6 months for 2 studies [88,91].

CC prognostic thresholds. In clinical practice and decision algorithms, the minimum marker level showing a prognostic value versus a defined adverse outcome is considered as the prognostic threshold of the marker. Nevertheless there is no general agreement on the definition of such a value. In most of studies, patients were subdivided according to CC level quantiles (quartiles, quintiles or tertiles) measured at study entry [83,84,87,88,90,93]. Thus the evaluation of predictive cut- off value of CC emerged from the calculation of a significant HR versus different events. One study [85]) relied on ROC analysis, to estimate the CC threshold for development of MACE, but did not declare the rule that was applied. Other studies used a ROC Curve analysis to determine optimal cut off point of CC level to predict cardiovascular events during follow up [86,91]. It was calculated by determining the CC level that provided the greatest sum of specificity and sensitivity.

Prognostic role. According to data from studies in *Group A* (CHD patients) CC levels higher than 1.3 mg/L were a significant risk factor for fatal and non fatal cardiovascular events with HR estimates ranging from 1.72 to 2.27. One study [91] evaluated that CC was prognostic for the combined endpoint of death and MI with an OR of 5.6(95% confidence interval, CI:1,9-16.3). Moreover this study showed that the prediction of the endpoint was improved by the combination of different clinical parameters: in particular cTn $\geq$ 0.1µg/L, measured within 2h from admission, abnormal ECG and CC accounted as continuous variable. As a matter of fact the AUC from the combination of cTnI and ECG was 0.73 and significantly increased to 0.80 when CC or NT-proBNP measurements was added to the model. For *Group B*, 2 studies found a significant prognostic role of CC with risk of death and recurrence of MI. In particular for CC levels $\geq$1.25 mg/dL, the risk of death was about 12 times greater than the one of patients with lower CC levels (HR= 11.7, (CI: 4.7-29.3))(88). In the second study [90], considering CC $\geq$1.01mg/dL as cut-off value, a HR of 4.07, (CI:2.16-7.66) was found. Concerning AMI, a different prognostic role of CC was found in the considered studies. One study [88] declared that CC was significantly associated to risk of subsequent MI, by univariate analysis. However after adjusting for other variables associated with the outcome, CC did not resulted significant (the HR was not reported in the paper). On the contrary, in the second study [90] a significant contribution was shown after adjustment, with a HR=1.95(CI:1.05-3.63). For another study [84], according to a composite endpoint of fatal and non fatal cardiac events, with a CC cut-off $\geq$ 0.93 mg/dL, a HR of 1.57(CI:1.04-2.49) was shown. For *Groups C and D* similar results were obtained considering MACE end-point: the study including STEMI and NSTEACS matched patients [85] showed a HR=9.43(CI: 4.0-21.8) for CC levels$\geq$1.05 mg/dL; in the study on STEMI [86] with a CC threshold $\geq$0.96 mg/dL a HR =2.17(CI:1.07-6.98) was obtained.

Discussion

6.0. Cystatin C: A Great Potential as Marker of Atherosclerosis, as Risk Factor for CVD and as Therapeutic Target

6.1. Cystatin C and C-Reactive Protein: Two Competing Markers for CVD Prediction in ACS Population

Several evidences, from basic experimental research and clinical observational case series, showed the critical role covered by the unbalance between CC and Cathepsin expression and release, in atherosclerosis progression and cardiovascular disease [47,54-63]. This equilibrium seems deeply conditioned by the cytokine network, that mainly inhibits the expression of CC and promotes the enhancement of cathepsin activity [1]. In particular, some cathepsins (S,K,L) exert the synergic activation of a wide spectrum of pathological mechanisms (unravelled in figure 1), resulting in the recruitment and stimulation of pro-inflammatory cells, ECM degradation and lipid accumulation in the vascular wall [2,3]. These mechanisms trigger the genesis of atheroma and simultaneously stimulate a spreading inflammatory process, by a positive feed-back, that further reinforces the progression of atherosclerosis. In turn, the inflammatory environment rich in cytokines (i.e. TGFB) stimulates the early release of CC in the blood flow [7,20]. Therefore, according to the available evidences, CC levels could early assess the inflammatory state and predict plaque destabilization [18]. It is noteworthy that inflammation is the leading pathological mechanism for the atherosclerotic process and contributes to all steps, from plaque initiation to final eventual rupture, thus promoting adverse cardiovascular events [94-95]. For this reason, in the setting of CVD, the most widely investigated predictive markers of atherosclerosis and cardiovascular disease are represented by the inflammatory ones. Particularly high-sensitivity C-reactive protein (hsCRP) is the main studied biomarker, since it is commonly measured by standardized immunoassays for clinical laboratory, and a consistent body of literature has shown that the marker high levels could predict risk for CVD, across different patient case series [96,97]. Nevertheless, in contrast to CC, it is yet unclear whether CRP could play a causal role in the pathophysiology of CVD [98-100]. The data, supporting the CRP contribute to atherothrombosis, are limited to studies on cultured cells and on animal models [98,99].

Thus, despite the great amount of studies and positive evidences on the predictive value of CRP, there is a need to introduce new biomarkers, characterized by a proven role in atherosclerotic pathways, with the final aim to disclose and exploit new therapeutic targets [100]. This is basic in ACS framework, where the search for biomarkers to gain more accurate risk classification aims to improve the choice of therapy in the secondary care. Thus, relying on the new pathological insights contributed by CC on atherosclerosis evolution, this marker could be really exploited for therapeutic purposes [12]. As a matter of fact, over the past decade advances in understanding the catalytic mechanism and fine structural features of CC as inhibitor of Cathepsins, made the design of protease inhibitors feasible for therapeutic purposes [105]. Several pharmaceutical companies became interested in the development of cystein protease inhibitors, that have yet shown promising results in treating osteoporosis and osteoarthritis [106,107]. Unfortunately no data are currently available in the cardiovascular framework.

On the other hand, despite the spreading evidences on prognostic value of CRP in cardiovascular disease, available data on the use of CRP as therapeutic target could be found only in one study on animal model [108]. The administration of a CRP inhibitor, to rats undergoing acute myocardial infarction, seemed to exert a cardioprotective effect. In addition, there are recent accumulating clinical evidences on the anti-inflammatory action of statins lowering plasma levels of hs-CRP, independently of LDL cholesterol [109]. However the net clinical benefit of hs-CRP reduction, and the possibility to employ this marker as therapeutic target should be yet explored [110]. Unfortunately this is mainly limited by the poor knowledge on the role of CRP in the pathways leading atherosclerotique progression. Thus, there is a real need to replace CRP with more specific inflammatory markers that play a direct role in atherogenesis and in its complications, as CC. [73]. Unfortunately, the amount of current clinical research available on the predictive value of CC in ACS patients, could not be compared to the wider one on CRP. Therefore, the present work, by analyzing available evidences, could disclose on the potential efficacy of CC measurement, with the final aim to encourage its introduction in the clinical practice and to exploit its therapeutic potential.

7.0. Evidences from Observational Studies on Cystatin C Prognostic Value in ACS Patients

7.1. Levels of Evidence and Quality of Study Design

Despite the interesting amount of evidences from basic research, showing the compelling role of CC in atherosclerosis evolution , there is yet a too low number of clinical studies on CC prognostic role in ACS. This could be due to the rather recent standardization and FDA approval of CC assays [111] that could have conditioned the pervasion of marker measurement in clinical laboratories. Moreover referring to Grade classification [82], these studies currently could provide only moderate evidence, mainly due to the poor quality of the study design. As a matter of fact, except for one study [84] most of evidences [85,86,88], relying on clinical case-series of specific ACS patients (NSTEACS, NSTEMI or STEMI patients), could be ascribed to level IV or V of GRADE, owing to a too low sample size. The remaining evidences could be of level II, as being sub-studies from clinical trials or from clinical studies based on prospective cohorts, characterized by larger sample size than previous one, but also by too heterogeneous case series [83-88,90,91,93]. To give a reliable assessment on the quality of evidence, in this framework, some issues concerning: 1) the importance of sample size estimation in the prognostic studies involving biomarkers in ACS and 2) the debated quality of evidences from these sub-studies, need to be further discussed.

7.2. Sample Size Estimation in Prognostic Studies Involving Biomarkers as CC

The studies here considered and generally on prognostic role of biomarkers, are often performed according to the availability of clinical data, with some gaps in the study design. There is no a clear specification of statistical hypothesis testing and of sample size estimation to assure adequate statistical power [112]. None of the studies in this work, reported on a sample size estimation, despite there is a real risk for invalidating the evidences, when obtained from an inadequate sample size [113]. Moreover, in studies based on survival data, the sample size estimation should account for different factors such as the length of follow-up and the prevalence of the risk factors in the case-series [114]. Concerning the first issue, it should be remarked that across the different studies considered, the duration of follow-up seems too variable for the same outcome (i.e. ranging from 6 months to 3 years for the recurrence of MACE or MI). This could doubt on the adequacy of this time length, that resulting too short could contrast with the experience from clinical practice in ACS patients, and further could invalidate the results. Concerning the second issue, it should be accounted that, in prognostic studies involving patients with cardiovascular disease, the investigation of the several risk factors (included as variables in the statistical model) mainly complicates the calculation of the sample size. Currently it is rather difficult to define the relationship between number of patients and number of variables that could be selected for a reliable prognostic modeling. Nevertheless, the rate between the number of events (i.e. death, recurrence of MI) and the number of prognostic variables included in the regression model of survival data, should be 10:1 [115]. Therefore, the two studies [85,86] with a small sample size and a ratio lower than the one previous reported, could be affected by instability in Cox regression models and thus provide unreliable evidences [115,116]. However the sample size estimation is seldom considered in observational longitudinal cohort studies, exploring the prognostic value of biomarkers.

7.3. Debatable Quality of Evidences from Sub-Studies

Several recent investigations on the prognostic value of biomarkers, as in CC case, are sub-studies derived from clinical trials or from prospective cohort studies with a different clinical end-point. These sub-studies might provide better clinical evidences than the ones obtained from small original prospective observational studies. As a matter of fact, the sub-studies generally consider large ACS case-series and theoretically should mirror the real clinical setting, with greater sample size than observational studies, and a rather unselected recruitment. Nevertheless these sub-studies could provide misleading evidences since their design is usually not suitable for an ancillary accurate evaluation of biomarkers [117]. As a matter of fact the too loose enrolment criteria, planned for the trial, could not account for potential interfering clinical conditions on marker levels. Furthermore, there is not a precise reference to the performance of marker measurement, in terms of time distance from acute event, number of detections and suitability of all patients to this evaluation. The lesson from literature on natriuretic peptides has taught that the evidences from the sub-studies, to assess the clinical value of the markers in ACS patients seem too rough [117-120]. Currently, it is widely accepted that perspective original studies with a planned evaluation of biomarkers, well documented in the method session of the design, tailored on the clinical relevance of marker variations around acute event, could really provide new insights in the ACS physiopathology and finally improve the quality of evidences.

7.4. Prognostic Value of CC and Statistical Issues

There is an interesting homogeneity of results from the studies considered in this review, providing a good final evidence on CC value as predictor of adverse fatal and non fatal cardiovascular events, in patients with ACS. However, according to the HRs, reported from two studies on NSTEACS [88,90], there is evidence of a relevant prognostic value versus cardiovascular death but not versus the recurrence of AMI. As previous anticipated, the latter result could be affected by the short follow up [88] and it seemed to be influenced by the adjustment for some potential confounders as age, gender, hospital site, BMI, HDL-cholesterol, history of diabetes, treatment with ACE inhibitors, and C Reactive Protein(CRP) levels, that are known to be associated to AMI recurrence [88,90]. Furthermore, it should be remarked that among the considered studies, few of them performed a statistical analyses that could fully explore the potential role of CC, by considering marker levels as a continuous variable in the multivariable regression function [83,85,88]. In most studies, CC levels were included in the models as dichotomized, according non homogeneous cut-off values [86,91], or grouped, according to the partition of CC marker levels in quantiles [83,84,87,88,90,93]. Some studies performed two models by considering CC both as continuous and as categorical variable. In the latter case, the patients were split into groups according to the median or to the quantiles of the distribution of CC levels (tertiles, quartiles or quintiles). However this approach could be rather debatable for different reasons. At first, the *a priori* assumption that patients in the highest quantiles could carry higher risk than the ones in the lowest quantile is groundless [121]. Moreover, this could provide a circular argumentation, since the aim of the analysis is to show whether the risk for adverse events could increase systematically as the level of a marker rises. However, in ACS primary care setting, a prognostic threshold of the marker is useful to classify the patients, according to risk decisional algorithms in the daily clinical practice. This threshold could be defined as the minimum marker level showing a prognostic value versus a defined adverse cardiovascular event.

The selection of a threshold level for the marker is troublesome since the same level is further used in computing P-values for: 1) the same marker, 2) displaying survival curves that compare subsets of patients respectively above and below the defined threshold, and 3) regression analyses involving the marker. It could be rather evident that, P-values, survival curves and regression coefficients resulting from these analyses are biased by pre-selection of the threshold level, since based on the same data [116,120,122]. The cut-off point could be used without introducing bias, only in the following cases: 1) when it could be inferred from other preliminary studies, aiming a sort of standardization of this threshold level, for a defined assay; or 2) when it could be assessed according to the distribution of patient marker levels, avoiding the use of clinical outcome data. Moreover, the partition of marker levels (and particularly the dichotomization) should discard relevant information carried by the marker [116]. To overcome this drawback, the marker levels should be included as a continuous variable in the multivariate analysis, under the assumption that the risk (log hazard ratio) increases linearly as the marker levels rise. This should be tested (for example with methods of regression splines) [115,123]. The reported approach, followed in three studies [83,85,88] could give evidence on the nature of the relation between marker levels and risk for adverse events, and thus it could provide utmost relevant information on the variation of the outcome according to the marker level of the patient.

Conclusions

Despite the too low number of evidences and their moderate level, there is a good agreement on the positive prognostic value of CC levels in ACS. These data are consistent with recent studies showing that also mild renal impairment is strongly associated with cardiovascular morbidity and mortality in patients with atherosclerotic disease and in the general population. This work should encourage to provide more reliable evidences than the one currently available, by improving the quality of the study design, by accounting for the relevant statistical issues reported and by investigating on more homogeneous case series of ACS. The ones investigated, in most of studies are extremely heterogeneous and there are no fairly data on STEMI and NSTEMI subsets. Particularly these patients could benefit from CC introduction in the clinical setting, since their therapeutic management after hospitalization is predominately gauged on risk classification and furthermore in the secondary prevention of myocardial infarction since new effective therapeutic targets are required.

Acknowledgments

The authors thank the "BANCA POPOLARE DI NOVARA" and the "FONDAZIONE DELLA COMUNITA' NOVARESE O.N.L.U.S" to have supported this publication and Mr Andrea Sirtori for the graphic support.

References

[1] Newman DJ. Cystatin C. *Ann. Clin. Biochem.* 2002; 39:89-104.

[2] Vray B, Hartmann S, Hoebeke J. Immunomodulatory properties of cystatins. *Cell Mol. Life Sci.* 2002;59:1503–12.

[3] Brown WM, Dziegielewska KM. Friends and relations of the cystatin superfamily—new members and their evolution. *Protein Sci.* 1997; 6:5–12.

[4] Chapman HA, Riese RJ, Shi GP.Emerging roles for cysteine proteases in human biology. *Annu. Rev. Physiol.* 1997; 59:63-88.

[5] Henriksen K, Tanko LB, Qvist P, Delmas PD, Christiansen C, Karsdal MA. Assessment of osteoclast number and function: application in the development of new and improved treatment modalities for bone diseases. *Osteoporos. Int.* 2007; 18:681-5.

[6] Cremers S, Garnero P.Biochemical markers of bone turnover in the clinical development of drugs for osteoporosis and metastatic bone disease: potential uses and pitfalls. *Drugs.* 2006; 66:2031-58.

[7] Leung-Tack J, Tavera C, Martinez J, Colle A. Neutrophil chemotactic activity is modulated by human cystatin C, an inhibitor of cysteine proteases. *Inflammation*. 1990 ;14:247-58.

[8] Sokol JP and Schiemann WP. Cystatin C Antagonizes Transforming Growth Factor β Signaling in Normal and Cancer Cells. *Mol. Cancer Res*. 2004; 2:183-95.

[9] Kos J, Werle B, Lah T, Brunner N. Cysteine proteinases and their inhibitors in extracellular fluids: markers for diagnosis and prognosis in cancer. *Int. J. Biol. Markers*. 2000; 15:84–9.

[10] 10.Villadangos JA, Bryant RA, Deussing J, Driessen C, Lennon-Duménil AM, Riese RJ, Roth W, Saftig P, Shi GP, Chapman HA, Peters C, Ploegh HL. Proteases involved in MHC class II antigen presentation. *Immunol. Rev*. 1999; 172:109-20.

[11] Gupta S, Singh RK, Dastidar S, Ray A.Cysteine cathepsin S as an immunomodulatory target: present and future trends. *Expert Opin. Ther. Targets*. 2008; 12:291-9.

[12] Lutgens SP, Cleutjens KB, Daemen MJ, Heeneman S. Cathepsin cysteine proteases in cardiovascular disease. *FASEB J*. 2007; 21:3029-41.

[13] Liu J, Sukhova GK, Sun JS, Xu WH, Libby P, Shi GP. Lysosomal cysteine proteases in atherosclerosis. *Arterioscler. Thromb Vasc. Biol.* 2004; 24:1359–66.

[14] Westhuyzen J.Cystatin C: a promising marker and predictor of impaired renal function. *Ann. Clin. Lab. Sci*. 2006; 36:387-94.

[15] Sukhova GK, Shi GP, Simon DI, Chapman HA, Libby P. Expression of the elastolytic cathepsins S and K in human atheroma and regulation of their production in smooth muscle cells. *J. Clin. Invest.* 1998; 102:576-83.

[16] Bengtsson E, To F, Håkansson K, Grubb A, Brånén L, Nilsson J, Jovinge S.Lack of the cysteine protease inhibitor cystatin C promotes atherosclerosis in apolipoprotein E-deficient mice. *Arterioscler. Thromb Vasc. Biol.* 2005; 25:2151-6.

[17] Shi GP, Sukhova GK, Grubb A, Ducharme A, Rhode LH, Lee RT, Ridker PM, Libby P, Chapman HA. Cystatin C deficiency in human atherosclerosis and aortic aneurysms. *J. Clin. Invest.* 1999; 104:1191–7.

[18] Sukhova GK, Zhang Y, Pan JH, Wada Y, Yamamoto T, Naito M, Kodama T, Tsimikas S, Witztum JL, Lu ML, Sakara Y, Chin MT, Libby P, Shi GP.Deficiency of cathepsin S reduces atherosclerosis in LDL receptor-deficient mice. *J. Clin. Invest.* 2003; 111:897-906.

[19] Sukhova GK, Wang B, Libby P, Pan JH, Zhang Y, Grubb A, Fang K, Chapman HA, Shi GP. Cystatin C deficiency increases elastic lamina degradation and aortic dilatation in apolipoprotein E-null mice. *Circ. Res*. 2005; 96:368-75.

[20] Leung-Tack J, Tavera C, Gensac MC, Martinez J, Colle A. Modulation of phagocytosis associated respiratory burst by human cystatin C: role of the N-terminal tetrapeptide Lys-Pro-Pro-Arg. *Exp. Cell Res.* 1990; 188:16-22.

[21] Knight EL, Verhave JC, Spiegelman D, Hillege HL, de Zeeuw D, Curhan GC, de Jong PE.Factors influencing serum cystatin C levels other than renal function and the impact on renal function measurement. *Kidney Int.* 2004; 65:1416-21.

[22] Shlipak MG, Katz R, Cushman M, Sarnak MJ, Stehman-Breen C, Psaty BM, Siscovick D, Tracy RP, Newman A, Fried L.Cystatin-C and inflammatory markers in the ambulatory elderly. *Am. J. Med.* 2005; 118:1416.

[23] Libby P, Ridker PM, Maseri A. Inflammation and atherosclerosis. *Circulation* 2002;105:1760-63.

[24] Grubb A, Björk J, Lindström V, Sterner G, Bondesson P, Nyman U.A. Cystatin C-based formula without anthropometric variables estimates glomerular filtration rate better than creatinine clearance using the Cockcroft-Gault formula. *Scand. J. Clin. Lab. Invest.* 2005; 65:153-62.

[25] Grubb A, Nyman U, Björk J, Lindström V, Rippe B, Sterner G, Christensson A.Simple cystatin C-based prediction equations for glomerular filtration rate compared with the modification of diet in renal disease prediction equation for adults and the Schwartz and the Counahan-Barratt prediction equations for children. *Clin. Chem.* 2005; 51:1420-31.

[26] Dharnidharka VR, Kwon C, Stevens G.Serum cystatin C is superior to serum creatinine as a marker of kidney function: a meta-analysis. *Am. J. Kidney Dis.* 2002 ;40:221-6.

[27] Hsu CY, Chertow GM, Curhan GC. Methodological issues in studying the epidemiology of mild to moderate chronic renal insufficiency. *Kidney Int.* 2002; 61: 1567-76.

[28] Levey AS, Bosch JP, Lewis JB, Greene T, Rogers N, Roth D. A more accurate method to estimate glomerular filtration rate from serum creatinine: a new prediction equation. Modification of Diet in Renal Disease Study Group. *Ann. Intern. Med.* 1999; 130:461-70.

[29] Shlipak MG, Praught ML, Sarnak MJ.Update on cystatin C: new insights into the importance of mild kidney dysfunction. *Curr. Opin. Nephrol. Hypertens.* 2006;15:270-5.

[30] Menon V, Shlipak MG, Wang X, Coresh J, Greene T, Stevens L, Kusek JW, Beck GJ, Collins AJ, Levey AS, Sarnak MJ. Cystatin C as a risk factor for outcomes in chronic kidney disease. *Ann. Intern. Med.* 2007; 147:19-27.

[31] Macdonald J, Marcora S, Jibani M, Roberts G, Kumwenda M, Glover R, Barron J, Lemmey A. GFR estimation using cystatin C is not independent of body composition. *Am. J. Kidney Dis.* 2006; 48:712-9.

[32] Stevens LA, Coresh J, Schmid CH, Feldman HI, Froissart M, Kusek J, Rossert J, Van Lente F, Bruce RD 3rd, Zhang YL, Greene T, Levey AS. Estimating GFR using serum cystatin C alone and in combination with serum creatinine: a pooled analysis of 3,418 individuals with CKD. *Am. J. Kidney Dis.* 2008; 5:395-406.

[33] Vupputuri S, Fox CS, Coresh J, Woodward M, Muntner P. Differential estimation of CKD using creatinine- versus cystatin C-based estimating equations by category of body mass index. *Am. J. Kidney Dis.* 2009; 53:993-1001.

[34] Tanaka A, Suemaru K, Araki H.A new approach for evaluating renal function and its practical application. *J. Pharmacol. Sci.* 2007; 10:1-5.

[35] Koenig W, Twardella D, Brenner H, Rothenbacher D.Plasma concentrations of cystatin C in patients with coronary heart disease and risk for secondary cardiovascular events: more than simply a marker of glomerular filtration rate. *Clin. Chem.* 2005; 51:321-7.

[36] Keller T, Messow CM, Lubos E, Nicaud V, Wild PS, Rupprecht HJ, Bickel C, Tzikas S, Peetz D, Lackner KJ, Tiret L, Münzel TF, Blankenberg S, Schnabel RB.Cystatin C and cardiovascular mortality in patients with coronary artery disease and normal or mildly reduced kidney function: results from the AtheroGene study. *Eur. Heart J.* 2009; 30:314-20.

[37] Windhausen F, Hirsch A, Fischer J, van der Zee PM, Sanders GT, van Straalen JP, Cornel JH, Tijssen JG, Verheugt FW, de Winter RJ; Invasive versus Conservative Treatment in Unstable Coronary Syndromes (ICTUS) Investigators.Cystatin C for

enhancement of risk stratification in non-ST elevation acute coronary syndrome patients with an increased troponin T. *Clin. Chem.* 2009; 55:1118-25.

[38] Gross ML, Adamczak M, Amann K, Ritz E. Mediators of inflammatory and ischemic renal damage: the role of neoangiogenesis. *J. Nephrol.* 2005; 18:513-20.

[39] Maruyama Y, Lindholm B, Stenvinkel P.Inflammation and oxidative stress in ESRD--the role of myeloperoxidase. *J. Nephrol.* 2004 ;17 Suppl 8:S72-6.

[40] Lu L, Pu LJ, Xu XW, Zhang Q, Zhang RY, Zhang JS, Hu J, Yang ZK, Lu AK, Ding FH, Shen J, Chen QJ, Lou S, Fang DH, Shen WF. Association of serum levels of glycated albumin, C-reactive protein and tumor necrosis factor-alpha with the severity of coronary artery disease and renal impairment in patients with type 2 diabetes mellitus. *Clin. Biochem.* 2007 ;40:810-6.

[41] Landray MJ, Wheeler DC, Lip GY, Newman DJ, Blann AD, McGlynn FJ, Ball S, Townend JN, Baigent C. Inflammation, endothelial dysfunction, and platelet activation in patients with chronic kidney disease: the chronic renal impairment in Birmingham (CRIB) study. *Am. J. Kidney Dis.* 2004; 43:244-53.

[42] Schalkwijk CG, Poland DC, van Dijk W, Kok A, Emeis JJ, Dräger AM, Doni A, van Hinsbergh VW, Stehouwer CD.Plasma concentration of C-reactive protein is increased in type I diabetic patients without clinical macroangiopathy and correlates with markers of endothelial dysfunction: evidence for chronic inflammation. *Diabetologia.* 1999; 42:351-7.

[43] Donahue RP, Stranges S, Rejman K, Rafalson LB, Dmochowski J, Trevisan M. Elevated cystatin C concentration and progression to pre-diabetes: the Western New York study. *Diabetes Care*. 2007; 30:1724-9.

[44] Servais A, Giral P, Bernard M, Bruckert E, Deray G, Isnard Bagnis C.Is serum cystatin-C a reliable marker for metabolic syndrome? *Am. J. Med.* 2008 ;121:426-32.

[45] Lee JG, Lee S, Kim YJ, Jin HK, Cho BM, Kim YJ, Jeong DW, Park HJ, Kim JE. Multiple biomarkers and their relative contributions to identifying metabolic syndrome. *Clin. Chim. Acta*. 2009; 408:50-5.

[46] Vigil L, Lopez M, Condés E, Varela M, Lorence D, Garcia-Carretero R, Ruiz J.Cystatin C is associated with the metabolic syndrome and other cardiovascular risk factors in a hypertensive population. *J. Am. Soc. Hypertens*. 2009 ; 3:201-9.

[47] Kestenbaum B, Rudser KD, de Boer IH, Peralta CA, Fried LF, Shlipak MG, Palmas W, Stehman-Breen C, Siscovick DS. Differences in kidney function and incident hypertension: the multi-ethnic study of atherosclerosis. *Ann. Intern. Med.* 2008; 148:501-8

[48] Moran A, Katz R, Smith NL, Fried LF, Sarnak MJ, Seliger SL, Psaty B, Siscovick DS, Gottdiener JS, Shlipak MG. Cystatin C concentration as a predictor of systolic and diastolic heart failure. *J. Card Fail.* 2008; 14:19-26.

[49] Willems D, Wolff F, Mekhali F, Gillet C. Cystatin C for early detection of renal impairment in diabetes. *Clin. Biochem*. 2009; 42:108-10.

[50] Shimizu A, Horikoshi S, Rinnno H, Kobata M, Saito K, Tomino Y. Serum cystatin C may predict the early prognostic stages of patients with type 2 diabetic nephropathy. J. *Clin. Lab. Anal.* 2003;17:164-7.

[51] Naruse H, Ishii J, Kawai T, Hattori K, Ishikawa M, Okumura M, Kan S, Nakano T, Matsui S, Nomura M, Hishida H, Ozaki Y. Cystatin C in acute heart failure without advanced renal impairment. *Am. J. Med.* 2009; 122:566-73.

[52] Watanabe S, Okura T, Kurata M, Irita J, Manabe S, Miyoshi K, Fukuoka T, Gotoh A, Uchida K, Higaki J. Valsartan reduces serum cystatin C and the renal vascular resistance in patients with essential hypertension. *Clin. Exp. Hypertens.* 2006; 28:451-61.

[53] Vasiljeva O, Reinheckel T, Peters C, Turk D, Turk V, Turk B. Emerging roles of cysteine cathepsins in disease and their potential as drug targets. *Curr. Pharm. Des.* 2007; 13:387-403.

[54] Parikh NI, Hwang SJ, Yang Q, Larson MG, Guo CY, Robins SJ, Sutherland P, Benjamin EJ, Levy D, Fox CS. Clinical correlates and heritability of cystatin C (from the Framingham Offspring Study). *Am. J. Cardiol.* 2008; 102:1194-8.

[55] Deo R, Sotoodehnia N, Katz R, Sarnak MJ, Fried LF, Chonchol M, Kestenbaum B, Psaty BM, Siscovick DS, Shlipak MG. Cystatin C and sudden cardiac death risk in the elderly. *Circ. Cardiovasc. Qual. Outcomes.* 2010; 3:159-64.

[56] Rifkin DE, Shlipak MG, Katz R, Fried LF, Siscovick D, Chonchol M, Newman AB, Sarnak MJ Rapid kidney function decline and mortality risk in older adults. *Arch. Intern. Med.* 2008 ; 168:2212-8.

[57] Shlipak MG, Katz R, Kestenbaum B, Fried LF, Siscovick D, Sarnak MJ. Clinical and subclinical cardiovascular disease and kidney function decline in the elderly. *Atherosclerosis.* 2009; 204:298-303.

[58] Shlipak MG, Katz R, Sarnak MJ, Fried LF, Newman AB, Stehman-Breen C, Seliger SL, Kestenbaum B, Psaty B, Tracy RP, Siscovick DS. Cystatin C and prognosis for cardiovascular and kidney outcomes in elderly persons without chronic kidney disease. *Ann. Intern. Med.* 2006; 145:237-46.

[59] Sarnak MJ, Katz R, Stehman-Breen CO, Fried LF, Jenny NS, Psaty BM, Newman AB, Siscovick D, Shlipak MG; Cardiovascular Health Study. Cystatin C concentration as a risk factor for heart failure in older adults. *Ann. Intern. Med.* 2005; 142:497-505.

[60] Kestenbaum B, Rudser KD, Shlipak MG, Fried LF, Newman AB, Katz R, Sarnak MJ, Seliger S, Stehman-Breen C, Prineas R, Siscovick DSKidney function, electrocardiographic findings, and cardiovascular events among older adults. *Clin. J. Am. Soc. Nephrol.* 2007; 2:501-8.

[61] Shlipak MG, Katz R, Fried LF, Jenny NS, Stehman-Breen CO, Newman AB, Siscovick D, Psaty BM, Sarnak MJ. Cystatin-C and mortality in elderly persons with heart failure. *J. Am. Coll. Cardiol.* 2005; 45:268-71.

[62] O'Hare AM, Newman AB, Katz R, Fried LF, Stehman-Breen CO, Seliger SL, Siscovick DS, Shlipak MG. Cystatin C and incident peripheral arterial disease events in the elderly: results from the Cardiovascular Health Study. *Arch. Intern. Med.* 2005; 165:2666-70.

[63] Shlipak MG, Wassel Fyr CL, Chertow GM, Harris TB, Kritchevsky SB, Tylavsky FA, Satterfield S, Cummings SR, Newman AB, Fried LF. Cystatin C and mortality risk in the elderly: the health, aging, and body composition study. *J. Am. Soc. Nephrol.* 2006; 17:254-61.

[64] Clerico A, Iervasi G, Pilo A. Turnover studies on cardiac natriuretic peptides:methodological, pathophysiological and therapeutical considerations. *Curr. Drug Metab.* 2000; 1:85–105.

[65] Ferraro S, Lupi A, Marano G, Rossi L, Ciardi L, Vendramin C, Bellomo G, Boracchi P, Bongo AS, Biganzoli E.Different patterns of NT-proBNP secretion in acute coronary syndromes. *Clin. Chim. Acta*. 2009; 402:176-81.

[66] Vittorini S, Clerico A.Cardiovascular biomarkers: increasing impact of laboratory medicine in cardiology practice. *Clin. Chem. Lab. Med*. 2008;46:748-63.

[67] Gerszten RE, Wang TJ. The search for new cardiovascular biomarkers. *Nature*. 2008;451:949-52.

[68] Pencina MJ, D'Agostino RB Sr, D'Agostino RB Jr and Vasan RS. Evaluating the added predictive ability of a new marker: from area under the ROC curve to reclassification and beyond. *Stat. Med*. 2008; 27:157-72; discussion 207-12.

[69] Ridker PM, Buring JE, Rifai N and Cook NR. Development and validation of improved algorithms for the assessment of global cardiovascular risk in women: the Reynolds Risk Score. *JAMA*. 2007; 297:611-9.

[70] Pepe M.S. The Statistical Evaluation of Medical Tests for Classification and Prediction. Oxford University Press, 2003.

[71] Kraaijeveld AO, de Jager SC, van Berkel TJ, Biessen EA, Jukema JW. Chemokines andatherosclerotic plaque progression: towards therapeutic targeting? *Curr. Pharm Des*. 2007; 13:1039-1052.

[72] Libby P. Current concept of the pathogenesis of the acute coronary syndromes. *Circulation* 2001; 104:365-372.

[73] Parikh SV, de Lemos JA. Biomarkers in cardiovascular disease: integrating pathophysiology into clinical practice. *Am. J. Med. Sci*. 2006; 332:186-97.

[74] Folsom AR, Chambless LE, Ballantyne CM, Coresh J, Heiss G, Wu KK, Boerwinkle E, Mosley TH Jr, Sorlie P, Diao G, Sharrett AR.An assessment of incremental coronary risk prediction using C-reactive protein and other novel risk markers: the atherosclerosis risk in communities study. *Arch Intern. Med*. 2006; 166:1368-73.

[75] Wang TJ, Gona P, Larson MG, Tofler GH, Levy D, Newton-Cheh C, Jacques PF, Rifai N, Selhub J, Robins SJ, Benjamin EJ, D'Agostino RB, Vasan RS. Multiple biomarkers for the prediction of first major cardiovascular events and death. *N. Engl. J. Med*. 2006; 355:2631-9.

[76] Parikh SV, de Lemos JA. Biomarkers in cardiovascular disease: integrating pathophysiology into clinical practice. *Am. J. Med. Sci*. 2006;332:186-197.

[77] Frangogiannis NG, Smith CW, Entman ML.The inflammatory response in myocardial infarction. *Cardiovasc. Res*. 2002; 53:31-47.

[78] Howes JM, Keen JN, Findlay JB, Carter AM. The application of proteomics technology tothrombosis research: the identification of potential therapeutic targets in cardiovascular disease. *Diab. Vasc. Dis. Res*. 2008;5:205-212.

[79] de Lemos JA, Lloyd-Jones DM.Multiple biomarker panels for cardiovascular risk assessment. *N. Engl. J. Med*. 2008; 358:2172-4.

[80] McCann CJ, Glover BM, Menown IB, Moore MJ, McEneny J, Owens CG, et al. Prognostic value of a multimarker approach for patients presenting to hospital with acute chest pain. *Am. J. Cardiol*. 2009; 103:22-8.

[81] Eggers KM, Dellborg M, Oldgren J, Swahn E, Venge P, Lindahl B. Risk prediction in chest pain patients by biochemical markers including estimates of renal function. *Int. J. Cardiol*. 2008; 128:207-13.

[82] Schünemann JH, Oxman AD, Brozek J, Glasziou P, Jaeschke R, Vist GE et al. Grading quality of evidence and strength of recommendations for diagnostic tests and strategies *BMJ*, 2008; 336:1106-10.

[83] Shlipak MG, Ix JH, Bibbins-Domingo K, Lin F, Whooley MA Biomarkers to predict recurrent cardiovascular disease: the Heart and Soul Study. *Am. J. Med.* 2008; 121:50-7.

[84] Taglieri N, Fernandez-Berges DJ, Koenig W, Consuegra-Sanchez L, Fernandez JM, Robles NR, Sánchez PL, Beiras AC, Orbe PM, Kaski JC; SIESTA Investigators. Plasma cystatin C for prediction of 1-year cardiac events in Mediterranean patients with non-ST elevation acute coronary syndrome. *Atherosclerosis*. 2010; 209:300-5.

[85] Kilic T, Oner G, Ural E, Yumuk Z, Sahin T, Bildirici U, Acar E, Celikyurt U, Kozdag G, Ural D. Comparison of the long-term prognostic value of cystatin C to other indicators of renal function, markers of inflammation and systolic dysfunction among patients with acute coronary syndrome. *Atherosclerosis*. 2009; 207:552-8.

[86] Ichimoto E, Jo K, Kobayashi Y, Inoue T, Nakamura Y, Kuroda N, Miyazaki A, Komuro I._Prognostic significance of cystatin C in patients with ST-elevation myocardial infarction. *Circ. J.* 2009; 73:1669-73.

[87] Ix JH, Shlipak MG, Chertow GM, Whooley MA. Association of cystatin C with mortality, cardiovascular events, and incident heart failure among persons with coronary heart disease: data from the Heart and Soul Study. *Circulation*. 2007; 115:173-9.

[88] Jernberg T, Lindahl B, James S, Larsson A, Hansson LO, Wallentin L. Cystatin C: a novel predictor of outcome in suspected or confirmed non-ST-elevation acute coronary syndrome. *Circulation*. 2004; 110:2342-8.

[89] Ge C, Ren F, Lu S, Ji F, Chen X, Wu X. Clinical prognostic significance of plasma cystatin C levels among patients with acute coronary syndrome. *Clin. Cardiol.* 2009; 32:644-8.

[90] Windhausen F, Hirsch A, Fischer J, van der Zee PM, Sanders GT, van Straalen JP, Cornel JH, Tijssen JG, Verheugt FW, de Winter RJ. Invasive versus Conservative Treatment in Unstable Coronary Syndromes (ICTUS) Investigators. Cystatin C for enhancement of risk stratification in non-ST elevation acute coronary syndrome patients with an increased troponin T. *Clin. Chem.* 2009; 55:1118-25.

[91] 91. Eggers KM, Dellborg M, Oldgren J, Swahn E, Venge P, Lindahl B. Risk prediction in chest pain patients by biochemical markers including estimates of renal function. *Int. J. Cardiol.* 2008; 128:207-13.

[92] García Acuña JM, González-Babarro E, Grigorian Shamagian L, Peña-Gil C, Vidal Pérez R, López-Lago AM, Gutiérrez Feijoó M, González-Juanatey JR. Cystatin C provides more information than other renal function parameters for stratifying risk in patients with acute coronary syndrome. *Rev. Esp. Cardiol.* 2009; 62:510-9.

[93] Koenig W, Twardella W, Brenner H and Rothenbancher D.Plasma Concentration of Cystatin C in Patients with Coronary Heart disease and Risk for Secondary Cardiovascular Events: More than simply a marker of Glomerular Filtration Rate. *Clin. Chem.* 2005; 51:321-7.

[94] Hansson GK. Inflammation, atherosclerosis, and coronary artery disease. *N. Engl. J. Med.* 2005; 352:1685-95.

[95] Ross R. Atherosclerosis is an inflammatory disease. *Am. Heart J.* 1999; 138:S419-20.

[96] Danesh J, Wheeler JG, Hirschfield GM, Eda S, Eiriksdottir G, Rumley A, Lowe GD, Pepys MB, Gudnason V. C-reactive protein and other circulating markers of inflammation in the prediction of coronary heart disease. *N. Engl. J. Med.* 2004; 350: 1387-97.

[97] Ridker PM, Paynter NP, Rifai N, Gaziano JM, Cook NR. C-reactive protein and parental history improve global cardiovascular risk prediction: the Reynolds Risk Score for men. *Circulation* 2008; 118:2243-51.

[98] Verma S, Devaraj S, Jialal I.Is C-reactive protein an innocent bystander or proatherogenic culprit? C-reactive protein promotes atherothrombosis. *Circulation.* 2006; 113:2135-50.

[99] Scirica BM, Morrow DA. Is C-reactive protein an innocent bystander or proatherogenic culprit? The verdict is still out. *Circulation.* 2006; 113:2128-34.

[100] Vasan RS.Biomarkers of cardiovascular disease: molecular basis and practical considerations. *Circulation* 2006; 113:2335-62.

[101] Klingenberg R, Hansson GK.Treating inflammation in atherosclerotic cardiovascular disease: emerging therapies. *Eur. Heart J.* 2009; 30:2838-44.

[102] Moubayed SP, Heinonen TM, Tardif JC.Anti-inflammatory drugs and atherosclerosis. *Curr. Opin. Lipidol.* 2007; 18:638-44.

[103] Montecucco F, Mach F.Update on statin-mediated anti-inflammatory activities in atherosclerosis. *Semin. Immunopathol.* 2009; 31:127-42.

[104] Libby P, Okamoto Y, Rocha VZ, Folco E.Inflammation in atherosclerosis: transition from theory to practice. *Circ. J.* 2010; 74:213-20.

[105] Roberts R.Lysosomal cysteine proteases: structure, function and inhibition of cathepsins. *Drug News Perspect.* 2005; 18:605-14.

[106] Abbenante G, Fairlie DP.Protease inhibitors in the clinic. *Med. Chem.* 2005; 1:71-104.

[107] Yasuda Y, Kaleta J, Brömme D.The role of cathepsins in osteoporosis and arthritis: rationale for the design of new therapeutics. *Adv. Drug Deliv. Rev.* 2005; 57:973-93.

[108] Pepys MB, Hirschfield GM, Tennent GA, Gallimore JR, Kahan MC, Bellotti V, Hawkins PN, Myers RM, Smith MD, Polara A, Cobb AJ, Ley SV, Aquilina JA, Robinson CV, Sharif I, Gray GA, Sabin CA, Jenvey MC, Kolstoe SE, Thompson D, Wood SP. Targeting C-reactive protein for the treatment of cardiovascular disease. *Nature* 2006; 440:1217-21.

[109] Ridker PM, Fonseca FA, Genest J, Gotto AM, Kastelein JJ, Khurmi NS, Koenig W, Libby P, Lorenzatti AJ, Nordestgaard BG, Shepherd J, Willerson JT, Glynn RJ; JUPITER Trial Study Group. Baseline characteristics of participants in the JUPITER trial, a randomized placebo-controlled primary prevention trial of statin therapy among individuals with low low-density lipoprotein cholesterol and elevated high-sensitivity C-reactive protein. *Am. J. Cardiol.* 2007; 100:1659-64.

[110] Mora S, Ridker PM. Justification for the Use of Statins in Primary Prevention: an Intervention Trial Evaluating Rosuvastatin (JUPITER)--can C-reactive protein be used to target statin therapy in primary prevention? *Am. J. Cardiol.* 2006; 97:33A-41A.

[111] Filler G, Bökenkamp A, Hofmann W, Le Bricon T, Martínez-Brú C, Grubb A. Cystatin C as a marker of GFR--history, indications, and future research. *Clin. Biochem.* 2005; 38:1-8.

[112] Simon R and Altman D.G. Statistical aspects of prognostic factor studies in oncology. *Br. J. Cancer*,1994; 69:979-985 .

[113] Freiman J.A., Chalmers, T.C., Smith H. AND Kuebler R.R. The importance of beta, the type II error, and sample size in the design and interpretation of the randomized controlled trial: survey of two sets of 'negative' trials. In Medical Uses of Statistics, 1992, 2nd edn. Bailar, J.C. and Mosteller, F. (eds) pp. 357-73. Books: Boston.

[114] Schoenfeld D. The asymptotic properties of nonparametric tests for comparing survival distributions. *Biometrika* 1981; 68: 316-9.

[115] Harrell F.E., Lee K.L., Matchar D.B. and Reichert T.A. Regression models for prognostic prediction: advantages, problems, and suggested solutions. *Cancer Treat. Rep.* 1985; 69:1071-7.

[116] Altman D.G. and Andersen P.K. Bootstrap investigation of the stability of a Cox regression model. *Stat. Med.* 1989; 8:771-83.

[117] Bonaca MP, Wiviott SD, Sabatine MS, Buros J, Murphy SA, Scirica BM. Hemodynamic significance of periprocedural myocardial injury assessed with N-terminal pro-B-type natriuretic peptide after percutaneous coronary intervention in patients with stable and unstable coronary artery disease (from the JUMBO-TIMI 26 trial). *Am. J. Cardiol.* 2007; 99:344-8.

[118] Wiviott SD, Cannon CP, Christopher P, Morrow DA, David A, Murphy SA. Differential expression of cardiac biomarkers by gender in patients with unstable angina/non-ST elevation myocardial infarction: a TACTICS-TIMI 18 (Treat Angina with Aggrastat and determine Cost of Therapy with an Invasive or Conservative Strategy – Thrombolysis In Myocardial Infarction 18) Substudy. *Circulation* 2004; 109: 580–6.

[119] Bjorklund E, Jernberg T, Johanson P, Venge P, Dellborg M, Wallentin L. ASSENT-2 and ASSENT-PLUS Study Groups. Admission N-terminal pro-brain natriuretic peptide and its interaction with admission troponin T and ST segment resolution for early risk stratification in ST elevation myocardial infarction. *Heart* 2006; 92:735-40.

[120] Tang WH, Steinhubl SR, Van Lente F, Brennan D, McErlean E, Maroo A, et al. Risk stratification for patients undergoing nonurgent percutaneous coronary intervention using Nterminal pro-B-type natriuretic peptide: a Clopidogrel for the Reduction of Events During Observation (CREDO) substudy. *Am. Heart J.* 2007; 153:36-41.

[121] Hilsenbeck S.G., Clark G.M. and McGuire W.L. Why do so many prognostic factors fail to pan out? *Breast Cancer Res. Treat.* 1992; 22: 197-206.

[122] Durrleman S. and Simon R. Flexible regression models with cubic splines. *Stat. Med.* 1989; 8:551-61.

Reviewed by: Aldo Clerico, Istituto fisiologia clinica Area della Ricerca CNR Pisa.

In: Cystatins: Protease Inhibitors ... ISBN: 978-1-61209-343-7
Editors: John B. Cohen and Linda P. Ryseck

Chapter 6

CURRENT APPROACHES FOR SENSITIVE DETECTION OF DRUG-INDUCED ACUTE KIDNEY INJURY

Yutaka Tonomura, Mitsunobu Matsubara and Takeki Uehara

Drug Safety Evaluation, Drug Developmental Research Laboratories, Shionogi and Co., Ltd., Toyonaka, Osaka, Japan
Division of Molecular Medicine, Center for Translational and Advanced Animal Research, Tohoku University School of Medicine, Aoba-ku, Sendai, Japan

ABSTRACT

The kidney is particularly vulnerable to various drugs. Early screening of drug-induced acute kidney injury (AKI) is therefore critical for its clinical management, leading to a better outcome of clinical treatment. For the pharmaceutical industry, drug-induced AKI is a major concern in the early stage of preclinical safety evaluations. One major limitation in early detection of AKI has been the low detection power of traditional biomarkers, such as creatinine and blood urea nitrogen. Recent advances in basic and clinical research have provided several valuable biomarkers for early detection of AKI for the preclinical safety evaluation of drugs, clinical trials, and early therapeutic intervention. Serum cystatin c (CysC) has been identified as an attractive alternative biomarker for estimation of the glomerular filtration rate. Additionally, several biomarkers, such as kidney injury molecule-1 (KIM-1), neutrophil gelatinase-associated lipocalin, interleukin-18 and liver type fatty acid binding protein, have been discovered and their usefulness has been evaluated in cross-sectional studies. Recently, the Predictive Safety Testing Consortium's Nephrotoxicity Working Group, a collaboration between biotech and pharmaceutical industries, the US Food and Drug Administration (FDA), the European Medicines Agency (EMEA) and academia, published a report concerning the qualification of seven urinary nephrotoxic biomarkers, including total protein, albumin, KIM-1, clusterin (CLU), β_2-microglobulin, CysC, and trefoil factor 3, for particular uses in regulatory decision-making. Furthermore, the International Life Sciences Institute Health and Environmental Sciences Institute reported an extensive data package on the four urinary nephrotoxic biomarkers glutathione S-transferase α (GSTα),

GSTμ, renal papillary antigen-1 and CLU to the FDA and the EMEA. This chapter describes the usefulness of these biomarkers with respect to clinical and preclinical usage and the possible mechanisms that underlie their alterations in serum and/or urine. Finally, we discuss the future view of AKI biomarkers.

INTRODUCTION

The morbidity of acute kidney injury (AKI) is associated with mortality. Recently, two different groups proposed criteria to define AKI [Bellomo *et al.*, 2004; Mehta *et al.*, 2007]. The RIFLE classification proposed by the Acute Quality Dialysis Initiative is based on serum creatinine (Cr) levels and urine output and defines the following categories: risk of renal dysfunction (R category), a 150% increase in serum Cr over baseline and a glomerular filtration rate (GFR) 25% lower than baseline or less than 0.5 mL/kg/h of urine output sustained for 6 hours; injury to the kidney (I category), a 200% increase in serum Cr over baseline and a GFR 50% lower than baseline or less than 0.5 mL/kg/h of urine output sustained for 12 hours; failure of kidney function (F category), a 300% increase in serum Cr over baseline (or not less than 4 mg/dL), a GFR 75% lower than baseline, or less than 0.3 mL/kg/h of urine output or anuria sustained for 24 or 12 hours, respectively; loss of kidney function (L category), requirement for dialysis for more than 4 weeks; and end-stage renal disease (E category), requirement for dialysis for more than 3 months. Thereafter, the Acute Kidney Injury Network (AKIN) removed GFR criteria and the L and E categories from the RIFLE classification and modified the R, I and F categories as stages I, II and III, respectively. In this revised classification, Stage I is similar to the R category but contains the additional definition of an increase in serum Cr that is not less than 0.3 mg/dL over baseline and AKI is defined as an increase in serum Cr over 48 hours rather than 7 days, as proposed by RIFLE.

AKI is a complication in up to 7% of all hospital admissions and in up to 25% of intensive care unit (ICU) admissions [de Mendonca *et al.*, 2000; Hou *et al.*, 1983; Nash *et al.*, 2002]. The mortality rate of AKI patients in the ICU currently remains high, and can reach up to 88% [Ympa *et al.*, 2005]. Moreover, a large number of survivors show persistent renal dysfunction [Nash *et al.*, 2002; Uchino *et al.*, 2005]. Furthermore, the kidney is a major drug target organ for the pharmaceutical industry, and nephrotoxicity can be fatal for drug research and development (R&D). Therefore, it is important to detect renal injury at an early stage in the hospital and pharmaceutical industries in order to facilitate prompt decisions for selection of the best therapy and to reduce the cost of drug R&D, respectively. Although the recent definition of AKI mentioned above might contribute to consensus recognition of AKI, AKIN and RIFLE criteria have several drawbacks due to the reference biomarkers serum Cr, GFR and urine output. In other words, these criteria might be not able to determine the injured site of nephron segment [Schrier, 2010], and the reference biomarkers show relatively late alteration after renal injury has occurred [Bagshow, 2010]. Thus, introduction of more site-specific and sensitive biomarkers into the definition of AKI would further improve the clinical outcome. Recently, several biomarkers, including kidney injury molecule-1 (KIM-1), neutrophil gelatinase-associated lipocalin (NGAL), interleukin-18 (IL-18) and liver type fatty acid binding protein (L-FABP), have been discovered and their usefulness has been evaluated using human subjects. Moreover, the Predictive Safety Testing Consortium (PSTC)'s

Nephrotoxicity Working Group, a collaboration of the biotech and pharmaceutical industries, the US Food and Drug Administration (FDA), the European Medicines Agency (EMEA) and academia, have published reports concerning the qualification of seven urinary nephrotoxic biomarkers, including total protein (uTP), albumin (uALB), KIM-1, clusterin (CLU), β_2-microglobulin (B2M), cystatin c (CysC), and trefoil factor 3 (TFF3), for particular uses in regulatory decision-making [Dieterle *et al.*, 2010; Ozer *et al.*, 2010; Vaidya *et al.*, 2010; Yu *et al.*, 2010]. Furthermore, the International Life Sciences Institute Health and Environmental Sciences Institute (ILSI-HESI) reported an extensive data package on the four urinary nephrotoxic biomarkers glutathione S-transferase α (GST-α), GST-μ, renal papillary antigen-1 (RPA-1) and CLU to the FDA and the EMEA. In this chapter, we discuss the possible mechanisms that underlie the alterations of these biomarkers in serum and/or urine and their utility in clinical and preclinical fields. Finally, we discuss the future of biomarkers for AKI.

Possible Mechanisms Underlying the Alteration of Each Biomarker and Biomarker Usefulness

Figure 1 and Table 1 summarize the site where each biomarker reflects the damage by AKI and the possible mechanisms underlying the alteration of each biomarker.

Cystatin C

Cystatins are inhibitors of cysteine proteases including cathepsins L, B and H. Cystatin c (CysC) is produced by all nucleated cells and is constantly released into the bloodstream. The low molecular weight (13 kDa) and positive charge of CysC allows serum CysC to be freely filtrated by glomeruli without the influence of glomerular size or the charge barrier in the kidney. Moreover, CysC has a further advantage over Cr in that the level of serum CysC is independent of age, sex, muscle mass and dehydration state, and there is no tubular secretion or extrarenal clearance of CysC [Herget-Rosenthal *et al.*, 2007; Ozer *et al.*, 2010; Takuwa *et al.*, 2002]. Thus, serum CysC has been used as an alternative biomarker to serum Cr and BUN, since serum CysC is superior to these traditional biomarkers for estimation of GFR in AKI [Herget-Rosenthal *et al.*, 2004; Madero *et al.*, 2006]. The concentration of CysC is very low in urine under normal renal conditions [Tenstad *et al.*, 1996] and increases in AKI. The mechanisms that underlie the elevation of low molecular weight urinary proteins such as CysC are explained by two theories. First, their reabsorption in proximal tubules may be inhibited by a large amount of high molecular weight proteins caused by abnormal transglomerular passage in glomerular injury [D'Amico and Bazzi, 2003]. Second, their tubular reabsorption may be impaired by treatment with several drugs and chemicals which directly injure proximal tubules [Jaafar *et al.*, 2009].

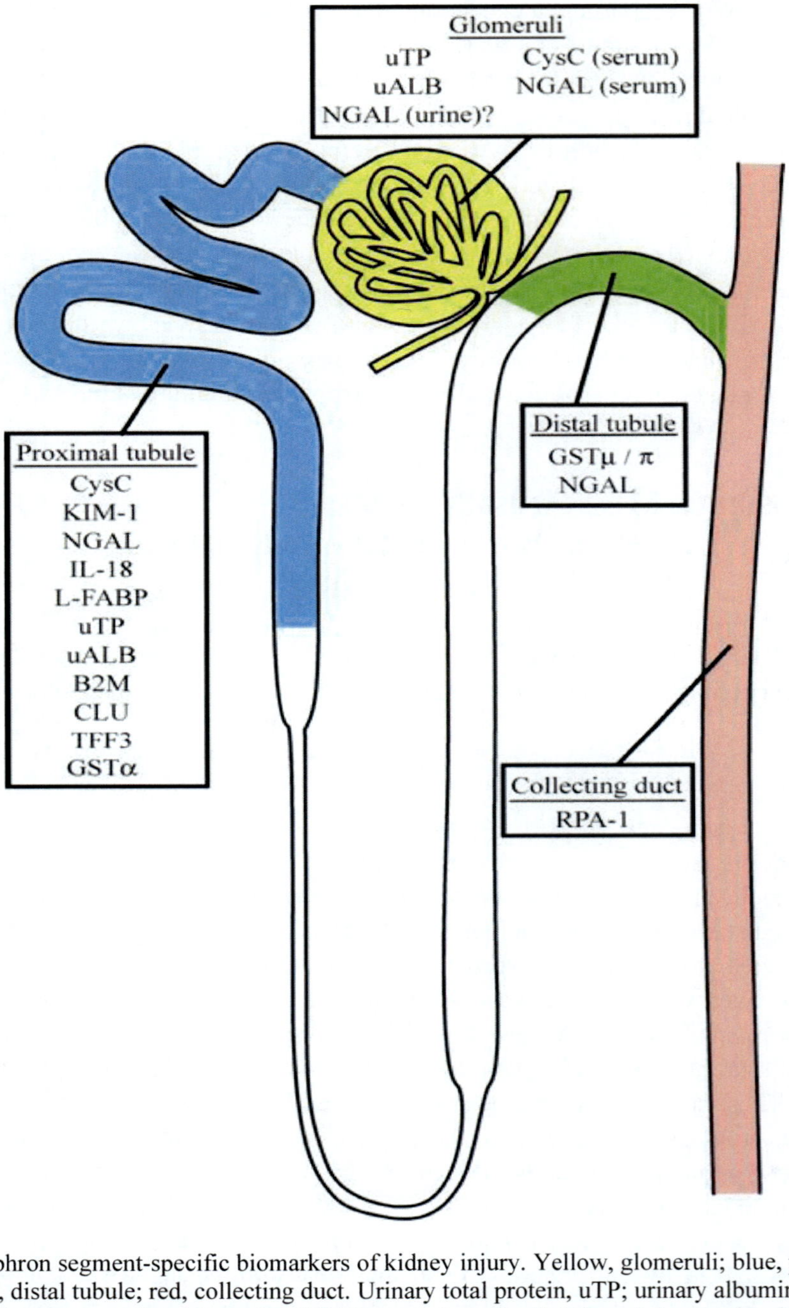

Figure 1. Nephron segment-specific biomarkers of kidney injury. Yellow, glomeruli; blue, proximal tubule; green, distal tubule; red, collecting duct. Urinary total protein, uTP; urinary albumin, uALB; neutrophil gelatinase-associated lipocalin, NGAL; cystatin c, CysC; kidney injury molecule-1, KIM-1; interleukin-18, IL-18; liver type fatty acid binding protein, L-FABP; β_2-microglobulin, B2M; clusterin, CLU; trefoil factor 3, TFF3; glutathione S-transferase, GST; renal papillary antigen-1, RPA-1.

Table 1. List of biomarkers for acute kidney injury

Parameter	Renal region	Possible mechanisms
CysC (urine)	Glomerulus	Abnormality of transglomerular passage
	Proximal tubule	Inhibition of reabsorption in renal tubule
CysC (serum)	Glomerular filtration	Accumulation in bloodstream
KIM-1	Proximal tubule	Release into extracellular space after induction
NGAL (urine)	Glomerulus	Abnormality of transglomerular passage
	Proximal tubule	Inhibition of reabsorption in renal tubule
	Distal tubule	Release into extracellular space after induction
NGAL (serum)	Glomerular filtration	Accumulation in bloodstream
IL-18	Proximal tubule	Leakage of pre-existing protein from cytosol
L-FABP	Proximal tubule	Inhibition of reabsorption in renal tubule
uTP	Glomerulus	Abnormality of transglomerular passage
	Proximal tubule	Inhibition of reabsorption in renal tubule
uALB	Glomerulus	Abnormality of transglomerular passage
	Proximal tubule	Inhibition of reabsorption in renal tubule
TFF3	Proximal tubule	Leakage of pre-existing protein from cytosol
CLU	Proximal tubule	Release into extracellular space after induction
B2M	Glomerulus	Abnormality of transglomerular passage
	Proximal tubule	Inhibition of reabsorption in renal tubule
GSTα	Proximal tubule	Leakage of pre-existing protein from cytosol
GSTμ / π	Distal tubule	Leakage of pre-existing protein from cytosol
RPA-1	Collecting duct	Leakage of pre-existing protein from cytosol

In a cross-sectional study, serum CysC was superior to serum Cr with respect to the detection of AKI as defined by the RIFLE criteria [Herget-Rosenthal *et al.*, 2004]. In patients at high risk of developing AKI, a 50% increase in serum CysC was observed 1 to 2 days earlier than that of serum Cr, and serum CysC was shown to detect AKI with high accuracy in receiver operating characteristics (ROC) analysis, in which the area under the curve (AUC) was 0.82 and 0.97 at 2 days and 1 day, respectively, before the increase in serum Cr. Additionally, in a preclinical study of several nephrotoxicants, serum CysC demonstrated high accuracy in ROC analysis based on the histopathological score [Ozer *et al.*, 2010]. In this study, the AUC for serum CysC was 0.72 in contrast to 0.57 and 0.66 for serum Cr and BUN, respectively. Therefore, serum CysC is a more useful biomarker for AKI compared to traditional serum biomarkers such as Cr and BUN.

Cross-sectional studies have been undertaken regarding urinary CysC levels in patients with renal injury. Urinary CysC was higher in patients with renal tubular disease diagnosed by pathological examination than in controls or in patients with glomerular disease diagnosed by serum Cr, serum CysC and Cr clearance [Conti *et al.*, 2006]. Another cross-sectional study, in which patients were classified as having tubulointerstitial or glomerular disease based on histopathological findings, suggested that the ratio of urinary CysC to urinary Cr was related to tubulointerstitial rather than to glomerular disease [Herget-Rosenthal *et al.*, 2007]. These two cross-sectional studies indicate that urinary CysC is a specific biomarker for renal tubular injury. By contrast, in a preclinical study, Dieterle and co-workers reported urinary CysC as a biomarker for glomerular injury and showed that the AUC of urinary CysC in glomerular injury ranged from 0.90 to 0.92, in contrast to 0.55 and 0.76 for serum Cr and BUN, respectively [Dieterle *et al.*, 2010]. Their study employed puromycin and doxorubicin, which induce glomerular injury. Urinary CysC was elevated in groups treated with both compounds, and correlated with glomerular histopathological severity. The combined data suggested a discrepancy in the meaning of increase in urinary CysC between clinical and preclinical results. However, both clinical and preclinical researchers commonly suggested that the cause of an increase in urinary CysC might be competitive inhibition of reabsorption in proximal tubules due to high molecular weight proteins accompanying abnormality transglomerular passage [Dieterle *et al.*, 2010; Herget-Rosenthal *et al.*, 2007]. Furthermore, it was also suggested that urinary CysC was increased by treatment with cisplatin and gentamicin, which are known as proximal tubular nephrotoxicants [Dieterle *et al.*, 2010; Tonomura *et al.*, 2010]. Thus, urinary CysC is thought to be a biomarker not only for reflecting both glomerular and proximal tubular injuries but also as a translational biomarker from animals to humans.

Kidney Injury Molecule-1

Kidney injury molecule-1 (KIM-1) is a type 1 transmembrane glycoprotein, also known as T-cell immunoglobulin mucin domains-1 and hepatitis A virus cellular receptor-1, with an immunoglobulin-like domain and a mucin-like domain that contain several putative *N*- and *O*-glycosylation sites. KIM-1 functions as a receptor for phosphatidylserine, which is induced on the cell surface by renal tubular injury and serves as an "eat-me" signal. The recognition of apoptotic tubular epithelial cells by KIM-1-expressing tubular epithelial cells induces the phagocytosis of injured tubular epithelial cells. Therefore, KIM-1 confers a phagocytic phenotype on non-myeloid cells and is considered to be associated with facilitation of the clearance of apoptotic debris from the tubular lumen [Ichimura *et al.*, 2008]. KIM-1 mRNA and protein are not detected in the normal kidney. When the kidney is damaged, KIM-1 mRNA is immediately and strongly expressed and is translated to its protein, which can be detected in urine. The reason why KIM-1 protein is detectable in urine, despite the fact that it is a transmembrane protein, is that its extracellular domain, termed the ectodomain, is cleaved by an undefined metallopeptidase and is subsequently shed into urine [Bailly *et al.*, 2002; Zhang *et al.*, 2007]. However, the role of the ectodomain of KIM-1 in renal tubular function remains unknown.

Urinary KIM-1 is used as a biomarker for AKI because of its rapid response to AKI. A cross-sectional study suggested that KIM-1 can specifically detect acute tubular necrosis (ATN; one of the symptoms of AKI, which is induced by several factors, such as drugs, chemicals or ischemia followed by reperfusion) that was confirmed by biopsy of patients who were diagnosed as AKI based on the elevation of serum Cr over baseline levels [Han *et al.*, 2002]. Another cross-sectional study also indicated the usefulness of urinary KIM-1 for detection of AKI following cardiopulmonary bypass [Liangos *et al.*, 2007]. In that study, urinary KIM-1 was the most accurate predictor of AKI of the urinary biomarkers, such as N-acetyl-β-D-glucosaminidase (NAG), in which the AUC of the ROC curve of KIM-1 was 0.78, whereas the AUC value of other biomarkers was lower. Several preclinical studies using different nephrotoxic rat models have also suggested that urinary KIM-1 can detect AKI with a higher accuracy than traditional biomarkers such as serum Cr, BUN and urinary NAG [Ichimura *et al.*, 2004; Tonomura *et al.*, 2010; Vaidya *et al.*, 2006; Vaidya *et al.*, 2010]. Thus,

KIM-1 is considered to be the best biomarker, not only for detection of AKI, but also as a translational biomarker from animals to humans.

Neutrophil Gelatinase-Associated Lipocalin

Neutrophil gelatinase-associated lipocalin (NGAL) is a cytosolic and secreted protein belonging to the lipocalin family [Mori and Nakao, 2007], and forms a complex with a siderophore and iron, and exerts iron-dependent biological activity. When the siderophore is mixed with iron, the NGAL complex transfers iron into cells. Conversely, when the siderophore is mixed without iron, the NGAL complex chelates iron [Li *et al.*, 2004; Yang *et al.*, 2002]. *In vivo*, the NGAL complex induces heme oxygenase-1 (HO-1) by transferring iron into epithelial cells in the proximal tubules. HO-1 is thought to protect epithelial cells from ischemia and nephrotoxicity by reducing the level of intracellular non-ferritin-bound iron. This type of iron is mislocalized iron that causes the mutagenization of molecules and is related to the induction of ATN. Recent research showing that renal injury is alleviated by treatment with the NGAL protein supports these theories of NGAL function [Mishra *et al.*, 2004; Mori *et al.*, 2005]. NGAL may also be useful as a biomarker for AKI because the expression of NGAL mRNA and protein in kidney is dramatically elevated in ischemia-reperfusion renal injury [Mishra *et al.*, 2003; Mori *et al.*, 2005]. It has been reported that the urinary NGAL protein level is increased in renal injury, which is caused by inhibition of its reabsorption into the renal tubule [Kuwabara *et al.*, 2009]. NGAL is known as a ligand for megalin, which is an endocytic receptor that is highly expressed in the brush border of the proximal tubule and functions in the reabsorption of several low molecular weight serum proteins [Birn and Christensen, 2006; Christensen *et al.*, 1999; Hvidberg *et al.* 2005]. Therefore, one reason why urinary NGAL is increased in renal injury may be because of renal reabsorptive dysfunction due to an abnormality in activity and/or a decrease of megalin. In addition, it is known that the *de novo* synthesis of NGAL in distal tubules is up-regulated during ureteral obstruction, which leads to an increase in urinary NGAL [Kuwabara *et al.*, 2009]. Therefore, the cause of urinary NGAL elevation is diverse. Serum NGAL is also elevated in AKI, which may be explained by two possible mechanisms. First, NGAL appears to be induced in the liver in renal injury and be released into the bloodstream, resulting in accumulation of NGAL in blood [Schmidt-Ott *et al.*, 2006]. Second, the decrease in GFR in AKI appears to contribute to a reduction in the clearance of serum NGAL through the kidney [Dent *et al.*, 2007].

In two cross-sectional studies [Haase-Fielitz *et al.*, 2009; Makris *et al.*, 2009], both serum and urinary NGAL predicted AKI with a higher accuracy than serum Cr; the AUCs of the ROCs of NGAL vs. serum Cr were 0.98 vs. 0.79 and 0.80 vs. 0.68, respectively, in each study. In addition to these two cross-sectional studies, numerous other cross-sectional studies have been conducted globally. However, these other studies showed various cutoff values due to the use of non-standardized methods. Thereafter, a meta-analysis was undertaken by collecting data from 19 clinical trials in order to set a unified cutoff value for AKI. In this meta-analysis, NGAL in blood or urine performed equally well, and its cutoff value and AUC for ROC were >150 ng/mL and 0.830, respectively, using standardized platforms [Haase *et al.*, 2009]. Preclinical studies using several AKI models suggested that urinary NGAL was more sensitive than conventional serum or urinary biomarkers (e.g., serum Cr and urinary

NAG) [Mishra *et al.*, 2004; Mori *et al.*, 2005; Tonomura *et al.*, 2010]. Thus, NGAL is a well-validated biomarker for AKI and a candidate translational biomarker.

Interleukin-18

Interleukin-18 (IL-18) is a proinflammatory cytokine and its activation requires the cleavage of the immature form at an Asp-X site by caspase-1, a proinflammatory caspase that can specifically activate IL-18 [Ghayur *et al.*, 1997; Fantuzzi *et al.*, 1998]. It has been reported that IL-18-neutralizing antiserum mitigates ischemic AKI *in vivo* [Melnikov *et al.*, 2001]. In contrast, it has been suggested that the IL-18 binding protein, which is an inhibitor of IL-18, did not show a protective effect against hypoxic membrane injury *in vitro* [Edelstein *et al.*, 2007]. Therefore, the overall mechanism that underlies IL-18-mediated ischemic AKI remains unclear, although IL-18 is partially associated with the degree of ischemic AKI. With regards to the origin of urinary IL-18, the immature form of IL-18 is expressed in intact proximal tubules, although IL-18 is barely detected in urine [Edelstein *et al.*, 2007; Melnikov *et al.*, 2001]. Once ischemic AKI occurs, the mature form of IL-18 is released into urine from the proximal tubule. Moreover, macrophages and blood monocytes are a second source of IL-18 [Obregon *et al.*, 2003], and IL-18 in blood may pass through the glomeruli due to its low molecular weight and may be detectable in urine under systemic inflammatory conditions. While it has been reported that urinary IL-18 is a useful biomarker for AKI in several cross-sectional studies [Parikh *et al.*, 2004; Parikh *et al.*, 2005; Parikh *et al.*, 2006; Washburn *et al.*, 2008], other cross-sectional studies suggested that the measurement of urinary IL-18 is not valuable for the prediction of AKI since urinary IL-18 may reflect systemic inflammation [Bulent Gul *et al.*, 2008; Haase *et al.*, 2008]. Therefore, the usefulness of urinary IL-18 is controversial and further studies are required.

Liver Type Fatty Acid Binding Protein

Fatty acid binding proteins (FABPs) are small cytoplasmic proteins. To date, nine types of FABPs have been identified [Zimmerman and Veerkamp, 2002]. FABPs can bind fatty acids (FA) and act as carrier proteins that bring FA from the outside into the inside of the cell. Intracellular FAs that are carried by FABPs are distributed to mitochondria, peroxisomes, the endoplasmic reticulum (ER) or the nucleus. In mitochondria and peroxisomes, FAs are used as a resource for β-oxidation [Wanders *et al.*, 2003]. In the ER, FAs are esterified into triglycerides or phospholipids [Frayn *et al.*, 2006]. In the nucleus, FAs act as second messengers in the transduction of external signals [Nunez, 1997]. Liver type fatty acid binding protein (L-FABP) is an FABP family member that is expressed in liver, intestine, kidney and stomach, and transports FA to the nucleus [Zimmerman and Veerkamp, 2002]. Oleic acid, a monounsaturated FA, promotes the distribution of L-FABP within the nucleus and its interaction with nucleic proteins [Lawrence *et al.*, 2000]. In addition, L-FABP interacts with the peroxisome proliferator-activated receptors (PPAR) α and γ, which play important roles in FA catabolism and in the regulation of adipose differentiation and adipogenesis, respectively [Amri *et al.*, 1995; Spiegelman and Flier, 1996]. The combined data suggest that L-FABP might be involved in informing the condition of FA metabolism from cytoplasm to nucleus. Moreover, it has been shown that L-FABP has a renoprotective effect in cisplatin-induced AKI and in ischemia-reperfusion renal injury [Negishi *et al.*, 2007; Sugaya *et al.*, 2005; Yamamoto *et al.*, 2007]. The mechanism that has been speculated to underlie its renoprotective effect is that L-FABP binds to adverse lipid peroxidative products in the proximal tubule and transfers them to the tubular lumen. In rat and human kidney, L-FABP protein is localized in proximal tubules and/or distal tubules, which are the major sites of oxidation in the kidney [Freeman *et al.*, 1986; Hohenleitner and Spitzer, 1961; Lee *et al.*, 1962; Maatman *et al.*, 1991; Maatman *et al.*, 1992]. In the rodents, L-FABP mRNA is hardly expressed in the kidney even though L-FABP protein can be detectable [Maatman *et al.*, 1992; Oyama *et al.*, 2005; Simon et al., 1993]. This discrepancy can be explained by the characteristics of L-FABP that L-FABP protein can be freely filtrated in renal glomeruli due to low-molecular-weight protein and is reabsorbed in proximal tubule via megalin [Oyama *et al.*, 2005]. Thus, almost all of the L-FABP protein detected in the kidney is thought to be reabsorbed protein and derived from circulation, and in the case of AKI, L-FABP can become detectable in urine due to two mechanisms: competitive inhibition of reabsorption by high molecular weight proteins accompanying abnormality of transglomerular passage; reabsorptive dysfunction due to direct damage to proximal tubules. On the other hand, the L-FABP protein is expressed in proximal tubules in the human kidney, and in ischemia-reperfusion renal injury, L-FABP in proximal tubules leaks into the tubular lumen [Yamamoto *et al.*, 2007]. Furthermore, in cardiopulmonary bypass-induced AKI, the elevation of urinary L-FABP could not be explained only by increased glomerular filtration because urinary L-FABP was observed to increase before the elevation of serum L-FABP [Portilla *et al.*, 2008]. Therefore, in humans, the source of urinary L-FABP is proximal tubular cells in addition to L-FABP which passes through glomeruli. Several cross-sectional studies have evaluated the usefulness of urinary L-FABP as a biomarker for AKI [Ferguson *et al.*, 2010; Nakamura *et al.*, 2006; Portilla *et al.*, 2008]. These studies indicated that L-FABP is a more sensitive and specific predictor of AKI than serum Cr. Ferguson and colleagues indicated that L-FABP was a better biomarker than NGAL, KIM-1, NAG or IL-18 in a study that included adult patients with AKI and heterogeneous controls [Ferguson *et al.*, 2010]. On the other hand, Portilla and colleagues suggested that urinary L-FABP was inferior to urinary NGAL with regard to the prediction of cardiopulmonary bypass-induced AKI in children. Therefore, to date, it remains controversial whether urinary L-FABP is useful for detection of AKI. There has been no preclinical study in which the usefulness of urinary L-FABP was investigated. Animal experiments will be required to determine if urinary L-FABP will be a useful translational biomarker.

Urinary Total Protein

The elevation of urinary total protein (uTP), so-called proteinuria, generally reflects an increase in glomerular permeability for albumin and other high molecular weight proteins in plasma [Orth and Ritz, 1998]. Proteinuria is classified into several types including tubular and overflow proteinuria [Mundel and Reiser *et al.*, 2010]. Low molecular weight proteins such as CysC and B2M, are freely filtrated by glomeruli and are almost completely reabsorbed by proximal tubules [Bergón *et al.*, 2002; D'Amico and Bazzi, 2003]. These low molecular weight proteins are not reabsorbed if there is tubular dysfunction. As a result, these proteins can be detected in urine. Overflow proteinuria occurs under myeloma conditions, in which the amount of filtered paraprotein exceeds the reabsorptive ability of proximal tubules, thereby leading to the detection of low molecular weight proteins in urine [Carroll and Temte, 2000]. In a healthy person, total excretion of uTP is less than 150 mg/day, and if a higher level of uTP is detected, then the presence of the above-mentioned diseases should be assumed [Mundel and Reiser *et al.*, 2010]. A large-scale preclinical evaluation of uTP has been carried out [Dieterle *et al.*, 2010]. In this study, uTP was a more accurate biomarker of glomerular injury than serum Cr or BUN, and the AUCs of the ROC curves were 0.86, 0.80 and 0.53, respectively. Since the usefulness of uTP has been adequately evaluated in animals and humans, uTP is considered to be a translational biomarker for renal injury.

Urinary Albumin

Albumin is a negatively charged protein that is produced predominantly in the liver and is the most abundant protein in blood. Although the functions of albumin are diverse, it is well-known that albumin modulates colloid osmotic pressure and serves as a carrier protein for molecules such as fatty acids, bilirubin and acidic xenobiotics. Intraglomerular plasma is filtrated in kidney glomeruli. Albumin is not filtrated very well under normal conditions, due to its size and to the charge-selective glomerular barrier [Mundel and Reiser *et al.*, 2010]. In the case of renal injury, two major mechanisms that underlie the increase in urinary albumin (uALB) are known. The first mechanism is an increased transglomerular passage of albumin due to the loss of charge restriction and/or size restriction of the glomerular capillary wall [D'Amico and Bazzi, 2003; Mundel and Reiser *et al.*, 2010]. The second mechanism is an impairment of reabsorption by the proximal tubule, where megalin is expressed (albumin is a ligand for megalin), due to dysfunction or toxic injury [Cui *et al.*, 1996; D'Amico and Bazzi, 2003]. In humans, the ratio of uALB to urinary Cr is less than 30 μg/mg Cr [Mattix *et al.*, 2002], and, if higher levels of uALB are detected, then glomerular and tubular injuries should be assumed. Two preclinical evaluations of uALB have been reported [Tonomura *et al.*, 2010; Yu *et al.*, 2010]. In both studies, uALB was evaluated as a biomarker for renal tubular injury and glomerular injury in several nephrotoxicant-induced rat models and showed high accuracy of prediction. The AUCs for ROC curves were 0.901-0.984 for renal tubular injury and 0.888 for glomerular injury. Since uALB has been adequately evaluated in animals and humans, uALB is also thought to be a translational biomarker for renal injury.

Clusterin

Clusterin (CLU) is a heterodimeric glycoprotein composed of α- and β-subunits whose precursor is encoded by a single gene. This precursor has a signal peptide at the amino terminal end and a cleavage site at the center [Jones and Jomary, 2002]. The mature form of CLU is associated with multiple functions including apoptosis, cell adhesion, oxidative stress and lipid transport [Buttyan *et al.*, 1989; Gelissen *et al.*, 1998; Gobé *et al.*, 1995; Nath *et al.*, 1994; Schwochau *et al.*, 1998; Silkensen *et al.*, 1995]. Moreover, in the kidney, CLU appears to have a protective effect against AKI for evidence of the mitigation of injury in renal tubular epithelial cells by treatment with CLU and the deterioration of ischemic-reperfusion renal injury in lack of CLU [Girton *et al.*, 2002; Zhou *et al.*, 2010]. In AKI, CLU mRNA is induced in the kidney and localized in renal tubular cell debris [Hidaka *et al.*, 2002]. Furthermore, because there was an increase in urinary CLU in tubular injury, but not in glomerular injury, urinary CLU has the potential to act as a specific biomarker for tubular injury [Hidaka *et al.*, 2002].

One cross-sectional study, which investigated whether urinary CLU is useful for discrimination of patients with or without kidney injury [Solichova *et al.*, 2007], indicated that urinary CLU is not useful for detection of kidney injury. However, in that study, the population of patients included various nephropathies: diabetic nephropathy, glomerulonephropathy and chronic kidney disease. Since urinary CLU is a specific biomarker for tubular injury [Hidaka *et al.*, 2002], urinary CLU may not have been a useful biomarker for that study. On the other hand, several preclinical investigations of the ability of urinary CLU to detect AKI [Aulitzky *et al.*, 1992; Dieterle *et al.*, 2010; Hoffmann *et al.*, 2010; Sieber *et al.*, 2009; Tonomura *et al.*, 2010], indicated that urinary CLU was elevated earlier than serum Cr and BUN. Thus, urinary CLU is considered to be more useful than traditional serum biomarkers of AKI. However, in comparison with other urinary biomarkers for AKI (e.g., KIM-1 and NGAL), CLU is considered to be of equal or lower usefulness for the accurate and early detection of AKI.

β_2-microglobulin

β_2-microglobulin (B2M) is a 12-kDa negatively charged protein that is expressed on the surface of most nucleated cells [Berggård and Bearn, 1968]. B2M is known to combine with the α subunit to form the major histocompatibility complex class I [Rask *et al.*, 1974]. B2M can be freely filtrated by glomeruli in the kidney due to its low molecular weight. After glomerular filtration, B2M is reabsorbed by the proximal tubular epithelial cells [Gauthier *et al.*, 1984; Portman *et al.*, 1986]. With respect to reabsorption in proximal tubules, there are two mechanisms mentioned above: competitive inhibition of reabsorption by high-molecular-weight proteins accompanying abnormality of transglomerular passage; reabsorptive dysfunction due to direct damage to proximal tubules. Thus, it is thought that urinary B2M is a biomarker for reflecting both glomerular and proximal tubular injuries. Several cross-sectional studies have evaluated whether urinary B2M is useful for detecting AKI [Aggarwal *et al.*, 2005; Bagshaw *et al.* 2007; Gordjani *et al.*, 1995; Mehta *et al.*, 1997]. These studies suggested that, although urinary B2M is more sensitive than serum Cr or BUN for detection of AKI, the detection power of urinary B2M is inferior to that of KIM-1 and enzymuria. Moreover, a preclinical study carried out by Tonomura and colleagues suggested that KIM-1 and other urinary biomarkers are more accurate than urinary B2M based on the AUCs of ROC curves with respect to proximal tubular injury [Tonomura *et al.*, 2010]. Conversely, one preclinical report indicates that urinary B2M is one of the most accurate biomarkers for the detection of glomerular injury [Dieterle *et al.*, 2010].

Trefoil Factor 3

Trefoil factor 3 (TFF3) is a small secretory protein that belongs to the trefoil factor family. After translation, the precursor is believed to be enzymatically modified and to act as a monomer and/or dimer [Chadwick *et al.*, 1997; Chinery *et al.*, 1993]. TFF3 is abundantly expressed in goblet cells in the intestine [Suemori *et al.*, 1991]. It is known that TFF3 has protective and regenerative effects upon intestinal injury [Mashimo *et al.*, 1996; Taupin and Podolsky, 2003]. In addition to the intestine, rat kidney also abundantly expresses TFF3 mRNA [Suemori *et al.*, 1991; Yu *et al.*, 2010.]. Considering the protective and regenerative effect of TFF3 in the intestine, TFF3 may therefore also act as a protective and regenerative factor in the kidney. There has only been one preclinical study of the biomarker ability of TFF3 [Yu *et al.*, 2010]. That study showed that TFF3 mRNA is abundantly expressed in the normal kidney, where it is mainly distributed at the corticomedullary junction consisting of the S3 segment of proximal tubules. Conversely, TFF3 mRNA expression is reduced in carbapenem A-induced AKI, whose nephrotoxicity is believed to be expressed in the S3 segment of proximal tubules [Ozer *et al.*, 2010]. As a result, urinary TFF3 protein is decreased. Thus, urinary TFF3 is a potential biomarker for proximal tubular injury. To date, the usefulness of urinary TFF3 as a biomarker for AKI has not been evaluated in humans. Thus, further study is required to determine if urinary TFF3 is a translational biomarker for AKI.

Glutathione S-transferases

Glutathione S-transferases (GSTs) form a family of homo- and heterodimeric proteins that are classified into several classes based on similarity in substrate specificity, inhibitor sensitivity and reactivity with specific antibodies [Mannervik *et al.*, 1985]. Functionally, GSTs catalyze the reaction of electrophilic compounds with reduced glutathione. Thus, they are important enzymes for detoxification of many xenobiotics [Dirr *et al.*, 1994; Ketterer *et al.*, 1983].

GSTα consists of the subunits 1, 2 and 8, which are mostly expressed in proximal tubules in humans and rats [Campbell *et al.*, 1991; Rozell *et al.*, 1993; Sundberg *et al.*, 1993]. In the rat, GSTμ consists of the subunits 3, 4 and 6, which are expressed in epithelial cells from the ascending thin limbs of the loop of Henle to the urinary tract. Subunit 3, also called GSTYb1, is also strongly expressed in distal tubules [Rozell *et al.*, 1993]. In humans, GSTπ consists of subunit 7 and is expressed in cells of the descending and ascending thin limbs of the loop of Henle, in distal tubules and in collecting ducts [Sundberg *et al.*, 1994]. These GSTs are localized in the cellular cytoplasm and are leaked from tubular cells during renal injury. They have therefore been used as biomarkers for renal injury of each nephron segment. In humans, the usefulness of urinary GSTα and GSTπ as biomarkers for the prediction of AKI was evaluated in two cross-sectional studies [Walshe *et al.*, 2009; Westhuyzen *et al.*, 2003]. In patients in the ICU, both GSTα and π could predict AKI with high accuracy (the AUCs for ROC were 0.893 and 0.929, respectively). On the other hand, urinary GSTπ, but not GSTα, was increased in patients with sepsis. However, its AUC for ROC indicated that GSTπ cannot predict AKI following sepsis [Walshe *et al.*, 2009]. The usefulness of urinary GSTα as a biomarker for AKI was evaluated in several preclinical studies [Bass *et al.*, 1979; Kharasch *et al.*, 1998; Ozer *et al.*, 2010; Tonomura *et al.*, 2010]. Urinary GSTα was considerably increased in cisplatin- and carbapenem A-induced AKI models, which exhibit tubular damage at the corticomedullary junction [Ozer *et al.*, 2010; Tonomura *et al.*, 2010]. In contrast to urinary GSTα, there has only been one preclinical evaluation of the usefulness of urinary GSTμ as a biomarker for distal tubular injury in AKI [Tonomura *et al.*, 2010]. In that study, urinary GSTμ was dramatically increased in a cisplatin-induced AKI model. Interestingly, there was no apparent pathological alteration in distal tubules in the AKI model, indicating that urinary GSTμ may be a more sensitive indicator of distal tubular AKI than pathological examination. However, further studies of urinary GSTμ as a biomarker of AKI are required. Regarding its ability to function as a translational biomarker, GSTα is a suitable translational biomarker for AKI because GSTα shows the same distribution in the nephron in animals and humans.

RPA-1

RPA-1, previously called Pap X 5C10, is an antibody against an antigen (PapA1) that is expressed in the collecting ducts of rat kidney, which is released into urine during renal papillary necrosis (RPN) [Falkenberg *et al.*, 1996; Hildebrand *et al.*, 1999; Price *et al.*, 2010]. Recent efforts to identify PapA1 suggested that PapA1 is a high molecular weight membrane-bound glycoprotein. However, the gene encoding PapA1, and the amino acid sequence of PapA1, remain unclear [Price *et al.*, 2010]. Nevertheless, the usage of the RPA-1 antigen as a biomarker for RPN is increasing based on the lack of other biomarkers for RPN [Hildebrand *et al.*, 1999; Price *et al.*, 2010]. In these studies, PapA1 was evaluated as a biomarker for RPN using several RPN rat models and these studies revealed a correlation between PapA1 levels and histopathological alterations in renal papillae. One report also raised the possibility that PapA1 could be up-regulated by proximal tubular nephrotoxicants [Zhang *et al.*, 2008]. However, it is possible that the observed induced expression of PapA1 was a false-positive due to the high background following immunohistochemical staining [Betton, 2008]. Although PapA1 is currently thought to be a biomarker for RPN, further study of its usefulness as a biomarker is required. Future use of PapA1 as a biomarker for RPN in humans depends on identification of the coding gene and on confirmation of specific expression of PapA1.

THE FUTURE OF BIOMARKERS FOR KIDNEY INJURY

Exosome-Derived Biomarkers

Exosomes are small membrane vesicles of endocytic origin that are released into the extracellular environment upon fusion of multivesicular bodies with the plasma membrane [van Niel *et al.*, 2006]. Exosomes are known to be released into the urine from every segment of the nephron [Pisitkun *et al.*, 2006; McKee *et al.*, 2000]. Furthermore, since exosomes contain various proteins, mRNA and microRNA [Gonzales *et al.*, 2009; Miranda *et al.*, 2010;

Pisitkun *et al.*, 2004], research has focused on exosomes as a new resource of biomarkers for renal injury. Indeed, the level of urinary exosomal proteins (e.g., activating transcription factor 3, Wilms Tumor 1, aquaporin-1, fetuin-A and Na-H exchanger isoform 3) are altered in AKI [du Cheyron *et al.*, 2003; Sonoda *et al.*, 2009; Zhou *et al.*, 2006; Zhou *et al.*, 2008]. Moreover, mRNA that is specifically expressed in each segment of the nephron can be detected in urinary exosomes [Miranda *et al.*, 2010]. Although a large-scale cross-sectional study has not yet been carried out, these exosomal biomarker candidates for AKI have the potential to lead to new applications for the detection and therapy of AKI. Thus, the usefulness of exosome-derived biomarker candidates should be evaluated in humans in the future.

Conclusion

The mechanisms underlying alterations in the level of biomarkers in serum and urine appear to be classified into two and four mechanisms, respectively. The increase and accumulation of biomarkers for renal injury in serum are due to (I) a reduction in GFR and (II) their release into the bloodstream after induction of their expression in several tissues. The alteration in urinary biomarker levels is due to (I) abnormalities in their transglomerular passage, (II) inhibition of their reabsorption in renal tubules, (III) their release into the extracellular space after induction of their protein expression and (IV) leakage of pre-existing protein from the cytosol. Consideration of the characteristics of each biomarker and their combinatorial measurement will further enhance their detection of AKI. Conversely, attention should be paid to the differences between biomarker characteristics in humans and animals. In addition, several biomarkers have not been evaluated with regards to whether their usefulness as a biomarker translates from animals to humans and vice versa. Thus, validation of biomarker candidates with respect to translational usage is of future importance. As a result of this validation, the usage of valuable biomarkers will increase and should lead to adequate therapy for AKI, which will improve the mortality of patients with AKI.

References

Aggarwal A, Kumar P, Chowdhary G, Majumdar S, Narang A. Evaluation of renal functions in asphyxiated newborns. *J. Trop. Pediatr.* 2005;51:295-9.

Amri EZ, Bonino F, Ailhaud G, Abumrad NA, Grimaldi PA. Cloning of a protein that mediates transcriptional effects of fatty acids in preadipocytes. Homology to peroxisome proliferator-activated receptors. *J. Biol. Chem.* 1995;270:2367-71.

Aulitzky WK, Schlegel PN, Wu DF, Cheng CY, Chen CL, Li PS, *et al.* Measurement of urinary clusterin as an index of nephrotoxicity. *Proc. Soc. Exp. Biol. Med.* 1992;199:93-6.

Bagshaw SM, Langenberg C, Haase M, Wan L, May CN, Bellomo R. Urinary biomarkers in septic acute kidney injury. *Intensive Care Med.* 2007;33:1285-96.

Bagshaw SM. Acute kidney injury: diagnosis and classification of AKI: AKIN or RIFLE? *Nat. Rev. Nephrol.* 2010;6:71-3.

Bailly V, Zhang Z, Meier W, Cate R, Sanicola M, Bonventre JV. Shedding of kidney injury molecule-1, a putative adhesion protein involved in renal regeneration. *J. Biol. Chem.* 2002; 277:39739-48.

Bass NM, Kirsch RE, Tuff SA, Campbell JA, Saunders JS. Radioimmunoassay measurement of urinary ligandin excretion in nephrotoxin-treated rats. *Clin. Sci.* 1979;56:419-26.

Bellomo R, Ronco C, Kellum JA, Mehta RL, Palevsky P; Acute Dialysis Quality Initiative workgroup. Acute renal failure - definition, outcome measures, animal models, fluid therapy and information technology needs: the Second International Consensus Conference of the Acute Dialysis Quality Initiative (ADQI) Group. *Crit. Care.* 2004; 8:R204-12.

Berggård I, Bearn AG. Isolation and properties of a low molecular weight beta-2-globulin occurring in human biological fluids. *J. Biol. Chem.* 1968;243:4095-103.

Bergón E, Granados R, Fernández-Segoviano P, Miravalles E, Bergón M. Classification of renal proteinuria: a simple algorithm. *Clin. Chem. Lab. Med.* 2002;40:1143-50.

Betton G. Immunohistochemical localization of RPA-1; comment on paper by Zhang *et al.*, Toxicol Pathol 36;397. *Toxicol. Pathol.* 2008;36:890.

Birn H, Christensen EI, Renal albumin absorption in physiology and pathology. *Kidney Int.* 2006;69:440-9.

Bulent Gul CB, Gullulu M, Oral B, Aydinlar A, Oz O, Budak F, *et al.* Urinary IL-18: a marker of contrast-induced nephropathy following percutaneous coronary intervention? *Clin. Biochem.* 2008;41:544-7.

Buttyan R, Olsson CA, Pintar J, Chang C, Bandyk M, Ng PY, *et al.* Induction of the TRPM-2 gene in cells undergoing programmed death. *Mol. Cell Biol.* 1989;9:3473-81.

Campbell JA, Corrigall AV, Guy A, Kirsch RE. Immunohistologic localization of alpha, mu, and pi class glutathione S-transferases in human tissues. *Cancer.* 1991;67:1608-13.

Carroll MF, Temte JL. Proteinuria in adults: a diagnostic approach. *Am. Fam. Physician.* 2000;62:1333-40.

Chadwick MP, Westley BR, May FE. Homodimerization and hetero-oligomerization of the single-domain trefoil protein pNR-2/pS2 through cysteine 58. *Biochem. J.* 1997;327:117-23.

Chinery R, Poulsom R, Elia G, Hanby AM, Wright NA. Expression and purification of a trefoil peptide motif in a beta-galactosidase fusion protein and its use to search for trefoil-binding sites. *Eur. J. Biochem.* 1993;212:557-63.

Christensen EI, Willnow TE. Essential role of megalin in renal proximal tubule for vitamin homeostasis. *J. Am. Soc. Nephrol.* 1999;10:2224-36.

Conti M, Moutereau S, Zater M, Lallali K, Durrbach A, Manivet P, *et al.* Urinary cystatin C as a specific marker of tubular dysfunction. *Clin. Chem. Lab. Med.* 2006;44:288-91.

Cui S, Verroust PJ, Moestrup SK, Christensen EI. Megalin/gp330 mediates uptake of albumin in renal proximal tubule. *Am. J. Physiol.* 1996;271:F900-7.

D'Amico G, Bazzi C. Pathophysiology of proteinuria. *Kidney Int.* 2003;63:809-5.

de Mendonça A, Vincent JL, Suter PM, Moreno R, Dearden NM, Antonelli M, *et al.* Acute renal failure in the ICU: risk factors and outcome evaluated by the SOFA score. *Intensive Care Med.* 2000;26:915–21.

Dent CL, Ma Q, Dastrala S, Bennett M, Mitsnefes MM, Barasch J, *et al.* Plasma neutrophil gelatinase-associated lipocalin predicts acute kidney injury, morbidity and mortality after pediatric cardiac surgery: a prospective uncontrolled cohort study. *Crit. Care*. 2007;11: R127.

Dieterle F, Perentes E, Cordier A, Roth DR, Verdes P, Grenet O, *et al*. Urinary clusterin, cystatin C, beta2-microglobulin and total protein as markers to detect drug-induced kidney injury. *Nat. Biotechnol*. 2010;28:463-9.

Dirr H, Reinemer P, Huber R. X-ray crystal structures of cytosolic glutathione S-transferases. Implications for protein architecture, substrate recognition and catalytic function. *Eur. J. Biochem*. 1994;220:645-61.

du Cheyron D, Daubin C, Poggioli J, Ramakers M, Houillier P, Charbonneau P, *et al*. Urinary measurement of Na+/H+ exchanger isoform 3 (NHE3) protein as new marker of tubule injury in critically ill patients with ARF. *Am. J. Kidney Dis*. 2003;42:497-506.

Edelstein CL, Hoke TS, Somerset H, Fang W, Klein CL, Dinarello CA, *et al*. Proximal tubules from caspase-1-deficient mice are protected against hypoxia-induced membrane injury. *Nephrol. Dial Transplant*. 2007;22:1052-61.

Falkenberg FW, Hildebrand H, Lutte L, Schwengberg S, Henke B, Greshake D, *et al*. Urinary antigens as markers of papillary toxicity. I. Identification and characterization of rat kidney papillary antigens with monoclonal antibodies. *Arch. Toxicol*. 1996;71:80-92.

Fantuzzi G, Puren AJ, Harding MW, Livingston DJ, Dinarello CA. Interleukin-18 regulation of interferon gamma production and cell proliferation as shown in interleukin-1beta-converting enzyme (caspase-1)-deficient mice. *Blood*. 1998;91:2118-25.

Ferguson MA, Vaidya VS, Waikar SS, Collings FB, Sunderland KE, Gioules CJ, *et al*. Urinary liver-type fatty acid-binding protein predicts adverse outcomes in acute kidney injury. *Kidney Int*. 2010;77:708-14.

Frayn KN, Arner P, Yki-Järvinen H. Fatty acid metabolism in adipose tissue, muscle and liver in health and disease. *Essays Biochem*. 2006;42:89-103.

Freeman DM, Chan L, Yahaya H, Holloway P, Ross BD. Magnetic resonance spectroscopy for the determination of renal metabolic rate in vivo. *Kidney Int*. 1986;30:35-42.

Gauthier C, Nguyen-Simonnet H, Vincent C, Revillard JP, Pellet MV. Renal tubular absorption of beta 2 microglobulin. *Kidney Int*. 1984;26:170-5.

Gelissen IC, Hochgrebe T, Wilson MR, Easterbrook-Smith SB, Jessup W, Dean RT, *et al*. Apolipoprotein J (clusterin) induces cholesterol export from macrophage-foam cells: a potential anti-atherogenic function? *Biochem J*. 1998;331:231-7.

Ghayur T, Banerjee S, Hugunin M, Butler D, Herzog L, Carter A, *et al*. Caspase-1 processes IFN-gamma-inducing factor and regulates LPS-induced IFN-gamma production. *Nature*. 1997; 386:619-23.

Girton RA, Sundin DP, Rosenberg ME. Clusterin protects renal tubular epithelial cells from gentamicin-mediated cytotoxicity. *Am. J. Physiol. Renal. Physiol*. 2002;282:F703-9.

Gobé GC, Buttyan R, Wyburn KR, Etheridge MR, Smith PJ. Clusterin expression and apoptosis in tissue remodeling associated with renal regeneration. *Kidney Int*. 1995; 47: 411-20.

Gonzales PA, Pisitkun T, Hoffert JD, Tchapyjnikov D, Star RA, Kleta R, *et al*. Large-scale proteomics and phosphoproteomics of urinary exosomes. *J. Am. Soc. Nephrol*. 2009;20: 363-79.

Gordjani N, Burghard R, Müller D, Mathäi H, Mergehenn G, Leititis JU, *et al*. Urinary excretion of adenosine deaminase binding protein in neonates treated with tobramycin. *Pediatr. Nephrol*. 1995;9:419-22.

Haase M, Bellomo R, Story D, Davenport P, Haase-Fielitz A. Urinary interleukin-18 does not predict acute kidney injury after adult cardiac surgery: a prospective observational cohort study. *Crit. Care*. 2008;12:R96.

Haase M, Bellomo R, Devarajan P, Schlattmann P, Haase-Fielitz A; NGAL Meta-analysis Investigator Group. Accuracy of neutrophil gelatinase-associated lipocalin (NGAL) in diagnosis and prognosis in acute kidney injury: a systematic review and meta-analysis. *Am. J. Kidney Dis*. 2009;54:1012-24.

Haase-Fielitz A, Bellomo R, Devarajan P, Story D, Matalanis G, Dragun D, *et al*. Novel and conventional serum biomarkers predicting acute kidney injury in adult cardiac surgery--a prospective cohort study. *Crit. Care Med*. 2009;37:553-60.

Han WK, Bailly V, Abichandani R, Thadhani R, Bonventre JV. Kidney Injury Molecule-1 (KIM-1): a novel biomarker for human renal proximal tubule injury. *Kidney Int*. 2002; 62:237-44.

Herget-Rosenthal S, Marggraf G, Hüsing J, Göring F, Pietruck F, Janssen O, *et al*. Early detection of acute renal failure by serum cystatin C. *Kidney Int*. 2004;66:1115-22.

Herget-Rosenthal S, van Wijk JA, Bröcker-Preuss M, Bökenkamp A. Increased urinary cystatin C reflects structural and functional renal tubular impairment independent of glomerular filtration rate. *Clin. Biochem*. 2007;40:946-51.

Hidaka S, Kränzlin B, Gretz N, Witzgall R. Urinary clusterin levels in the rat correlate with the severity of tubular damage and may help to differentiate between glomerular and tubular injuries. *Cell Tissue Res*. 2002;310:289-96.

Hildebrand H, Rinke M, Schlüter G, Bomhard E, Falkenberg FW. Urinary antigens as markers of papillary toxicity. II: Application of monoclonal antibodies for the determination of papillary antigens in rat urine. *Arch Toxicol*. 1999;73:233-45.

Hoffmann D, Adler M, Vaidya VS, Rached E, Mulrane L, Gallagher WM, *et al*. Performance of novel kidney biomarkers in preclinical toxicity studies. *Toxicol. Sci*. 2010;116:8-22.

Hohenleitner FJ, Spitzer JJ. Changes in plasma free fatty acid concentrations on passage through the dog kidney. *Am. J. Physiol*. 1961;200:1095-8.

Hou SH, Bushinsky DA, Wish JB, Cohen JJ, Harrington JT. Hospital-acquired renal insufficiency: a prospective study. *Am. J. Med*. 1983;74:243-8.

Hvidberg V, Jacobsen C, Strong RK, Cowland JB, Moestrup SK, Borregaard N. The endocytic receptor megalin binds the iron transporting neutrophil-gelatinase-associated lipocalin with high affinity and mediates its cellular uptake. *FEBS Lett*. 2005;579: 773-7.

Ichimura T, Hung CC, Yang SA, Stevens JL, Bonventre JV. Kidney injury molecule-1: a tissue and urinary biomarker for nephrotoxicant-induced renal injury. *Am. J. Physiol. Renal. Physiol*. 2004;286:F552-63.

Ichimura T, Asseldonk EJ, Humphreys BD, Gunaratnam L, Duffield JS, Bonventre JV. Kidney injury molecule-1 is a phosphatidylserine receptor that confers a phagocytic phenotype on epithelial cells. *J. Clin. Invest*. 2008;118:1657-68.

Jaafar A, Séronie-Vivien S, Malard L, Massip P, Chatelut E, Tack I. Urinary cystatin C can improve the renal safety follow-up of tenofovir-treated patients. AIDS. 2009;23:257-9.

Jones SE, Jomary C. Clusterin. *Int. J. Biochem. Cell Biol*. 2002;34:427-31.

Ketterer B, Coles B, Meyer DJ. The role of glutathione in detoxication. *Environ Health Perspect*. 1983;49:59-69.

Kharasch ED, Hoffman GM, Thorning D, Hankins DC, Kilty CG. Role of the renal cysteine conjugate beta-lyase pathway in inhaled compound A nephrotoxicity in rats. *Anesthesiology.* 1998;88:1624-33.

Kuwabara T, Mori K, Mukoyama M, Kasahara M, Yokoi H, Saito Y, *et al.* Urinary neutrophil gelatinase-associated lipocalin levels reflect damage to glomeruli, proximal tubules, and distal nephrons. *Kidney Int.* 2009;75:285-94.

Lawrence JW, Kroll DJ, Eacho PI. Ligand-dependent interaction of hepatic fatty acid-binding protein with the nucleus. *J. Lipid Res.* 2000;41:1390-401.

Lee JB, Vance VK, Cahill GF Jr. Metabolism of C14-labeled substrates by rabbit kidney cortex and medulla. *Am. J. Physiol.* 1962;203:27-36.

Li JY, Ram G, Gast K, Chen X, Barasch K, Mori K, *et al.* Detection of intracellular iron by its regulatory effect. *Am. J. Physiol. Cell Physiol.* 2004;287:C1547-59.

Liangos O, Perianayagam MC, Vaidya VS, Han WK, Wald R, Tighiouart H, *et al.* Urinary N-acetyl-beta-(D)-glucosaminidase activity and kidney injury molecule-1 level are associated with adverse outcomes in acute renal failure. *J. Am. Soc. Nephrol.* 2007;18: 904-12.

Maatman RG, Van Kuppevelt TH, Veerkamp JH. Two types of fatty acid-binding protein in human kidney. Isolation, characterization and localization. *Biochem. J.* 1991;273:759-66.

Maatman RG, van de Westerlo EM, van Kuppevelt TH, Veerkamp JH. Molecular identification of the liver- and the heart-type fatty acid-binding proteins in human and rat kidney. Use of the reverse transcriptase polymerase chain reaction. *Biochem. J.* 1992; 288:285-90.

Madero M, Sarnak MJ, Stevens LA. Serum cystatin C as a marker of glomerular filtration rate. *Curr. Opin. Nephrol. Hypertens.* 2006;15:610-6.

Makris K, Markou N, Evodia E, Dimopoulou E, Drakopoulos I, Ntetsika K, *et al.* Urinary neutrophil gelatinase-associated lipocalin (NGAL) as an early marker of acute kidney injury in critically ill multiple trauma patients. *Clin. Chem. Lab. Med.* 2009;47:79-82.

Mannervik B, Alin P, Guthenberg C, Jensson H, Tahir MK, Warholm M, *et al.* Identification of three classes of cytosolic glutathione transferase common to several mammalian species: correlation between structural data and enzymatic properties. *Proc. Natl. Acad. Sci. USA* 1985;82:7202-6.

Mashimo H, Wu DC, Podolsky DK, Fishman MC. Impaired defense of intestinal mucosa in mice lacking intestinal trefoil factor. *Science* 1996;274:262-5.

Mattix HJ, Hsu CY, Shaykevich S, Curhan G. Use of the albumin/creatinine ratio to detect microalbuminuria: implications of sex and race. *J. Am. Soc. Nephrol.* 2002;13:1034-9.

McKee JA, Kumar S, Ecelbarger CA, Fernández-Llama P, Terris J, Knepper MA. Detection of Na(+) transporter proteins in urine. *J. Am. Soc. Nephrol.* 2000;11:2128-32.

Mehta KP, Ali US, Shankar L, Tirthani D, Ambadekar M. Renal dysfunction detected by beta-2 microglobulinuria in sick neonates. *Indian Pediatr.* 1997;34:107-11.

Mehta RL, Kellum JA, Shah SV, Molitoris BA, Ronco C, Warnock DG, *et al.* Acute Kidney Injury Network. Acute Kidney Injury Network: report of an initiative to improve outcomes in acute kidney injury. *Crit. Care.* 2007;11:R31.

Melnikov VY, Ecder T, Fantuzzi G, Siegmund B, Lucia MS, Dinarello CA, *et al.* Impaired IL-18 processing protects caspase-1-deficient mice from ischemic acute renal failure. *J. Clin. Invest.* 2001;107:1145-52.

Miranda KC, Bond DT, McKee M, Skog J, Păunescu TG, Da Silva N, *et al.* Nucleic acids within urinary exosomes/microvesicles are potential biomarkers for renal disease. *Kidney Int.* 2010;78:191-9.

Mishra J, Ma Q, Prada A, Mitsnefes M, Zahedi K, Yang J, *et al.* Identification of neutrophil gelatinase-associated lipocalin as a novel early urinary biomarker for ischemic renal injury. *J. Am. Soc. Nephrol.* 2003;14:2534-43.

Mishra J, Mori K, Ma Q, Kelly C, Barasch J, Devarajan P. Neutrophil gelatinase-associated lipocalin: a novel early urinary biomarker for cisplatin nephrotoxicity. *Am. J. Nephrol.* 2004;24:307-15.

Mori K, Lee HT, Rapoport D, Drexler IR, Foster K, Yang J, *et al.* Endocytic delivery of lipocalin-siderophore-iron complex rescues the kidney from ischemia-reperfusion injury. *J. Clin. Invest.* 2005;115:610-21.

Mori K, Nakao K. Neutrophil gelatinase-associated lipocalin as the real-time indicator of active kidney damage. *Kidney Int.* 2007;71:967-70.

Mundel P, Reiser J. Proteinuria: an enzymatic disease of the podocyte? *Kidney Int.* 2010;77:571-80.

Nakamura T, Sugaya T, Node K, Ueda Y, Koide H. Urinary excretion of liver-type fatty acid-binding protein in contrast medium-induced nephropathy. *Am. J. Kidney Dis.* 2006; 47:439-44.

Nash K, Hafeez A, Hou S. Hospital-acquired renal insufficiency. *Am. J. Kidney Dis.* 2002; 39:930-6.

Nath KA, Dvergsten J, Correa-Rotter R, Hostetter TH, Manivel JC, Rosenberg ME. Induction of clusterin in acute and chronic oxidative renal disease in the rat and its dissociation from cell injury. *Lab. Invest.* 1994;71:209-18.

Negishi K, Noiri E, Sugaya T, Li S, Megyesi J, Nagothu K, *et al.* A role of liver fatty acid-binding protein in cisplatin-induced acute renal failure. *Kidney Int.* 2007;72:348-58.

Nunez EA. Fatty acids involved in signal cross-talk between cell membrane and nucleus. *Prostaglandins Leukot Essent Fatty Acids.* 1997;57:429-34.

Obregon C, Dreher D, Kok M, Cochand L, Kiama GS, Nicod LP. Human alveolar macrophages infected by virulent bacteria expressing SipB are a major source of active interleukin-18. *Infect Immun.* 2003;71:4382-8.

Orth SR, Ritz E. The nephrotic syndrome. *N. Engl. J. Med.* 1998;338:1202-11.

Oyama Y, Takeda T, Hama H, Tanuma A, Iino N, Sato K, *et al.* Evidence for megalin-mediated proximal tubular uptake of L-FABP, a carrier of potentially nephrotoxic molecules. *Lab. Invest.* 2005;85:522-31.

Ozer JS, Dieterle F, Troth S, Perentes E, Cordier A, Verdes P, *et al.* A panel of urinary biomarkers to monitor reversibility of renal injury and a serum marker with improved potential to assess renal function. *Nat. Biotechnol.* 2010;28:486-94.

Parikh CR, Jani A, Melnikov VY, Faubel S, Edelstein CL. Urinary interleukin-18 is a marker of human acute tubular necrosis. *Am. J. Kidney Dis.* 2004;43:405-14.

Parikh CR, Abraham E, Ancukiewicz M, Edelstein CL. Urine IL-18 is an early diagnostic marker for acute kidney injury and predicts mortality in the intensive care unit. *J. Am. Soc. Nephrol.* 2005;16:3046-52.

Parikh CR, Mishra J, Thiessen-Philbrook H, Dursun B, Ma Q, Kelly C, *et al.* Urinary IL-18 is an early predictive biomarker of acute kidney injury after cardiac surgery. *Kidney Int.* 2006; 70:199-203.

Pisitkun T, Shen RF, Knepper MA. Identification and proteomic profiling of exosomes in human urine. Proc Natl Acad Sci U S A. 2004;101:13368-73.Pisitkun T, Johnstone R, Knepper MA. Discovery of urinary biomarkers. *Mol. Cell Proteomics*. 2006;5:1760-71.

Portilla D, Dent C, Sugaya T, Nagothu KK, Kundi I, Moore P, *et al.* Liver fatty acid-binding protein as a biomarker of acute kidney injury after cardiac surgery. *Kidney Int.* 2008;73:465-72.

Portman RJ, Kissane JM, Robson AM. Use of beta 2 microglobulin to diagnose tubulo-interstitial renal lesions in children. *Kidney Int.* 1986;30:91-8.

Price SA, Davies D, Rowlinson R, Copley CG, Roche A, Falkenberg FW, *et al.* Characterization of renal papillary antigen 1 (RPA-1), a biomarker of renal papillary necrosis. *Toxicol. Pathol.* 2010;38:346-58.

Rask L, Lindblom JB, Peterson PA. Subunit structure of H-2 alloantigens. *Nature*. 1974; 249:833-4.

Rozell B, Hansson HA, Guthenberg C, Tahir MK, Mannervik B. Glutathione transferases of classes alpha, mu and pi show selective expression in different regions of rat kidney. *Xenobiotica.* 1993;23:835-49.

Schmidt-Ott KM, Mori K, Kalandadze A, Li JY, Paragas N, Nicholas T, *et al.* Neutrophil gelatinase-associated lipocalin-mediated iron traffic in kidney epithelia. *Curr. Opin. Nephrol. Hypertens*. 2006;15:442-9.

Schrier RW. ARF, AKI, or ATN? *Nat. Rev. Nephrol*. 2010;6:125.

Schwochau GB, Nath KA, Rosenberg ME. Clusterin protects against oxidative stress in vitro through aggregative and nonaggregative properties. *Kidney Int*. 1998;53:1647-53.

Sieber M, Hoffmann D, Adler M, Vaidya VS, Clement M, Bonventre JV, *et al*. Comparative analysis of novel noninvasive renal biomarkers and metabonomic changes in a rat model of gentamicin nephrotoxicity. *Toxicol. Sci*. 2009;109:336-49.

Silkensen JR, Skubitz KM, Skubitz AP, Chmielewski DH, Manivel JC, Dvergsten JA, *et al.* Clusterin promotes the aggregation and adhesion of renal porcine epithelial cells. *J. Clin. Invest.* 1995;96:2646-53.

Simon TC, Roth KA, Gordon JI. Use of transgenic mice to map cis-acting elements in the liver fatty acid-binding protein gene (Fabpl) that regulate its cell lineage-specific, differentiation-dependent, and spatial patterns of expression in the gut epithelium and in the liver acinus. *J. Biol. Chem*. 1993 Aug 25;268(24):18345-58.

Solichova P, Karpisek M, Ochmanova R, Hanulova Z, Humenanska V, Stejskal D, *et al.* Urinary clusterin concentrations--a possible marker of nephropathy? Pilot study. Biomed Pap Med Fac Univ Palacky Olomouc Czech Repub. 2007;151:233-6.

Sonoda H, Yokota-Ikeda N, Oshikawa S, Kanno Y, Yoshinaga K, Uchida K, *et al*. Decreased abundance of urinary exosomal aquaporin-1 in renal ischemia-reperfusion injury. *Am. J. Physiol. Renal. Physiol*. 2009;297:F1006-16.

Spiegelman BM, Flier JS. Adipogenesis and obesity: rounding out the big picture. *Cell.* 1996;87:377-89.

Suemori S, Lynch-Devaney K, Podolsky DK. Identification and characterization of rat intestinal trefoil factor: tissue- and cell-specific member of the trefoil protein family. *Proc. Natl. Acad. Sci. USA* 1991;88:11017-21.

Sugaya T, Noiri E, Yamamoto T, Doi K, Negishi K, Kamijo A, *et al*. L-type fatty acid biding protein (L-FABP) ameliorates renal ischemia reperfusion injury (I/R) in human L-FABP transgenic mice. *Nephrology*. 2005;10:A133

Sundberg AG, Nilsson R, Appelkvist EL, Dallner G. Immunohistochemical localization of alpha and pi class glutathione transferases in normal human tissues. *Pharmacol. Toxicol.* 1993;72:321-31.

Sundberg AG, Appelkvist EL, Bäckman L, Dallner G. Quantitation of glutathione transferase-pi in the urine by radioimmunoassay. *Nephron.* 1994;66:162-9.

Takuwa S, Ito Y, Ushijima K, Uchida K. Serum cystatin-C values in children by age and their fluctuation during dehydration. *Pediatr Int.* 2002;44:28-31.

Taupin D, Podolsky DK. Trefoil factors: initiators of mucosal healing. Nat Rev Mol Cell Biol. 2003 Sep;4(9):721-32. Review. Erratum in: *Nat. Rev. Mol. Cell Biol.* 2003;4:819.

Tenstad O, Roald AB, Grubb A, Aukland K. Renal handling of radiolabelled human cystatin C in the rat. *Scand. J. Clin. Lab. Invest.* 1996;56:409-14.

Tonomura Y, Tsuchiya N, Torii M, Uehara T. Evaluation of the usefulness of urinary biomarkers for nephrotoxicity in rats. *Toxicology.* 2010;273:53-9.

Uchino S, Kellum JA, Bellomo R, Doig GS, Morimatsu H, Morgera S, *et al.* Acute renal failure in critically ill patients: a multinational, multicenter study. *JAMA.* 2005;294:813-8.

Vaidya VS, Ramirez V, Ichimura T, Bobadilla NA, Bonventre JV. Urinary kidney injury molecule-1: a sensitive quantitative biomarker for early detection of kidney tubular injury. *Am. J. Physiol. Renal. Physiol.* 2006;290:F517-29.

Vaidya VS, Ozer JS, Dieterle F, Collings FB, Ramirez V, Troth S, *et al.* Kidney injury molecule-1 outperforms traditional biomarkers of kidney injury in preclinical biomarker qualification studies. *Nat. Biotechnol.* 2010;28:478-85.

van Niel G, Porto-Carreiro I, Simoes S, Raposo G. Exosomes: a common pathway for a specialized function. *J. Biochem.* 2006;140:13-21.

Walshe CM, Odejayi F, Ng S, Marsh B. Urinary glutathione S-transferase as an early marker for renal dysfunction in patients admitted to intensive care with sepsis. *Crit. Care Resusc.* 2009;11:204-9.

Wanders RJ, van Roermund CW, Visser WF, Ferdinandusse S, Jansen GA, van den Brink DM, *et al.* Peroxisomal fatty acid alpha- and beta-oxidation in health and disease: new insights. *Adv. Exp. Med. Biol.* 2003;544:293-302.

Washburn KK, Zappitelli M, Arikan AA, Loftis L, Yalavarthy R, Parikh CR, *et al.* Urinary interleukin-18 is an acute kidney injury biomarker in critically ill children. *Nephrol. Dial Transplant.* 2008;23:566-72.

Westhuyzen J, Endre ZH, Reece G, Reith DM, Saltissi D, Morgan TJ. Measurement of tubular enzymuria facilitates early detection of acute renal impairment in theintensive care unit. *Nephrol. Dial Transplant.* 2003;18:543-51.

Yamamoto T, Noiri E, Ono Y, Doi K, Negishi K, Kamijo A, *et al.* Renal L-type fatty acid--binding protein in acute ischemic injury. *J. Am. Soc. Nephrol.* 2007;18:2894-902.

Yang J, Goetz D, Li JY, Wang W, Mori K, Setlik D, *et al.* An iron delivery pathway mediated by a lipocalin. *Mol. Cell.* 2002;10:1045-56.

Ympa YP, Sakr Y, Reinhart K, Vincent JL. Has mortality from acute renal failure decreased? A systematic review of the literature. *Am. J. Med.* 2005;118:827-32.

Yu Y, Jin H, Holder D, Ozer JS, Villarreal S, Shughrue P, *et al.* Urinary biomarkers trefoil factor 3 and albumin enable early detection of kidney tubular injury. *Nat. Biotechnol.* 2010;28:470-7.

Zhang Z, Humphreys BD, Bonventre JV. Shedding of the urinary biomarker kidney injury molecule-1 (KIM-1) is regulated by MAP kinases and juxtamembrane region. *J. Am. Soc. Nephrol.* 2007;18:2704-14.

Zhang J, Brown RP, Shaw M, Vaidya VS, Zhou Y, Espandiari P, *et al.* Immunolocalization of Kim-1, RPA-1, and RPA-2 in kidney of gentamicin-, mercury-, or chromium-treated rats: relationship to renal distributions of iNOS and nitrotyrosine. *Toxicol. Pathol.* 2008; 36:397-409.

Zhou H, Pisitkun T, Aponte A, Yuen PS, Hoffert JD, Yasuda H, *et al.* Exosomal Fetuin-A identified by proteomics: a novel urinary biomarker for detecting acute kidney injury. *Kidney Int.* 2006;70:1847-57.

Zhou H, Cheruvanky A, Hu X, Matsumoto T, Hiramatsu N, Cho ME, *et al.* Urinary exosomal transcription factors, a new class of biomarkers for renal disease. *Kidney Int.* 2008; 74:613-21.

Zhou W, Guan Q, Kwan CC, Chen H, Gleave ME, Nguan CY, *et al.* Loss of clusterin expression worsens renal ischemia-reperfusion injury. *Am. J. Physiol. Renal Physiol.* 2010; 298:F568-78.

Zimmerman AW, Veerkamp JH. New insights into the structure and function of fatty acid-binding proteins. *Cell Mol. Life Sci.* 2002;59:1096-116.

In: Cystatins: Protease Inhibitors …
Editors: John B. Cohen and Linda P. Ryseck
ISBN: 978-1-61209-343-7

Chapter 7

CYSTATIN C AND ACUTE KIDNEY INJURY

M. Guillouet, C. Guennegan, M. Coat, A. Khalifa, Z. Alavi, F. Lion, R. Deredec, C. C. Arvieux, and G. Gueret*

Pôle anesthésie réanimation, Centre Hospitalier Universitaire, Brest, France
Service de chirurgie cardiaque, thoracique et vasculaire, Centre Hospitalier Universitaire, Brest, France
Inserm CIC 0502, CHU Brest, France

ABSTRACT

Acute kidney injury (AKI) is defined as an abrupt and sustained decrease in kidney function. There are a lot of definitions in the literature, which explain the large variations in the reported incidence. It is well recognized for its impact on the outcome of patients, as it increases morbidity and mortality. Diagnosis of AKI is always difficult, particularly in the early stage of the disease. Numerous definitions and parameters have been used but the gold standard in clinical practice remains the creatinine clearance. Recently, in order to develop early biomarkers, cystatin C (CysC) was proposed. CysC is a protease inhibitor produced in a constant manner by nucleated cells. This molecule is passively filtrated by the glomerule and quite completely catabolized in the proximal tubules. It has the advantage of not being influenced by age, sex, race or muscular mass. Its excretion increases after reversible and mild dysfunction and may not necessarily be associated with persistent or irreversible damage. However, it is primilarly a sensitive marker of reduction in glomerular filtration rate but it cannot differentiate between different types of AKI. Some studies demonstrated the superiority of CysC over plasma creatinine while other studies did not, depending on the population. On the other hand, urinary CysC seems superior to conventional and new plasma markers, but its main disadvantage is its instability in the urine samples. In conclusion, we can use CysC like an additional argument of renal failure but it does not seem to be superior to plasma creatinine in the early diagnosis of AKI in all situations.

---

* Corresponding author: Gildas Gueret, Pôle Anesthésie Réanimation, Centre Hospitalier Universitaire, 29609 Brest, France. Phone: 33 2 98 34 74 29; Fax: 33 2 98 34 78 10; Email: gildas.gueret@chu-brest.fr.

INTRODUCTION

Acute renal injury (AKI) is a common complication in hospitalized patients and its incidence is increasing [1-3]. Mortality and morbidity rates remain high despite technical advances in patient management. Furthermore, it represents an important part in hospital costs. An ideal biomarker that can diagnose AKI during early stage would help physician to treat patients and improve the evolution of the disease. Indeed, its aim is to reduce the delay in initiating therapy. At the time being, serum creatinine and creatinine clearance remain the gold standard to evaluate renal function but researches try to find other biomarkers. Among them, Cystatin C (CysC) has been proposed. Actually, it can certainly be used as another tool to diagnose AKI but its superiority compared to creatinine is still discussed.

Incidence of Acute Kidney Injury

AKI is a common complication in hospital. Its incidence deeply varies according to the cohort and the diagnosis criteria [4, 5]. For example, AKI is attributed to sepsis in 11 % to 64 % of the cases in ICU [6, 7] and its incidence varies from 3 % to 60% [8] in postoperative cardiac surgery patients. The incidence was observed to be greater in men and increases with age [9].

Physiopathology of Acute Kidney Injury

Indeed, tubular damage, glomerular disease, interstitial disease and acute tubular necrosis have been described. The mechanism implicated was observed to be dependant of the department of hospitalization. Tubular damage is the most frequent pathology, and can be attributed either to nephrotoxic or ischemic mechanisms [10]. Shock and resuscitation, major vascular and cardiac surgery and kidney transplantation leads to ischemia/reperfusion injury [11]. Renal toxicity of antibiotics, antifungals, antivirals, radiotherapy and cytostatic agents is another important factor leading to kidney injury. Inflammatory reaction [8] and reactive oxygen species [12] also play an important role. AKI could be iatrogenic to certain medications or the result of dehydration and hypovolemia. Obstructive kidney injury is observed in case of nephritic lithiasis or cancer. However, sepsis remains the main cause of AKI [13].

Prognosis of Acute Kidney Injury

AKI is associated with a high mortality depending on the severity of the disease [14]. Twelve percent of the surviving patients need renal replacement therapy at 3 and 6 month [15]. Studies have demonstrated that a post-operative estimated glomerular filtration rate of less than 60 ml/min/1,73 m^2 is associated with the worst 5-years survival [16]. Even a small increase in creatinine level is associated with a worse outcome, in cardiac surgery [17,18] as in ICU [19]. However, the role of early diagnosis of AKI remains to be determined.

Treatment of Acute Kidney Injury

Prevention of AKI is the first intervention in critically ill patients [20]. It includes controlled fluids resuscitation, maintained adequate arterial blood pressure and cardiac output using inotropes/vasopressors if necessary. Diuretics should not be used.

Renal replacement therapy (RRT) is the major therapeutic tool. Uchino et al. [21], with the Beginning and Ending Supportive Therapy for the Kidney (BEST Kidney) Investigators, found that among a cohort of 29 269 ICU patients admitted to 54 ICUs in 23 countries, 4.2% of patients were treated with RRT for AKI. Despite advances in technical therapy like continuous renal replacement therapy and biocompatible membrane, mortality of AKI remains high [22]. Retrospective and observational studies including AKI patients showed that late dialysis was associated with a worse outcome [23-25], but this point needs to be confirmed by a randomized control study [26].

Hospital stay is always longer thus increasing the risk to develop other complications, particularly nosocomial infections. It also increases hospital costs.

The major risk is to develop a chronic kidney disease or end stage kidney disease but even a mild injury in kidney function can have an important effect on outcome [4]. According to different studies, end-stage kidney disease appears in 8.3 % to 50% of patients who suffered from AKI [21,27,28].

Detecting AKI as early as possible appears to be the solution to improve patient prognosis outcomes and reduce hospital costs. As clinical evaluation has already been improved with the RIFLE classification, new early biomarkers of AKI are needed.

Classical Diagnosis of Acute Kidney Injury

Traditional biomarkers for the detection of AKI are urine output, creatinine, urea, and measured or calculated creatinine clearance.

Early diagnosis of postoperative AKI is difficult, particularly after cardiac surgery. Although the role of early diagnosis of AKI remains to be determined, clinical trials may however be useful to establish effective protocols for possible therapeutic strategies.

The first obstacle in the diagnosis AKI is the existence of numerous conflicting definitions and parameters currently in use. Definition of acute renal injury differs among studies. The society of Thoracic surgeons' definition is at least a 2-fold creatinine rise to a value exceeding 2 mg/dL and/or new dialysis. Increase in creatinine levels up to 20 % [18], 25 % [29-31], 30 % [32], 50 % [33-36] or 100 % of preoperative values were also used to define AKI. Other definitions are: postoperative serum creatinine level above 177 μmol/L with a preoperative-to-postoperative increase above 62 μmol/L [37] or increase in serum creatinine level greater than 0.5 mg/dL or a 25% increase from baseline [31]. Most of the studies included 2 groups (with or without AKI), but some studies included different groups according to the degree of the creatinine rise [8,38-40]. In cardiac surgery, the duration of clinical observation of patients varied from 2 to 7 days after surgery [29,41,42].

Urine output is widely measured in operating rooms and intensive care units (ICU) with indwelling catheters but it is more of an hemodynamic indicator than of a clearance one [43]. Urinary output more than 1 ml/kg/h is not an outstanding indicator of renal function after cardiopulmonary bypass (CPB) whereas a low urinary output during CPB has been associated with postoperative AKI [44]. AKI and oligouria are not always associated and could exist independently. Furthermore, therapeutics like diuretics and vasopressors are confounders of the urine output.

Creatinine is the product of the breakdown of creatine in skeletal muscle and the subsequent liver metabolism of creatine to form creatinine. It is released in the plasma at a constant rate and filtered by the glomerulus without tubular reabsorption. Then, if glomerular filtration is deficient, blood levels of creatinine rise with a non-linear, inverse relationship. Furthermore, a loss of 50% of the glomerular filtration rate (GFR) is necessary to detect an abnormal level of creatinine. The plasma creatinine dosage provides insufficient clinical data, particularly in older patients. Furthermore, serum creatinine concentration is greatly influenced by changes in muscle mass which varies between patients, race, age, sex, drugs, muscle metabolism, protein intake and tubular secretion, and is lower in conditions of prolonged bed rest [45]. In AKI, serum creatinine is even a poorer index of kidney function because the patients are not stable, and serum creatinine changes occur after the onset of renal injury. Another point is that tubular secretion of creatinine can overestimate the GFR from 10% to 50% [46]. Creatinemia can remain normal, even in cases of serious damage to the renal function [47]. In spite of these limitations, the changes of plasma creatinine are still widely used [18,29,39,40,48-50]. In order to evaluate the intensity of the kidney injury, RIFLE classification has been recently proposed [43,51,52]. It stands for the increasing severity classes Risk, Injury, and Failure, and the two outcome classes Loss and End-stage kidney disease. Creatinine increase is one of the main criteria used in the RIFLE classification [43,51-53].

Many methods exist to evaluate the GFR, but inulin clearance and isotopic methods (125 Iode p aminohippurate, 51Cr-EDTA) [54] remain the methods of reference. They are not routinely used because of their cost, the difficulty of implementation and the availability of equipment.

Creatinine clearance measurement aims to evaluate GFR and it is widely used. Its clinical performance is reduced for values below 20 ml/min. It needs a stable state during 24 hours. When measured creatinine clearance is used, urinary creatinine and urinary output must be obtained. The measurement of creatinine clearance has been proposed for shorter periods than 24 hours [55-61], but a minimum collection during four to eight hours is recommended [55,62] as the measured urinary volume depends on the quality of the urinary sample. The renal excretion of creatinine can vary throughout the day [63]. Goldberg et al. observed large daily variations of urinary creatinine excretion leading to unreliability of urine collections [64], and O'Riordan et al. have shown that more than 50% of the variations of creatinine clearance could not be explained by GFR modifications [47]. Measured creatinine clearance is not reliable in the early post-operative period after cardiac surgery with CPB [62]. Indeed, sepsis itself reduces the production of creatinine which blunts the increase of the biomarker and limits the early detection of AKI [65].

Estimated GFR was first developed by Cockcroft et al [66]. In this work, creatinine calculated clearance was more sensitive than creatinine for renal injury diagnosis. After CPB, the MDRD formula [67] seems more reliable than the Cockcroft and Gault formula [62].

Urea is a product of protein metabolism. Its level is inversely related to GFR but its production and clearance are not constant. Changes in circulatory volume, protein intake and gastrointestinal bleeding modify its plasma concentration. Furthermore, its clearance is not constant and almost 50 % of the filtered urea may be reabsorbed in the tubules.

Thus, traditional biomarkers do not detect injury early enough and this leads to delayed implementation. In order to solve these problems, new technologies have facilitated detection of prospective new biomarkers of AKI.

What Is an Ideal Biomarker of AKI

At the moment, there is no ideal biomarker to diagnose AKI but the criteria of such a marker are known. Indeed, it should have biological and physicochemical properties.

Biological properties

- Rapid and reliable response to injury
- Highly sensitive for AKI with a wide dynamic range and cutoff values
- Etiologic specificity (given multifactorial etiology of AKI)
- Its level should correlate with injury severity
- Its level should provide prognostic information
- Applicable across different populations

Physicochemical properties

- Its level should be unaffected by drugs or other endogenous substances
- Stable across a wide range of temperature and pH
- Easily measured in urine or serum
- Rapid, reliable, and inexpensive measurement using standardized assay platforms

The Need for New Biomarkers

New biomarkers, which can diagnose AKI at an early stage, seem promising. They include plasma and urinary markers (α1 microglobulin, Gro-alpha, meprin, β2 microglobulin).

Urinary excretion of brush border tubular proximal cell glycoproteins during ischemia has often been used as a marker of infra-clinical kidney injuries [54,68-70].

The N Acetyl β D glucosaminidas (NAG) is a lysosomial enzyme specific to proximal tubular cells which is neither secreted nor reabsorbed on a tubular level. The NAG urinary/creatinine report is a sensitive and specific marker of tubular injury [68].

The brush border antigen adenosin desaminase binding protein (ADBP) glycoprotein of the surface of tubular proximal cells, plasma and urinary β2 microglobulin and α1 microglobulin are low molecular weight molecules which are filtered by the glomerulus and reabsorbed at 95% by the renal tubules in normal conditions (excepting tubular injury). These are sensitive markers for tubular failure and have long been used in experimental studies [71] and in cardiac surgery [72].

The retinol binding protein (RBP) is also a low molecular weight glycoprotein that is reabsorbed by the renal tubules and used as an urinary marker of tubular injury [73], and could be a more practical and reliable index of proximal tubular function than β2 microglobulin [74].

The neutrophil gelatinase-associated lipocalcin (NGAL) normally exerts protective bacteriostatic and antioxidant effects involving iron transport and is thought to act as an iron scavenger and growth factor [75]. It is generated by ischemic renal tubular cells and appears in the urine and in the plasma of cardiac surgery patients in less than six hours [76,77]. The two hours NGAL urine collection levels correlate with severity and duration of AKI, length of stay in hospital, dialysis requirement and death [33]. Mishra et al have also shown an early increase of urinary NGAL concentration in children with AKI after pediatric cardiac surgery [36]. As a general rule, a concentration >150 ng/mL can identify patients at high risk for AKI, and a level >350 ng/mL, those at high risk for renal replacement therapy.

Kidney injury molecule-1 (KIM-1) is a transmembrane protein that is over expressed in proximal tubules of patients with established acute renal injury. This marker seems to be more specific to nephrotoxic or ischemic kidney injury and is not affected by chronic kidney disease or urinary tract infections. Thus it may be important for differentiating between various subtypes of acute kidney injury. The increase of KIM-1 in urinary concentration is delayed by 12-24h after the insult. However, in conjunction with other sensitive biomarkers it may add specificity to early acute kidney injury diagnosis.

IL-18 also represents a promising candidate. It is induced and cleaved in the proximal tubule and is detected in urine following experimental AKI [78]. It can be used in predicting AKI within six hours after cardiac surgery therefore, differentiating it from other types of renal injury [79]. The latter is still being discussed [80]. Nevertheless, it can also be considered as a non-specific marker of inflammation requiring further evaluation.

Liver-type fatty acid binding protein (L-FABP) is expressed in various organs including kidney. Its function is the cellular uptake of fatty acids (FA) from plasma and the promotion of intracellular FA metabolism. L-FABP can then prevent oxidation of free FA and cellular injury by oxidative stress. It can be filtered by glomeruli and reabsorbed in the proximal tubule cells therefore increasing the proximal tubular cell injury. Urine L-FABP has showed great potential for early and accurate detection of histological and functional decline in both nephrotoxin-induced and ischemia-reperfusion injury in mice [81]. In a further study, higher urine L-FABP levels differentiated patients with septic shock from those with severe sepsis, and those with AKI from healthy controls [82]. However it appears that its level rises later than that of NGAL.

Finally, among these new biomarkers, the CysC evaluated in several populations seems interesting.

Cystatin C, a New Biomarker

Description of the Cystatin C

CysC is a non glycosylate 13kD basic protein of 122 amino acids. It is a member of the cystatin family of cysteine protease inhibitors. Its production by all nucleated cells is constant. It is filtered by the glomeruli and completely reabsorbed and catabolized in the proximal tubules but not secreted. Its physiological function is the inhibition of lysosomial and cysteine proteases and thus the protection against tissue destruction. It can be detected in all body fluids and was first detected in the cerebrospinal fluid [83].

Its secretion is unaffected by muscle mass, age until fifty years, sex, race and hydration level. Due to its constant rate of production, its serum level can only be attributed to GFR [84]. Urinary concentration of CysC is normally very low except in case of tubulo-interstitial disease or tubular dysfunction [85]. It was proposed as GFR marker in 1979 [86].

One of its interesting aspects is that it can not be removed by the standard hemodialyzer. Thus, it could be used during peritoneal dialysis and intermittent hemodialysis [87]. A study also showed that continuous veno-venous hemofiltration (CVVH) was unlikely to influence serum concentrations of CysC significantly, which suggests that it can be used to monitor residual renal function during CVVH [88]. CysC can be measured either in plasma or in urine.

Evaluation of the Kidney Function by CysC

GFR can be evaluated by CysC through its changes in serum and urine levels.

This is achieved by Particle-enhanced turbidimetric immunoassay (PETIA) or Particle-enhanced nephelometric immunoassay (PENIA) methods. The former can sometimes under evaluate measurements in case of bilirubine level > 100 mg/l [89] or triglycerid level > 15 g/l [90]. The latter was proposed by Finney in 1997 and seems not to be influenced by other biological values [91]. The two methods provide automated results in five minutes [92]. Half mL of serum sample is necessary.

Because of the non circadian dependence of its release, one sample is sufficient. Upper urine concentration of CysC is a sign of kidney tubular damage, independently of changes in GFR [85]. It can increase by 200 folds.

Normal values are increased when age is over 50-60 years. Reference values of CysC are 0,70-1,21 mg/L before 50 years and 0,84-1,55 mg/L after 50 years but there is no standardization.

Glomerular Filtration Rate Estimating Equations for CysC

There is no validated equation to estimate the GFR with CysC value. Several equations have been proposed in adults as well as in children [93]. However, the bias and the precision differ between the different equations. Another problem is the lack of reproducibility between different methods for CysC measurement [93,94].

Serum CysC

The advantage of a serum marker for diagnosis of AKI is that it can be used even if the patient presents severe oliguria.

The majority of the studies compared CysC to the actual gold standard creatinine. Results and conclusions are discussed. Some authors agree with an increase in CysC in AKI 24-48 hours earlier than creatinine and found that it allows to observe and to follow up the progression of the disease [95]. However it can not be considered as a specific marker of AKI with tubular lesion but rather an early marker of impaired glomerular filtration [96].

Critical Care Patients

Early recognition of AKI in ICU remains a critical problem, with a rising incidence and a high mortality rate. The accurate diagnosis of AKI is a major issue in critically ill patients, in whom the renal function is in an unsteady state. Therefore, the validity of creatinine-based baseline assessment measures is reduced.

A study on critical care patients with normal baseline GFR rate found an earlier increase in CysC compared to creatinine [97]. This is supported by other studies. Villa et al. showed that serum CysC is a better marker of GFR than serum creatinine in unstable, critically ill patients [98]. In critically ill children, Herrero-Morin et al. found that serum CysC was a better marker than serum creatinine to detect AKI [99]. Their explanation was that GFR can change rapidly in critically ill children, but changes in serum creatinine take more time.

Finally, Royakkers et al. made an overwiew of the literature on the subject and concluded that in ICU patients, serum CysC seems to be an early and adequate endogenous marker for renal dysfunction. Even with mild reductions in GFR, it is a better predictor for the development of renal failure than plasma creatinine [100].

Furthermore, CysC is correlated with mortality in ICU, independently of renal function [101].

Patients Who Underwent Cardiac Surgery

Patients who underwent cardiac surgery have a high risk of AKI [8,62] compared to other surgery. This is mainly due to the use of extracorporeal circulation, hemodynamics disorders during and after the operation and postoperative inflammatory response [8,102]. Depending on the definition of acute renal failure, the incidence varies across studies between 2.5% and 60% [8,103]. Necessity for dialysis occurs in approximately 1% to 2%.

Haase-fielitz et al. found that CysC was an early biomarker of AKI but it was still inferior to NGAL which is another new biomarker [104]. This difference was more important in patients without preoperative renal impairment. Nevertheless, the two new biomarkers are good predictors for AKI (severity and duration), renal replacement therapy and hospital mortality. Another study on patients with cardiopulmonary bypass found a significant increase in CysC value in the first and the fifth postoperative days compared to creatinine level [105]. On the contrary, Gueret et al. did not demonstrate the superiority of CysC compared to creatinine as an early biomarker [62]. Koyner et al. , in their study, confirmed the above by concluding that CysC was not a useful predictor to determine AKI within the first six hours following surgery [30]. In addition, Heise et al. showed that GFR values <60 ml/min/1.73 m2 were detected with equal effectiveness using creatinine or CysC. Whereas for the detection of GFR <90 ml/min/1.73 m, the area under the serum creatinine's curve was significantly wider [106].

In fact, it seems that plasma CysC is not a useful early AKI biomarker in the setting of cardiac surgery perhaps due to the effects of perioperative hemodilution on plasma biomarker concentrations.

Transplanted Patients

Delayed graft function after kidney transplantation can also be predicted by CysC but later than other biomarkers [77]. Saudi's study agrees with this and concludes that serum CysC may be used as a marker of renal function after one-week post kidney transplantation [107]. On the contrary, after liver transplantation, postoperative CysC predicted AKI earlier and more accurately than creatinine [108]. Furthermore the same study found that preoperative serum CysC value was predictive of postoperative AKI. When using CysC formula, Qutb et al. found that the best correlation, the highest precision and the least bias [109]. In pediatric kidney transplant recipients, estimation of GFR yields lower values when using CysC rather than serum creatinine [110].

AKI after Contrast Application

After contrast application, serum CysC level increases earlier than that of creatinine [111]. Bachorzewska-Gajewska et al. found that serum CysC level increased at 8 and 24 hours while that of creatinine remained unchanged [112]. More recently, Briguori et al. showed that, in patients with chronic kidney disease, CysC seems to be a reliable marker for the early diagnosis and prognosis of contrast-induced AKI [113]. Its value increases within 24 hours after contrast injection.

Elderly Patients

CysC was tested as a marker of AKI in this population and the results did not show superiority but a complementarity of this dosage with the glomerular filtration rate estimated by the Modification in Diet in Renal Disease (MDRD) [114].

AKI in Chronic Kidney Disease

Studies tend to show that CysC is a better biomarker in early diagnosis of renal impairment in chronic kidney disease. Mussap et al. [115] concluded that CysC may be considered as an alternative and more accurate serum marker than serum creatinine .in differentiating type II diabetic patients, with reduced GFR, from those with normal GFR. Likewise, in patients with various types of glomerulonephritis, CysC serum concentration is well correlated with inulin clearance [116].

Urinary Cystatin C

Results are controversial in regard to the time of the dosage and the criteria used to define AKI. Furthermore, as CysC can be degraded by proteolytic enzymes present in urine (or other biological fluids), some authors have claimed that measurement of urinary CysC (uCysC) would suffer from a lack of reliability or even be impossible to assess [84]. Its stability must be improved by a protease inhibitory cocktail, otherwise, after 3 days at room temperature, a slight decrease of 6–12% is observed [117]. Another drawback is that 24-h urine collection is difficult to do, therefore spot urine samples are used. However, the influence of extreme urination volumes should be studied.

Critical Care Patients

It was demonstrated that in ICU patients, uCysC was independently associated with AKI and mortality even when adjusted for covariates: age, gender, hypotension, APACHE II, plasma CysC, plasma and urinary creatinine [118]. However, when patients with sepsis were excluded, it was not a good predictor of AKI. Indeed, uCysC increases in sepsis even without renal failure. Thus, the effect of AKI and sepsis are additive on uCysC but their time course is different, so uCysC cannot be a constant predictor of AKI. Another problem is that sepsis reduces production of creatinine [65], so when creatinine-based definitions of AKI are applied, it would reduce the sensitivity of uCysC as a marker of AKI.

Patients Who Underwent Cardiac Surgery

Koyner et al. [119] showed that level of uCysC six hours after cardiac surgery was highly predictive of AKI. Furthermore, the increase of the biomarker was proportional to the severity of AKI. In this study, AKI was defined by a peak increase of plasma creatinine greater than 25% from preoperative baseline or the need of renal replacement therapy within 3 days after the surgery. On the contrary, Heise et al. found no significant difference of uCysC concentration between patients with or without AKI defined by the Acute Kidney Injury Network [120]. Liangos et al. compared the performance of six candidate urinary biomarkers [121]: N-acetyl-β-(D)-glucosaminidase (NAG), NGAL, interleukin 18 (IL-18), CysC, and α-1 microglobulin, measured 2 hours following cardiopulmonary bypass (CPB). When measured two hours post CPB, uCysC did not allow to diagnose early AKI. Rentsch et al. found the same results [122].

Controversy about Cystatin C

It was long supposed that inflammatory conditions unaltered the rate of the CysC but a study showed the opposite in an elderly ambulatory population without chronic kidney disease [123]. Studying specifically C-reactive protein which is a marker of inflammation, Stevens et al [124] showed that higher CRP was associated with higher CysC levels.

Otherwise, CRP is increased after extracorporeal circulation [8], so AKI diagnosis with CysC in this case is still discussed [62]. According to that, Akerfeldt et al. chose to study the influence of inflammatory response on CysC after orthopedic surgery which is associated with an important inflammatory response [125]. They found a significant relation between plasma CysC and CRP in the preoperative period, but not after surgery (day 4th and day 30th).

Data exist concerning medication by corticosteroids and elevation of serum CysC, but the subject is still being discussed. A recent study reported that adult kidney transplant patients had a significantly higher amount of CysC than non transplanted adult with kidney disease - but no reason was found [126]. In adult renal transplant patients, Risch et al. showed an association in a dose-dependent manner between glucocorticoid medication and increased CysC, leading to systematic underestimation of GFR [127]. The phenomenon is attribuated to a transcription of the CysC gene induced by promoter mediated glucocorticoid [128]. Thus it seems that glucocorticoid dosage must be included in the analysis of AKI by CysC.

Thyroid function can also influence CysC as hyperthyroid state rises CysC level. It was shown that this did not result in a decrease in glomerular clearance but in an increased synthesis of the biomarker [129].

Knight et al. [130] specifically studied the influence of other factors than renal function on CysC and demonstrated not only older age, male gender, and greater weight, and height but also current cigarette smoking and higher CRP levels were independently associated with higher serum CysC levels after adjusting for creatinine clearance.

Furthermore, this value is higher in patients with acute coronary syndrome. The G73A polymorphism of the CST3 gene affects the plasma CysC level and leads to a change in the CysC/ creatinine ratio [131].

As CysC is implicated in the atherosclerosis process, atherosclerosis in itself may confound results of estimated kidney function. Indeed, patients with peripheral arterial diseases have an increased level of CysC, even if results are adjusted with covariates like IL-6 or CRP [132]. The evoked mechanism is an imbalance between the expression of cathepsines and their endogenous inhibitor CysC [133].

Finally, CysC measurement is more expensive than routine creatinine one. The cost is an important point to be taken into account nowadays.

Conclusion

Glomerular filtration rate is generally accepted as the best indicator of renal function. Its markers must be able to detect kidney disease and particularly the earliest most treatable stages. Early recognition of AKI would certainly prevent patients from progressing to end-stage renal failure. As no reliable marker is early enough to achieve the diagnosis, CysC was proposed. Both plasma and urinary concentrations where studied. Serum CysC seems to be an early marker of AKI except in the settings of cardio thoracic surgery where results of studies are controversial. Urinary CysC has not been studied enough and seems to be reliable only in critical care patients. One of the major drawbacks of uCysC sampling is that patients with AKI are sometimes in complete anuria where urine sampling is impossible. Furthermore, tubular marker concentration depends on urinary output and fluid balance.

The major issues in concluding the superiority of a biomarker over another one are: the differences in the criteria used to define AKI and the differences in the time course of kidney damage corresponding to samples.

Furthermore, considering the fact that measuring this new biomarker is more expensive than that of creatinine and the influence of other factors on its base level, it should actually be reserved for certain categories of patients and should be associated with other biomarkers measurements for early diagnosis of AKI.

REFERENCES

[1] Shusterman N, Strom BL, Murray TG, Morrison G, West SL, Maislin G: Risk factors and outcome of hospital-acquired acute renal failure. Clinical epidemiologic study. *Am. J. Med.* 1987; 83: 65-71

[2] Liano F, Pascual J: Epidemiology of acute renal failure: a prospective, multicenter, community-based study. Madrid Acute Renal Failure Study Group. *Kidney Int.* 1996; 50: 811-8

[3] Waikar SS, Curhan GC, Wald R, McCarthy EP, Chertow GM: Declining mortality in patients with acute renal failure, 1988 to 2002. *J. Am. Soc. Nephrol.* 2006; 17: 1143-50

[4] Hoste EA, Clermont G, Kersten A, Venkataraman R, Angus DC, De Bacquer D, Kellum JA: RIFLE criteria for acute kidney injury are associated with hospital mortality in critically ill patients: a cohort analysis. *Crit. Care* 2006; 10: R73

[5] Cruz DN, Bolgan I, Perazella MA, Bonello M, de Cal M, Corradi V, Polanco N, Ocampo C, Nalesso F, Piccinni P, Ronco C: North East Italian Prospective Hospital Renal Outcome Survey on Acute Kidney Injury (NEiPHROS-AKI): targeting the problem with the RIFLE Criteria. *Clin. J. Am. Soc. Nephrol.* 2007; 2: 418-25

[6] Hoste EA, Lameire NH, Vanholder RC, Benoit DD, Decruyenaere JM, Colardyn FA: Acute renal failure in patients with sepsis in a surgical ICU: predictive factors, incidence, comorbidity, and outcome. *J. Am. Soc. Nephrol.* 2003; 14: 1022-30

[7] Parmar A, Langenberg C, Wan L, May CN, Bellomo R, Bagshaw SM: Epidemiology of septic acute kidney injury. *Curr. Drug Targets* 2009; 10: 1169-78

[8] Gueret G, Lion F, Guriec N, Arvieux J, Dovergne A, Guennegan C, Bezon E, Baron R, Carre J, Arvieux C: Acute renal failure after cardiac surgery with cardiopulmonary bypass is associated with plasmatic IL6 increase. *Cytokine* 2008; 45: 92-8

[9] Hsu CY, McCulloch CE, Fan D, Ordonez JD, Chertow GM, Go AS: Community-based incidence of acute renal failure. *Kidney Int.* 2007; 72: 208-12

[10] Thadhani R, Pascual M, Bonventre JV: Acute renal failure. *N. Engl. J. Med.* 1996; 334: 1448-60

[11] Versteilen AM, Di Maggio F, Leemreis JR, Groeneveld AB, Musters RJ, Sipkema P: Molecular mechanisms of acute renal failure following ischemia/reperfusion. *Int. J. Artif. Organs.* 2004; 27: 1019-29

[12] Haase M, Haase-Fielitz A, Bellomo R: Cardiopulmonary bypass, hemolysis, free iron, acute kidney injury and the impact of bicarbonate. *Contrib. Nephrol.* 2010; 165: 28-32

[13] Ali T, Khan I, Simpson W, Prescott G, Townend J, Smith W, Macleod A: Incidence and outcomes in acute kidney injury: a comprehensive population-based study. *J. Am. Soc. Nephrol.* 2007; 18: 1292-8

[14] Ostermann M, Chang RW: Acute kidney injury in the intensive care unit according to RIFLE. *Crit. Care Med.* 2007; 35: 1837-43; quiz 1852

[15] Delannoy B, Floccard B, Thiolliere F, Kaaki M, Badet M, Rosselli S, Ber CE, Saez A, Flandreau G, Guerin C: Six-month outcome in acute kidney injury requiring renal replacement therapy in the ICU: a multicentre prospective study. *Intensive Care Med.* 2009; 35: 1907-15

[16] Brown JR, Cochran RP, MacKenzie TA, Furnary AP, Kunzelman KS, Ross CS, Langner CW, Charlesworth DC, Leavitt BJ, Dacey LJ, Helm RE, Braxton JH, Clough RA, Dunton RF, O'Connor GT: Long-term survival after cardiac surgery is predicted by estimated glomerular filtration rate. *Ann. Thorac. Surg.* 2008; 86: 4-11

[17] Anderson RJ, O'Brien M, MaWhinney S, VillaNueva CB, Moritz TE, Sethi GK, Henderson WG, Hammermeister KE, Grover FL, Shroyer AL: Mild renal failure is associated with adverse outcome after cardiac valve surgery. *Am. J. Kidney Dis.* 2000; 35: 1127-34

[18] Ryckwaert F, Boccara G, Frappier JM, Colson PH: Incidence, risk factors, and prognosis of a moderate increase in plasma creatinine early after cardiac surgery. *Crit. Care Med.* 2002; 30: 1495-8

[19] Chertow GM, Burdick E, Honour M, Bonventre JV, Bates DW: Acute kidney injury, mortality, length of stay, and costs in hospitalized patients. *J. Am. Soc. Nephrol.* 2005; 16: 3365-70

[20] Joannidis M, Druml W, Forni LG, Groeneveld AB, Honore P, Oudemans-van Straaten HM, Ronco C, Schetz MR, Woittiez AJ: Prevention of acute kidney injury and protection of renal function in the intensive care unit. Expert opinion of the Working Group for Nephrology, ESICM. *Intensive Care Med.* 2010; 36: 392-411

[21] Uchino S, Kellum JA, Bellomo R, Doig GS, Morimatsu H, Morgera S, Schetz M, Tan I, Bouman C, Macedo E, Gibney N, Tolwani A, Ronco C: Acute renal failure in critically ill patients: a multinational, multicenter study. *JAMA* 2005; 294: 813-8

[22] Groeneveld AB, Tran DD, van der Meulen J, Nauta JJ, Thijs LG: Acute renal failure in the medical intensive care unit: predisposing, complicating factors and outcome. *Nephron* 1991; 59: 602-10

[23] Liu KD, Himmelfarb J, Paganini E, Ikizler TA, Soroko SH, Mehta RL, Chertow GM: Timing of initiation of dialysis in critically ill patients with acute kidney injury. *Clin. J. Am. Soc. Nephrol.* 2006; 1: 915-9

[24] Bagshaw SM, Uchino S, Bellomo R, Morimatsu H, Morgera S, Schetz M, Tan I, Bouman C, Macedo E, Gibney N, Tolwani A, Oudemans-van Straaten HM, Ronco C, Kellum JA: Timing of renal replacement therapy and clinical outcomes in critically ill patients with severe acute kidney injury. *J. Crit. Care* 2009; 24: 129-40

[25] Iyem H, Tavli M, Akcicek F, Buket S: Importance of early dialysis for acute renal failure after an open-heart surgery. *Hemodial Int.* 2009; 13: 55-61

[26] Seabra VF, Balk EM, Liangos O, Sosa MA, Cendoroglo M, Jaber BL: Timing of renal replacement therapy initiation in acute renal failure: a meta-analysis. *Am. J. Kidney Dis.* 2008; 52: 272-84

[27] Bagshaw SM, Laupland KB, Doig CJ, Mortis G, Fick GH, Mucenski M, Godinez-Luna T, Svenson LW, Rosenal T: Prognosis for long-term survival and renal recovery in critically ill patients with severe acute renal failure: a population-based study. *Crit. Care* 2005; 9: R700-9

[28] Bell M, Granath F, Schon S, Ekbom A, Martling CR: Continuous renal replacement therapy is associated with less chronic renal failure than intermittent haemodialysis after acute renal failure. *Intensive Care Med.* 2007; 33: 773-80

[29] Tuttle KR, Worrall NK, Dahlstrom LR, Nandagopal R, Kausz AT, Davis CL: Predictors of ARF after cardiac surgical procedures. *Am. J. Kidney Dis.* 2003; 41: 76-83

[30] Koyner JL, Bennett MR, Worcester EM, Ma Q, Raman J, Jeevanandam V, Kasza KE, Connor MF, Konczal DJ, Trevino S, Devarajan P, Murray PT: Urinary cystatin C as an early biomarker of acute kidney injury following adult cardiothoracic surgery. *Kidney Int.* 2008

[31] Barrett BJ, Parfrey PS: Prevention of nephrotoxicity induced by radiocontrast agents. *N. Engl. J. Med.* 1994; 331: 1449-50

[32] Provenchere S, Plantefeve G, Hufnagel G, Vicaut E, De VC, Lecharny JB, Depoix JP, Vrtovsnik F, Desmonts JM, Philip I: Renal dysfunction after cardiac surgery with normothermic cardiopulmonary bypass: incidence, risk factors, and effect on clinical outcome. *Anesth. Analg.* 2003; 96: 1258-64

[33] Bennett M, Dent CL, Ma Q, Dastrala S, Grenier F, Workman R, Syed H, Ali S, Barasch J, Devarajan P: Urine NGAL predicts severity of acute kidney injury after cardiac surgery: a prospective study. *Clin. J. Am. Soc. Nephrol.* 2008; 3: 665-73

[34] Dent CL, Ma Q, Dastrala S, Bennett M, Mitsnefes MM, Barasch J, Devarajan P: Plasma neutrophil gelatinase-associated lipocalin predicts acute kidney injury, morbidity and mortality after pediatric cardiac surgery: a prospective uncontrolled cohort study. *Crit. Care* 2007; 11: R127

[35] Parikh CR, Mishra J, Thiessen-Philbrook H, Dursun B, Ma Q, Kelly C, Dent C, Devarajan P, Edelstein CL: Urinary IL-18 is an early predictive biomarker of acute kidney injury after cardiac surgery. *Kidney Int.* 2006; 70: 199-203

[36] Mishra J, Dent C, Tarabishi R, Mitsnefes MM, Ma Q, Kelly C, Ruff SM, Zahedi K, Shao M, Bean J, Mori K, Barasch J, Devarajan P: Neutrophil gelatinase-associated lipocalin (NGAL) as a biomarker for acute renal injury after cardiac surgery. *Lancet* 2005; 365: 1231-8

[37] Mangano CM, Diamondstone LS, Ramsay JG, Aggarwal A, Herskowitz A, Mangano DT: Renal dysfunction after myocardial revascularization: risk factors, adverse outcomes, and hospital resource utilization. The Multicenter Study of Perioperative Ischemia Research Group. *Ann. Intern. Med.* 1998; 128: 194-203

[38] Brown JR, Cochran RP, Dacey LJ, Ross CS, Kunzelman KS, Dunton RF, Braxton JH, Charlesworth DC, Clough RA, Helm RE, Leavitt BJ, Mackenzie TA, O'Connor GT: Perioperative increases in serum creatinine are predictive of increased 90-day mortality after coronary artery bypass graft surgery. *Circulation* 2006; 114: I409-13

[39] Kuitunen A, Vento A, Suojaranta-Ylinen R, Pettila V: Acute renal failure after cardiac surgery: evaluation of the RIFLE classification. *Ann. Thorac. Surg.* 2006; 81: 542-6

[40] Lassnigg A, Schmid ER, Hiesmayr M, Falk C, Druml W, Bauer P, Schmidlin D: Impact of minimal increases in serum creatinine on outcome in patients after cardiothoracic

surgery: do we have to revise current definitions of acute renal failure? *Crit. Care Med.* 2008; 36: 1129-37

[41] Burns KE, Chu MW, Novick RJ, Fox SA, Gallo K, Martin CM, Stitt LW, Heidenheim AP, Myers ML, Moist L: Perioperative N-acetylcysteine to prevent renal dysfunction in high-risk patients undergoing cabg surgery: a randomized controlled trial. *Jama* 2005; 294: 342-50

[42] Adabag AS, Ishani A, Koneswaran S, Johnson DJ, Kelly RF, Ward HB, McFalls EO, Bloomfield HE, Chandrashekhar Y: Utility of N-acetylcysteine to prevent acute kidney injury after cardiac surgery: a randomized controlled trial. *Am. Heart J.* 2008; 155: 1143-9

[43] Bellomo R, Ronco C, Kellum JA, Mehta RL, Palevsky P: Acute renal failure - definition, outcome measures, animal models, fluid therapy and information technology needs: the Second International Consensus Conference of the Acute Dialysis Quality Initiative (ADQI) Group. *Crit. Care* 2004; 8: R204-12

[44] Gueret G, Lion F, Guriec N, Arvieux J, Dovergne A, Guennegan C, Bezon E, Baron R, Carre J, Arvieux C: Acute renal failure after cardiac surgery with cardiopulmonary bypass is associated with plasmatic IL6 increase. Cytokine 2008; accepted for publication

[45] Bagshaw SM, Gibney RT: Conventional markers of kidney function. *Crit. Care Med.* 2008; 36: S152-8

[46] Rodrigo E, de Francisco AL, Escallada R, Ruiz JC, Fresnedo GF, Pinera C, Arias M: Measurement of renal function in pre-ESRD patients. *Kidney Int. Suppl.* 2002: 11-7

[47] O'Riordan SE, Webb MC, Stowe HJ, Simpson DE, Kandarpa M, Coakley AJ, Newman DJ, Saunders JA, Lamb EJ: Cystatin C improves the detection of mild renal dysfunction in older patients. *Ann. Clin. Biochem.* 2003; 40: 648-55

[48] Andersson LG, Ekroth R, Bratteby LE, Hallhagen S, Wesslen O: Acute renal failure after coronary surgery--a study of incidence and risk factors in 2009 consecutive patients. *Thorac. Cardiovasc. Surg.* 1993; 41: 237-41

[49] Bove T, Calabro MG, Landoni G, Aletti G, Marino G, Crescenzi G, Rosica C, Zangrillo A: The incidence and risk of acute renal failure after cardiac surgery. *J. Cardiothorac. Vasc. Anesth.* 2004; 18: 442-5

[50] Barret B.J. PPS: Prevention of nephrotoxicity induced by radiocontrast agents. *N. Engl. J. Med.* 1994; 331: 1449-1450

[51] Bagshaw SM, Uchino S, Cruz D, Bellomo R, Morimatsu H, Morgera S, Schetz M, Tan I, Bouman C, Macedo E, Gibney N, Tolwani A, Oudemans-van Straaten HM, Ronco C, Kellum JA: A comparison of observed versus estimated baseline creatinine for determination of RIFLE class in patients with acute kidney injury. *Nephrol. Dial Transplant* 2009; 24: 2739-44

[52] D'Onofrio A, Cruz D, Bolgan I, Auriemma S, Cresce GD, Fabbri A, Ronco C: RIFLE criteria for cardiac surgery-associated acute kidney injury: risk factors and outcomes. *Congest Heart Fail* 2010; 16 Suppl 1: S32-6

[53] Srisawat N, Hoste EE, Kellum JA: Modern classification of acute kidney injury. *Blood Purif.* 2010; 29: 300-7

[54] Lema G, Meneses G, Urzua J, Jalil R, Canessa R, Moran S, Irarrazaval MJ, Zalaquett R, Orellana P: Effects of extracorporeal circulation on renal function in coronary surgical patients. *Anesth. Analg.* 1995; 81: 446-51

[55] Cherry RA, Eachempati SR, Hydo L, Barie PS: Accuracy of short-duration creatinine clearance determinations in predicting 24-hour creatinine clearance in critically ill and injured patients. *J. Trauma* 2002; 53: 267-71

[56] Herget-Rosenthal S, Kribben A, Pietruck F, Ross B, Philipp T: Two by two hour creatinine clearance--repeatable and valid. *Clin. Nephrol.* 1999; 51: 348-54

[57] Markantonis SL, Agathokleous-Kioupaki E: Can two-, four- or eight-hour urine collections after voluntary voiding be used instead of twenty-four-hour collections for the estimation of creatinine clearance in healthy subjects? *Pharm World Sci.* 1998; 20: 258-63

[58] Nichols L: Can 12-hour creatinine clearances be substituted for 24-hour creatinine clearances in monitoring for nephrotoxicity from cancer chemotherapy? A pilot study and preliminary data. *Am. J. Clin. Pathol.* 1988; 90: 373-4

[59] O'Connell MB, Wong MO, Bannick-Mohrland SD, Dwinell AM: Accuracy of 2- and 8-hour urine collections for measuring creatinine clearance in the hospitalized elderly. *Pharmacotherapy* 1993; 13: 135-42

[60] Sladen RN, Endo E, Harrison T: Two-hour versus 22-hour creatinine clearance in critically ill patients. *Anesthesiology* 1987; 67: 1013-6

[61] Wilson RF, Soullier G: The validity of two-hour creatinine clearance studies in critically ill patients. *Crit. Care Med.* 1980; 8: 281-4

[62] Gueret G, Kiss G, Bezon E, Lion F, Fourmont C, Corre O, Vaillant C, Carre J, Arvieux C: [Evaluation of the renal function in cardiac surgery with CPB: role of the cystatin C and the calculated creatinine clearance]. *Ann. Fr. Anesth. Reanim.* 2007; 26: 412-7

[63] Cho MM, Yi MM: Variability of daily creatinine excretion in healthy adults. *Hum. Nutr. Clin. Nutr.* 1986; 40: 469-72

[64] Goldberg TH, Finkelstein MS: Difficulties in estimating glomerular filtration rate in the elderly. *Arch. Intern. Med.* 1987; 147: 1430-3

[65] Doi K, Yuen PS, Eisner C, Hu X, Leelahavanichkul A, Schnermann J, Star RA: Reduced production of creatinine limits its use as marker of kidney injury in sepsis. *J. Am. Soc. Nephrol.* 2009; 20: 1217-21

[66] Cockcroft DW, Gault MH: Prediction of creatinine clearance from serum creatinine. *Nephron* 1976; 16: 31-41

[67] Levey AS, Bosch JP, Lewis JB, Greene T, Rogers N, Roth D: A more accurate method to estimate glomerular filtration rate from serum creatinine: a new prediction equation. Modification of Diet in Renal Disease Study Group. *Ann. Intern. Med.* 1999; 130: 461-70

[68] Jorres A KO, Hess S, Farke S, Gahl GM, Muller C, Djurup R.: Urinary excretion of thromboxane and markers for renal injury in patients undergoing cardiopulmonary bypass. *Artif Organs* 1994; 18: 565-9

[69] Westhuyzen J, McGiffin DC, McCarthy J, Fleming SJ: Tubular nephrotoxicity after cardiac surgery utilising cardiopulmonary bypass. *Clin. Chim. Acta* 1994; 228: 123-32

[70] Gormley SM, McBride WT, Armstrong MA, Young IS, McClean E, MacGowan SW, Campalani G, McMurray TJ: Plasma and urinary cytokine homeostasis and renal dysfunction during cardiac surgery. *Anesthesiology* 2000; 93: 1210-6; discussion 5A

[71] Hall PW, 3rd, Ricanati ES: Renal handling of beta-2-microglobulin in renal disorders: with special reference to hepatorenal syndrome. *Nephron* 1981; 27: 62-6

[72] Fernandez F, de Miguel MD, Barrio V, Mallol J: Beta-2-microglobulin as an index of renal function after cardiopulmonary bypass surgery in children. *Child Nephrol. Urol.* 1988; 9: 326-30

[73] Sumeray M, Robertson C, Lapsley M, Bomanji J, Norman AG, Woolfson RG: Low dose dopamine infusion reduces renal tubular injury following cardiopulmonary bypass surgery. *J. Nephrol.* 2001; 14: 397-402

[74] Bernard AM, Moreau D, Lauwerys R: Comparison of retinol-binding protein and beta 2-microglobulin determination in urine for the early detection of tubular proteinuria. *Clin. Chim. Acta* 1982; 126: 1-7

[75] Schmidt C, Hocherl K, Schweda F, Kurtz A, Bucher M: Regulation of renal sodium transporters during severe inflammation. *J. Am. Soc. Nephrol.* 2007; 18: 1072-83

[76] Wagener G, Jan M, Kim M, Mori K, Barasch JM, Sladen RN, Lee HT: Association between increases in urinary neutrophil gelatinase-associated lipocalin and acute renal dysfunction after adult cardiac surgery. *Anesthesiology* 2006; 105: 485-91

[77] Lebkowska U, Malyszko J, Lebkowska A, Koc-Zorawska E, Lebkowski W, Malyszko JS, Kowalewski R, Gacko M: Neutrophil gelatinase-associated lipocalin and cystatin C could predict renal outcome in patients undergoing kidney allograft transplantation: a prospective study. *Transplant Proc.* 2009; 41: 154-7

[78] Melnikov VY, Faubel S, Siegmund B, Lucia MS, Ljubanovic D, Edelstein CL: Neutrophil-independent mechanisms of caspase-1- and IL-18-mediated ischemic acute tubular necrosis in mice. *J. Clin. Invest* 2002; 110: 1083-91

[79] Parikh CR, Jani A, Melnikov VY, Faubel S, Edelstein CL: Urinary interleukin-18 is a marker of human acute tubular necrosis. *Am. J. Kidney Dis.* 2004; 43: 405-14

[80] Haase M, Bellomo R, Story D, Davenport P, Haase-Fielitz A: Urinary interleukin-18 does not predict acute kidney injury after adult cardiac surgery: a prospective observational cohort study. *Crit. Care* 2008; 12: R96

[81] Negishi K, Noiri E, Doi K, Maeda-Mamiya R, Sugaya T, Portilla D, Fujita T: Monitoring of urinary L-type fatty acid-binding protein predicts histological severity of acute kidney injury. *Am. J. Pathol.* 2009; 174: 1154-9

[82] Nakamura T, Sugaya T, Koide H: Urinary liver-type fatty acid-binding protein in septic shock: effect of polymyxin B-immobilized fiber hemoperfusion. *Shock* 2009; 31: 454-9

[83] Clausen J: Proteins in normal cerebrospinal fluid not found in serum. *Proc. Soc. Exp. Biol. Med.* 1961; 107: 170-2

[84] Grubb A: Diagnostic value of analysis of cystatin C and protein HC in biological fluids. *Clin. Nephrol.* 1992; 38 Suppl 1: S20-7

[85] Herget-Rosenthal S, van Wijk JA, Brocker-Preuss M, Bokenkamp A: Increased urinary cystatin C reflects structural and functional renal tubular impairment independent of glomerular filtration rate. *Clin. Biochem.* 2007; 40: 946-51

[86] Lofberg H, Grubb AO: Quantitation of gamma-trace in human biological fluids: indications for production in the central nervous system. *Scand. J. Clin. Lab. Invest* 1979; 39: 619-26

[87] Hoek FJ, Korevaar JC, Dekker FW, Boeschoten EW, Krediet RT: Estimation of residual glomerular filtration rate in dialysis patients from the plasma cystatin C level. *Nephrol Dial Transplant* 2007; 22: 1633-8

[88] Baas MC, Bouman CS, Hoek FJ, Krediet RT, Schultz MJ: Cystatin C in critically ill patients treated with continuous venovenous hemofiltration. *Hemodial Int*. 2006; 10 Suppl 2: S33-7
[89] Kyhse-Andersen J, Schmidt C, Nordin G, Andersson B, Nilsson-Ehle P, Lindstrom V, Grubb A: Serum cystatin C, determined by a rapid, automated particle-enhanced turbidimetric method, is a better marker than serum creatinine for glomerular filtration rate. *Clin. Chem.* 1994; 40: 1921-6
[90] Newman DJ, Thakkar H, Edwards RG, Wilkie M, White T, Grubb AO, Price CP: Serum cystatin C measured by automated immunoassay: a more sensitive marker of changes in GFR than serum creatinine. *Kidney Int.* 1995; 47: 312-8
[91] Finney H, Newman DJ, Gruber W, Merle P, Price CP: Initial evaluation of cystatin C measurement by particle-enhanced immunonephelometry on the Behring nephelometer systems (BNA, BN II). *Clin. Chem*. 1997; 43: 1016-22
[92] Herget-Rosenthal S, Feldkamp T, Volbracht L, Kribben A: Measurement of urinary cystatin C by particle-enhanced nephelometric immunoassay: precision, interferences, stability and reference range. *Ann. Clin. Biochem*. 2004; 41: 111-8
[93] Tidman M, Sjostrom P, JonesI: A comparison of GFR estimating formulae based upon s-cystatin C and s-creatinine and a combination of the two. *Nephrol Dial Transplant* 2008; 23: 154-60
[94] Li J, Dunn W, Breaud A, Elliott D, Sokoll LJ, Clarke W: Analytical performance of 4 automated assays for measurement of cystatin C. *Clin. Chem.* 2010; 56: 1336-9
[95] Lisowska-Myjak B: Serum and urinary biomarkers of acute kidney injury. *Blood Purif;* 29: 357-65
[96] Coca SG, Yalavarthy R, Concato J, Parikh CR: Biomarkers for the diagnosis and risk stratification of acute kidney injury: a systematic review. *Kidney Int.* 2008; 73: 1008-16
[97] Herget-Rosenthal S, Marggraf G, Husing J, Goring F, Pietruck F, Janssen O, Philipp T, Kribben A: Early detection of acute renal failure by serum cystatin C. *Kidney Int.* 2004; 66: 1115-22
[98] Villa P, Jimenez M, Soriano MC, Manzanares J, Casasnovas P: Serum cystatin C concentration as a marker of acute renal dysfunction in critically ill patients. *Crit. Care* 2005; 9: R139-43
[99] Herrero-Morin JD, Malaga S, Fernandez N, Rey C, Dieguez MA, Solis G, Concha A, Medina A: Cystatin C and beta2-microglobulin: markers of glomerular filtration in critically ill children. *Crit. Care* 2007; 11: R59
[100] Royakkers AA, van Suijlen JD, Hofstra LS, Kuiper MA, Bouman CS, Spronk PE, Schultz MJ: Serum cystatin C-A useful endogenous marker of renal function in intensive care unit patients at risk for or with acute renal failure? *Curr. Med. Chem.* 2007; 14: 2314-7
[101] Bell M, Granath F, Martensson J, Lofberg E, Ekbom A, Martling CR: Cystatin C is correlated with mortality in patients with and without acute kidney injury. *Nephrol Dial Transplant* 2009; 24: 3096-102
[102] Zanardo G, Michielon P, Paccagnella A, Rosi P, Calo M, Salandin V, Da Ros A, Michieletto F, Simini G: Acute renal failure in the patient undergoing cardiac operation. Prevalence, mortality rate, and main risk factors. *J. Thorac. Cardiovasc. Surg*. 1994; 107: 1489-95

[103] Novis BK, Roizen MF, Aronson S, Thisted RA: Association of preoperative risk factors with postoperative acute renal failure. *Anesth. Analg.* 1994; 78: 143-9

[104] Haase-Fielitz A, Bellomo R, Devarajan P, Story D, Matalanis G, Dragun D, Haase M: Novel and conventional serum biomarkers predicting acute kidney injury in adult cardiac surgery--a prospective cohort study. *Crit. Care Med.* 2009; 37: 553-60

[105] Felicio ML, Andrade RR, Castiglia YM, Silva MA, Vianna PT, Martins AS: Cystatin C and glomerular filtration rate in the cardiac surgery with cardiopulmonary bypass. *Rev. Bras Cir. Cardiovasc.* 2009; 24: 305-11

[106] Heise D, Waeschle RM, Schlobohm J, Wessels J, Quintel M: Utility of cystatin C for assessment of renal function after cardiac surgery. *Nephron. Clin. Pract.* 2009; 112: c107-14

[107] Geramizadeh B, Azarpira N, Ayatollahi M, Rais-Jalali GA, Aghdai M, Yaghoobi R, Banihashemi M, Malekpour Z, Malek-Hosseini SA: Value of serum cystatin C as a marker of renal function in the early post kidney transplant period. *Saudi J. Kidney Dis. Transpl.* 2009; 20: 1015-7

[108] Hei ZQ, Li XY, Shen N, Pang HY, Zhou SL, Guan JQ: Prognostic values of serum cystatin C and beta2 microglobulin, urinary beta2 microglobulin and N-acetyl-beta-D-glucosaminidase in early acute renal failure after liver transplantation. *Chin. Med. J.* (Engl) 2008; 121: 1251-6

[109] Qutb A, Syed G, Tamim HM, Al Jondeby M, Jaradat M, Tamimi W, Al Ghamdi G, Al Qurashi S, Flaiw A, Hejaili F, Al Sayyari AA: Cystatin C-based formula is superior to MDRD, Cockcroft-Gault and Nankivell formulae in estimating the glomerular filtration rate in renal allografts. *Exp. Clin. Transplant* 2009; 7: 197-202

[110] Franco MC, Nagasako SS, Machado PG, Nogueira PC, Pestana JO, Sesso R: Cystatin C and renal function in pediatric kidney transplant recipients. *Braz. J. Med. Biol. Res.* 2009; 42: 1225-9

[111] Rickli H, Benou K, Ammann P, Fehr T, Brunner-La Rocca HP, Petridis H, Riesen W, Wuthrich RP: Time course of serial cystatin C levels in comparison with serum creatinine after application of radiocontrast media. *Clin. Nephrol.* 2004; 61: 98-102

[112] Bachorzewska-Gajewska H, Malyszko J, Sitniewska E, Malyszko JS, Pawlak K, Mysliwiec M, Lawnicki S, Szmitkowski M, Dobrzycki S: Could neutrophil-gelatinase-associated lipocalin and cystatin C predict the development of contrast-induced nephropathy after percutaneous coronary interventions in patients with stable angina and normal serum creatinine values? *Kidney Blood Press Res.* 2007; 30: 408-15

[113] Briguori C, Visconti G, Rivera NV, Focaccio A, Golia B, Giannone R, Castaldo D, De Micco F, Ricciardelli B, Colombo A: Cystatin C and contrast-induced acute kidney injury. *Circulation* 2010; 121: 2117-22

[114] Carbonnel C, Seux V, Pauly V, Oddoze C, Roubicek C, Larue JR, Thirion X, Soubeyrand J, Retornaz F: [Estimation of the glomerular filtration rate in elderly inpatients: comparison of four methods]. *Rev. Med. Interne* 2008; 29: 364-9

[115] Mussap M, Dalla Vestra M, Fioretto P, Saller A, Varagnolo M, Nosadini R, Plebani M: Cystatin C is a more sensitive marker than creatinine for the estimation of GFR in type 2 diabetic patients. *Kidney Int.* 2002; 61: 1453-61

[116] Hayashi T, Nitta K, Hatano M, Nakauchi M, Nihei H: The serum cystatin C concentration measured by particle-enhanced immunonephelometry is well correlated

with inulin clearance in patients with various types of glomerulonephritis. *Nephron* 1999; 82: 90-2
[117] M, Moutereau S, Zater M, Lallali K, Durrbach A, Manivet P, Eschwege P, Loric S: Urinary cystatin C as a specific marker of tubular dysfunction. *Clin. Chem. Lab. Med.* 2006; 44: 288-91
[118] Nejat M, Pickering JW, Walker RJ, Westhuyzen J, Shaw GM, Frampton CM, Endre ZH: Urinary cystatin C is diagnostic of acute kidney injury and sepsis, and predicts mortality in the intensive care unit. *Crit. Care;* 14: R85
[119] Koyner JL, Bennett MR, Worcester EM, Ma Q, Raman J, Jeevanandam V, Kasza KE, O'Connor MF, Konczal DJ, Trevino S, Devarajan P, Murray PT: Urinary cystatin C as an early biomarker of acute kidney injury following adult cardiothoracic surgery. *Kidney Int.* 2008; 74: 1059-69
[120] Heise D, Rentsch K, Braeuer A, Friedrich M, Quintel M: Comparison of urinary neutrophil glucosaminidase-associated lipocalin, cystatin C, and alpha(1)-microglobulin for early detection of acute renal injury after cardiac surgery. *Eur. J. Cardiothorac. Surg.*
[121] Liangos O, Tighiouart H, Perianayagam MC, Kolyada A, Han WK, Wald R, Bonventre JV, Jaber BL: Comparative analysis of urinary biomarkers for early detection of acute kidney injury following cardiopulmonary bypass. *Biomarkers* 2009; 14: 423-31
[122] Heise D, Rentsch K, Braeuer A, Friedrich M, Quintel M: Comparison of urinary neutrophil glucosaminidase-associated lipocalin, cystatin C, and alpha(1)-microglobulin for early detection of acute renal injury after cardiac surgery. *Eur. J. Cardiothorac. Surg.* 2010
[123] Keller CR, Odden MC, Fried LF, Newman AB, Angleman S, Green CA, Cummings SR, Harris TB, Shlipak MG: Kidney function and markers of inflammation in elderly persons without chronic kidney disease: the health, aging, and body composition study. *Kidney Int.* 2007; 71: 239-44
[124] Stevens LA, Schmid CH, Greene T, Li L, Beck GJ, Joffe MM, Froissart M, Kusek JW, Zhang YL, Coresh J, Levey AS: Factors other than glomerular filtration rate affect serum cystatin C levels. *Kidney Int.* 2009; 75: 652-60
[125] Akerfeldt T, Helmersson J, Larsson A: Postsurgical inflammatory response is not associated with increased serum cystatin C values. *Clin. Biochem*; 43: 1138-40
[126] Hermida J, Romero R, Tutor JC: Relationship between serum cystatin C and creatinine in kidney and liver transplant patients. *Clin. Chim. Acta* 2002; 316: 165-70
[127] Risch L, Herklotz R, Blumberg A, Huber AR: Effects of glucocorticoid immunosuppression on serum cystatin C concentrations in renal transplant patients. *Clin. Chem.* 2001; 47: 2055-9
[128] Bjarnadottir M, Grubb A, Olafsson I: Promoter-mediated, dexamethasone-induced increase in cystatin C production by HeLa cells. *Scand. J. Clin. Lab. Invest.* 1995; 55: 617-23
[129] Karawajczyk M, Ramklint M, Larsson A: Reduced cystatin C-estimated GFR and increased creatinine-estimated GFR in comparison with iohexol-estimated GFR in a hyperthyroid patient: a case report. *J. Med. Case Reports* 2008; 2: 66
[130] Knight EL, Verhave JC, Spiegelman D, Hillege HL, de Zeeuw D, Curhan GC, de Jong PE: Factors influencing serum cystatin C levels other than renal function and the impact on renal function measurement. *Kidney Int.* 2004; 65: 1416-21

[131] Noto D, Cefalu AB, Barbagallo CM, Pace A, Rizzo M, Marino G, Caldarella R, Castello A, Pernice V, Notarbartolo A, Averna MR: Cystatin C levels are decreased in acute myocardial infarction: effect of cystatin C G73A gene polymorphism on plasma levels. *Int. J. Cardiol.* 2005; 101: 213-7

[132] Arpegard J, Ostergren J, de Faire U, Hansson LO, Svensson P: Cystatin C--a marker of peripheral atherosclerotic disease? *Atherosclerosis* 2008; 199: 397-401

[133] Chapman HA, Riese RJ, Shi GP: Emerging roles for cysteine proteases in human biology. *Annu. Rev. Physiol.* 1997; 59: 63-88

In: Cystatins: Protease Inhibitors …
Editors: John B. Cohen and Linda P. Ryseck.
ISBN: 978-1-61209-343-7

Chapter 8

CYSTATIN C, ATHEROSCLEROSIS AND LIPID-LOWERING THERAPY BY STATINS

T. A. Korolenko, M. S. Cherkanova, E. A. Gashenko, T. P. Johnston, and I. Yu. Bravve

Institute of Physiology, Siberian Branch of the Russian Academy of Medical Sciences, Novosibirsk, Russia

Division of Pharmaceutical Sciences, School of Pharmacy, University of Missouri-Kansas City, Kansas City, USA

ABSTRACT

The search for new serum markers of aging, atherosclerosis, and predictors of cardiovascular emergencies is important in contemporary society. Recently, new non-lipid markers of atherosclerosis and predictors of cardiovascular events were introduced. These markers are related to inflammation and macrophage stimulation, such as cystatin C, matrix metalloproteases (MMPs), and chitotriosidase. Increased serum cystatin C concentration, an alternative measure of renal function, is now suggested as a strong predictor of cardiovascular events. We compared new non-lipid atherosclerosis indexes with common inflammatory (hs-CRP) and lipid markers in elderly persons and patients with atherosclerosis and ischemic heart disease (IHD) who have undergone coronary bypass surgery.

Cystatins are known to be very potent endogenous inhibitors of cysteine proteases of the papain superfamily. They form equimolar, tight, and reversible complexes with human cysteine proteases (cathepsins B, H, K, L and S), and express different cellular functions as proteins (cell proliferation, degradation of extracellular matrix, etc.). In humans, cystatin C, localized predominantly in the extracellular space, was used for early detection of impaired kidney function, as well as a marker in several inflammatory and tumor diseases. The question of whether cystatin C is an atherogenic or protective protein, as well as its possible role as a marker or predictor in IHD is still not clear.

Using an ELISA method, we have shown that in healthy persons aged 25-45, cystatin C concentration is the highest in cerebrospinal fluid and much lower in urine and especially in bile (cerebrospinal fluid>saliva>serum>urine>bile). A similar distribution was shown for procathepsin B, which is an enzymatically inactive precursor of the mature cysteine protease cathepsin B: cerebrospinal fluid>saliva>serum>urine. However, the relative concentration of cystatin C in serum was higher (~100-fold) than in urine, as compared to the procathepsin B level (~ 5-fold).

The serum cystatin C concentration was shown to increase significantly in elderly persons of 45-65 years old with a high risk of IHD (with normal serum and urine creatinine level), and especially in patients of the same age with atherosclerosis and IHD before coronary bypass surgery. The elevated serum level of cystatin C, MMPs, and chitotriosidase activity were observed in an elderly group with atherosclerosis and IHD, and lipid-lowering therapy by statins significantly decreased the hs-CRP concentration and the activity of MMPs, , but not the cystatin C level. After coronary bypass surgery, there was a rapid and marked increase in hs-CRP and a mild increase in cystatin C concentration. Moreover, 30 days, and one year after surgery, cystatin C levels were still elevated.

Protease inhibitors are generally regarded as atheroprotective, because proteases participate in matrix degradation, a process regarded mostly as atherogenic. The cystatin C most likely has its origin in different cell types present in aortic lesions, such as SMCs, endothelial cells, and macrophages, known to produce cystatin C. The role of cystatin C compared to hs-CRP and the activity of MMPs, as possible predictors of cardiovascular events is discussed.

INTRODUCTION

The search for new serum markers of aging, atherosclerosis, and predictors of cardiovascular events is important in contemporary society (Anderson, 2005; Brown, Bittner, 2008; Evrin et al., 2005), especially in regions and cities with a high mortality due to cardiovascular pathology (Novosibirsk and Novosibirsk Region). Recently, new non-lipid markers of atherosclerosis and predictors of cardiovascular events related to inflammation and macrophage stimulation were introduced, for example, cystatin C (Seronie-Vivien et al., 2008), several types of matrix metalloproteases (MMPs) (Nagase et al., 2006), and chitotriosidase (Malaguarnera, 2006; Kurt et al., 2007).

Increased serum cystatin C concentrations, used earlier primarily as an alternative measure of renal function, is now suggested by some authors as a strong predictor of cardiovascular events (Naruse et al., 2009). We compared new non-lipid atherosclerosis indexes with common inflammatory (hs-CRP) and lipid markers in elderly persons and patients with atherosclerosis and ischemic heart disease (IHD) who have undergone coronary bypass surgery, with a special attention to the role of cystatin C.

Cystatins are known to be very potent endogenous inhibitors of cysteine proteases of the papain superfamily (Turk et al., 2002). They form equimolar, tight, and reversible complexes with human cysteine proteases (cathepsins B, H, K, L and S), and express different cellular functions as proteins (cell proliferation, degradation of extracellular matrix, etc.) (Kepler, 2006). In humans, cystatin C, localized predominantly in the extracellular space, was used for early detection of impaired kidney function (Mussap, Plebani, 2004; Chew et al., 2008), as well as a marker in several inflammatory and tumor diseases (Mussap, Plebani, 2004; Korolenko et al., 2008). The question of whether cystatin C is an atherogenic or protective protein, as well as its possible role as a marker or predictor in IHD, is still not clear.

Using an ELISA method, we have shown that in biological fluids of healthy persons aged 25-45, the cystatin C concentration is the highest in cerebrospinal fluid and much lower in urine, and especially in bile (cerebrospinal fluid>saliva>serum>urine>bile). A similar distribution was shown for procathepsin B, which is an enzymatically inactive precursor of the mature cysteine protease cathepsin B: cerebrospinal fluid>saliva>serum>urine. However, the relative concentration of cystatin C in serum was higher (~100-fold) than in urine, as compared to the procathepsin B level (~ 5-fold). This is most likely a result of cystatin C molecules being able to cross the glomerular membrane due to their low molecular weight.

According to data obtained in this work, the serum cystatin C concentration was shown to increase significantly in elderly persons of 45-65 years old with a high risk of IHD (with normal serum and urine creatinine level), especially in patients of the same age with atherosclerosis and IHD before coronary bypass surgery. The elevated serum level of cystatin C, MMPs, and chitotriosidase activity were observed in the elderly group with atherosclerosis and IHD, and lipid-lowering therapy by statins significantly decreased the hs-CRP concentration and the activity of MMPs, but not the cystatin C level. Early after coronary bypass surgery, there was a rapid and significant increase in hs-CRP and a mild decrease in the cystatin C concentration, followed by elevation of the cystatin C level as noted before the operation. Moreover, 30 days, and one year after the operation, the cystatin C levels were still elevated, and statin treatment had no effect on this index.

Protease inhibitors are generally regarded as atheroprotective, because proteases participate in matrix degradation, a process regarded mostly as atherogenic. The cystatin C most likely has its origin in different cell types present in aortic lesions, such as SMCs, endothelial cells, and macrophages, known to produce cystatin C. The role of cystatin C, compared to hs-CRP and the activity of MMPs, as possible predictors of cardiovascular events is discussed.

Cystatin C: Structure and Biological Functions

The term "cystatin" was proposed by A. Barrett for designation of homologous proteins of the same superfamily (Turk et al., 2002). According to the classification of cystatins based on the homology of amino acid composition (similarity more than 50%), three main types of cystatin families have been recognized in mammals. The single domain cystatins (a) are referred to as *type 1* (cystatin A is also known as stefin and cystatin B) and *type 2* (the best studied cystatin C and other recently discovered cystatins D, S, SN, SA, E. F, G and H). Complex cystatins (b) *(type 3)* contain several domains of type 2 cystatins and include low- and high-molecular weight kininogens, fetuin, and glycoproteins (Turk et al., 2002; Keppler, 2006).

Cystatin C is preferentially an extracellular (and partially cytoplasmic) basic low-molecular weight protein of 13343- 13359 Da and an effective inhibitor of endogenous cysteine proteases. It belongs to the type 2 cystatin family (Seronie-Vivien et al., 2008). Cystatin C is encoded by the CST3 gene, and mutations in this gene are associated with development of some inherited diseases in humans (Turk et al., 2002). The cystatin C gene is referred to as "housekeeping genes", and nucleus containing cells synthesize cystatin C at a constant rate (Warfel et al., 1987; Seronie-Vivien et al., 2008; Mussap, Plebani, 2004).

Cell *functions* of cystatin C are not fully understood and represent an intensively studied subject. Cystatin C is known to be involved in the regulation of degradation of extracellular matrix (ECM) proteins, processing and presentation of antigens, processes of cell differentiation and proliferation, and also aging (Keppler, 2006; Chew et al., 2008). In experiments, it has been shown that cystatin C *knockout* mice do not demonstrate marked changes, including vascular impairments (expected and typical for atherosclerosis) (Keppler, 2006).

Serum cystatin C concentration are now more often used as an alternative measure of kidney function, especially in patients with hypertension and diabetes, because that index is less affected by age, sex, or muscle mass, and is a more sensitive indicator of early renal dysfunction than creatinine-based estimations of glomerular filtration rate (Bokenkamp et al., 2006).

We compared cystatin C and changes in some new non-lipid atherosclerosis indices with inflammatory (hs-CRP) and common lipid markers – serum cholesterol and triglyceride concentrations - in elderly persons and patients with atherosclerosis treated by cardio bypass surgery.

Serum of 26 healthy males aged 20-45 (donors, aged 31.4±6.5 years ← different font size for "years"; change to match rest of the text , group 1, low risk of CVD) and 25 elderly individuals, aged 50-65 years old, with arterial hypertension, without ischemic heart disease (IHD), non-treated (high risk of CVD, 56.8±2.9 years old, group 2) and treated by statins (simvastatin, 20-40 mg per day, group 3) and a group 4 of 32 patients with atherosclerosis and IHD, 56.5±6.97 years old, documented with coronarographic study before the cardio bypass surgical operation (Novosibirsk Regional Cardiovascular Center) were evaluated. Patients with type 2 diabetes were excluded from the investigation. All human specimens were obtained with the informed consent of patients according to a protocol approved by the Medical Ethical Committee Recommendations of the Institute of Physiology RAMS (Novosibirsk, Russia).

Serum was obtained after centrifugation of blood samples at 3000 x g for 15 min at 5° C and stored at –70° C until analysis for total cholesterol using commercially-available kits *Novochol* (Vector-Best, Russia), triglycerides (TG) – *Triglycerides-Novo* (Vector-Best, Russia), at 490 nm and 546 nm, respectively, using a photometer 5010 (Robert Riele, Germany). The concentration of HDL-cholesterol was measured with the commercial kits *HDL–Cholesterol-Novo* and *Novochol* (Vector-Best, Russia). Photometry was conducted using a half-automatic photometer 5010 with a flow-through thermostatted cuvette (Robert Riele, Germany) at 490 nm. All samples were provided under the control of commercially-available test serums (Lyonorm U and Lyonorm P, *Lachema*, Czechia). Increased cholesterol and TG levels were shown in the elderly group and patients with atherosclerosis (Figure1). Statins (simvastatin) treatment of elderly patients was followed by significant lowering of both cholesterol and TG concentrations (Figure 1).

Serum creatinine concentrations were measured by a kinetic colorimetric method using a commercial kit *Creatinin-Novo-A* (Vector-Best, Russia) using the photometer Screen Master (Hospitex Diagnostics, Switzerland) at 500 nm. The high-sensitivity C-reactive protein (hs-CRP) concentration was measured by commercial kit (Biosystems, Spain). The cystatin C concentration was determined with a commercial ELISA kit for quantitative determination of human cystatin C (BioVendor, Czechia) and the plate reader was a Multiscan Ex Termo Electron Co., Finland; human procathepsin B concentration was determined by a commercial R&D ELISA kit, USA; the activity of MMP and chitotriosidase - by fluorescent methods against MCA-Pro-Leu-Gly-Leu-DpA-Ala-Arg-NH_2 (American Peptide Co., USA) (Knight et al., 1992; Nagase, 1994) and MUF-β-D-N,N',N''-triacetylchitotrioside (Sigma) as substrates, respectively (Malaguarnera, 2006), where MCA denotes methylcoumarylamide and MUF – methylumbelliferyl. Extinction was measured using a fluorescent spectrophotomer, Shimadzu RF-530101(PC) S, Japan (at 360 and 445 nm for chitotriosidase and at 328 and 393 nm for MMP). Statistical analysis was performed using the SPSS-9 program. Data were deemed statistically significant at $p<0.05$. The Spearman test was used for correlation calculations.

We have shown that *healthy* persons of *low risk* of CVD development (group 1) were characterized by a normal lipid profile: the total serum cholesterol 4.37±0.11 mmol/L, high-density lipoprotein (HDL)-cholesterol 1.40±0.05 mmol/L, low-density lipoprotein (LDL)-cholesterol 2.40±0.08 mmol/L, triglycerides (TG. 1.30±0.08 mmol/L). These results confirmed data obtained earlier by other investigators (Chew et al., 2008). Two of 26 persons in this group smoked.

Group 2 had a *high* risk of CVD, with arterial hypertension (without IHD), *not* treated by statins, and was characterized by hyperlipidemia with an increased level of total serum cholesterol 6.06±0.12 and TG 1.92±0.09 mmol/L (Figure 1), decreased HDL-cholestrol 1.19±0,05 mmol/L, and increased LDL-cholesterol 4.00±0.09 mmol/L,).

Simvastatin treatment (group 3) was followed by a significant hypolipidemic effect: total serum cholesterol decreased to 4.55±0.06 mmol/L, TG 1.45±0.04 mmol/l; HDL-cholesterol 1.48±0.03 mmol/l, and LDL-cholesterol 2.41±0.09 mmol/L,).

In figure 1 below, you need to say in elderly persons and patiens, not elder

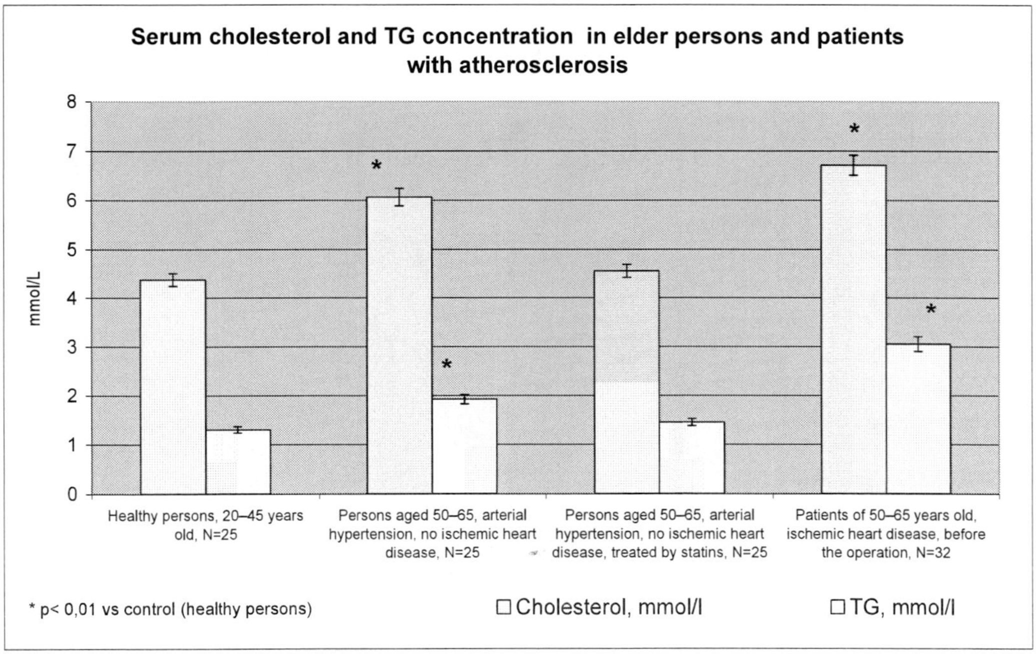

Figure 1.

Table 1. Serum hs-CRP concentration, chitotriosidase and MMP activity of persons with a high risk of CVD and patients with atherosclerosis

Groups, the number of persons	hs-CRP, mg/L	Chitotriosidase activity, nmol MUF/mL per hour	MMP activity, µmol MCA/L per hour
1.Healthy persons, 20 –45 years old, N=26	3.24 ± 0.08	158.6 ± 8.40	45.7±1.48
2. Persons aged 50 – 65, arterial hypertension, no ischemic heart disease, N=25	6.09 ± 0.07 $P_{1\text{-}2} < 0.001$	261.84 ± 14,80 $P_{1\text{-}2} < 0.001$	88.5 ± 4.16 $P_{1\text{-}2} < 0.001$
3. Persons aged 50 – 65, arterial hypertension, no ischemic heart disease, *treated by statins*, N=25	4.71 ± 0,07 $P_{1\text{-}3} < 0{,}001$ $P_{2\text{-}3} < 0.001$	399.34 ± 8.35 $P_{1\text{-}3} < 0.001$	65.8 ± 3.64 $P_{1\text{-}3} < 0.001$ $P_{2\text{-}3} < 0.001$
4. Patients of 50 – 65 years old, ischemic heart disease, before the operation, N=32	7.21 ± 0.10 $P_{1\text{-}4} < 0.001$ $P_{2\text{-}4} < 0.01$	380.50 ± 15.60 $P_{1\text{-}4} < 0.001$ $P_{2\text{-}4} < 0.001$	101.9 ± 7.91 $P_{1\text{-}4} < 0.001$ $P_{2\text{-}4} < 0.01$

Abbreviations: MUF- methylumbellyferyl; MCA- methylcoumarylamide.

Table 2. Serum cystatin C, creatinine concentration and GFR in persons with a high risk of CVD development and patients with atherosclerosis

Groups, the number of persons	Cystatin C, ng/mL	Creatinine, mmol/L	Glomerular filtration rate (GFR), ml/min
1 Healthy persons, 20 –45 years old, N= 26	846.0 ± 124.1	0.089± 0.015	131.5 ± 7.6
2. Persons aged 50 – 65, arterial hypertension, no ischemic heart disease, N= 15	1704.5 ± 235,0 $P_{1\text{-}2} < 0.05$	0.096 ± 0.012	117.3 ± 10.8
3. Persons aged 50 – 65, arterial hypertension, no ischemic heart disease, *treated by statins*, N=18	1687.9 ± 292.3 $P_{1\text{-}3} < 0.05$	0.090 ± 0.014	119.1 ± 9.6
4. Patients of 50 – 65 years old, ischemic heart disease, before the operation, N= 17	1824.7 ± 256.9 $P_{1\text{-}4} < 0.05$	0.092 ± 0.014	111.8 ±14.9

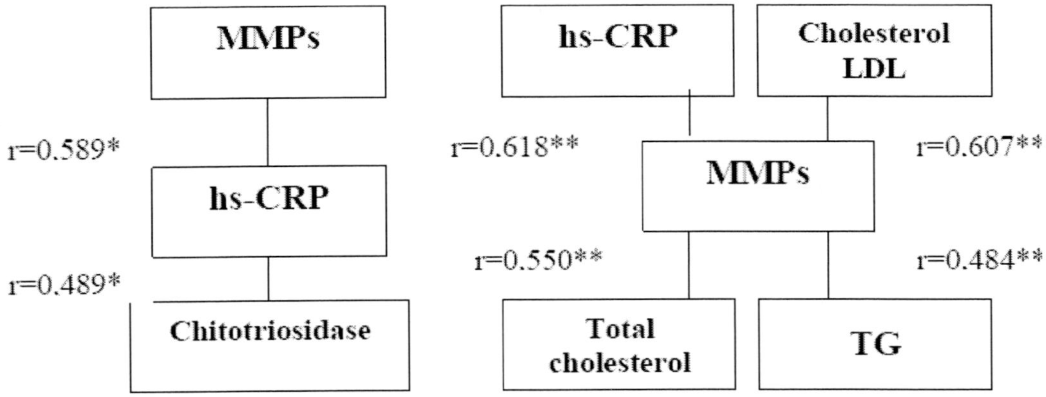

(*p < 0.05, **p <0.01).

Schema. Correlations between serum cholesterol, TG, hs-CRP, chitotriosidase, and MMPs activity in persons with a high risk of ischemic heart disease and patients with ischemic heart disease (before simvastatin treatment).

Before cardiosurgery patients (group 4) with IHD and arterial hypertension revealed hyperlipidemia: increased concentrations of total serum cholesterol up to 6.71±0.11 mmol/L, TG 3.24±0.18 mmol/L, decreased level of HDL-cholesterol 1.06±0.03 mmol/L, and with an increased concentration of LDL-cholesterol up to 4.29±0.08 mmol/L. Most patients in this group (87%) smoked.

Serum *hs-CRP* concentration increased both in elderly persons and patients with atherosclerosis (groups 2 and 4, Table 1). Similar results were obtained previously by other investigators (Evrin et al., 2005). The highest hs-CRP concentration was noted in patients with atherosclerosis before cardiosurgery (group 4, Table 1). Statin (simvastatin, 20-40 mg per day) treatment of elderly persons (group 3) significantly decreased the hs-CRP concentration, but not to the level of healthy young persons with a low risk of CVD (Table 1).

In patients with IHD, the following correlations were revealed: between serum MMP activity and hs-CRP ($r = 0.618$, $P<0.01$), TG ($r = 0.484$, $P<0.01$), LDL-cholesterol ($r = 0.607$, $P<0.01$), and total cholesterol ($r=0.550$, $p< 0.01$) (see Schema). A correlation was found between changes in serum MMPs and all lipid markers of atherosclerosis studied in our work.

The serum *cystatin C* concentration increased both in elderly persons and patients with atherosclerosis before cardiosurgery to the same degree (groups 2 and 4, Table 2).

Statin treatment did not significantly influence serum cystatin C concentration (Tables 2, 3), which was elevated both in treated (group 3) and non-treated patients (group 2). According to commonly used methods for renal function assessment, there was no change of glomerular filtration rate (GFR) and serum creatinine level in all groups studied (Table 2). Renal function is an important index of atherosclerosis, and serum cystatin C is a novel measure of GFR and also a predictor of cardiovascular events (Loew et al., 2005; Naruse et al., 2009). Increased cystatin C concentration, an alternative measure of renal function, is now accepted as a strong predictor of cardiovascular events, provided one has data on early disturbances of renal function.

The total MMP *activity* assay using the enzymatic fluorescent method provided information on several MMP types, including gelatinases (MMP-2 and MMP-9), stromelysins (MMP-3, -10, -11) playing the important role in plaque instability (Yang et al., 2001; Nagase,

1994). Changes of serum MMP *activity* in early atherosclerosis have not been studied thoroughly at the present time. In our work, MMP *activity* was shown to increase in elderly, group 2, as compare to young persons (Table 1). Statin treatment decreased the MMP activity in patients of group 3, as compared to the group 2, which were not treated by statins (Table 3). The significant elevation of MMP activity was revealed in patients with atherosclerosis before the operation (Table 1). Thus, both aging and atherosclerosis development were characterized by increased serum MMP activity, which was higher in group 4 patients with atherosclerosis and IHD. Patients in this group (4) with atherosclerosis and with IHD showed positive correlations of MMP activity with the hs-CRP ($r = 0.589$, $p < 0.05$), and hs-CRP with chitoriosidase activity ($r=0.489$, $p< 0.05$) before cardio surgery.

Serum *chitotriosidase* activity increased in elderly persons with hypertension and a high risk of CVD (Table 1) and statin treatment had no lowering effect on enzyme activity (there was even a tendency for this value to increase). As compared to elderly group 2 patients (not-treated by statins), significantly more elevated chitotriosidase activity was noted in patients with atherosclerosis (Table 1, group 4). A correlation was shown between chitotriosidase activity and TG concentrations ($r= -0.508$, $p<0.05$), and hs-CRP ($r =-o.683$, $p<0.05$).

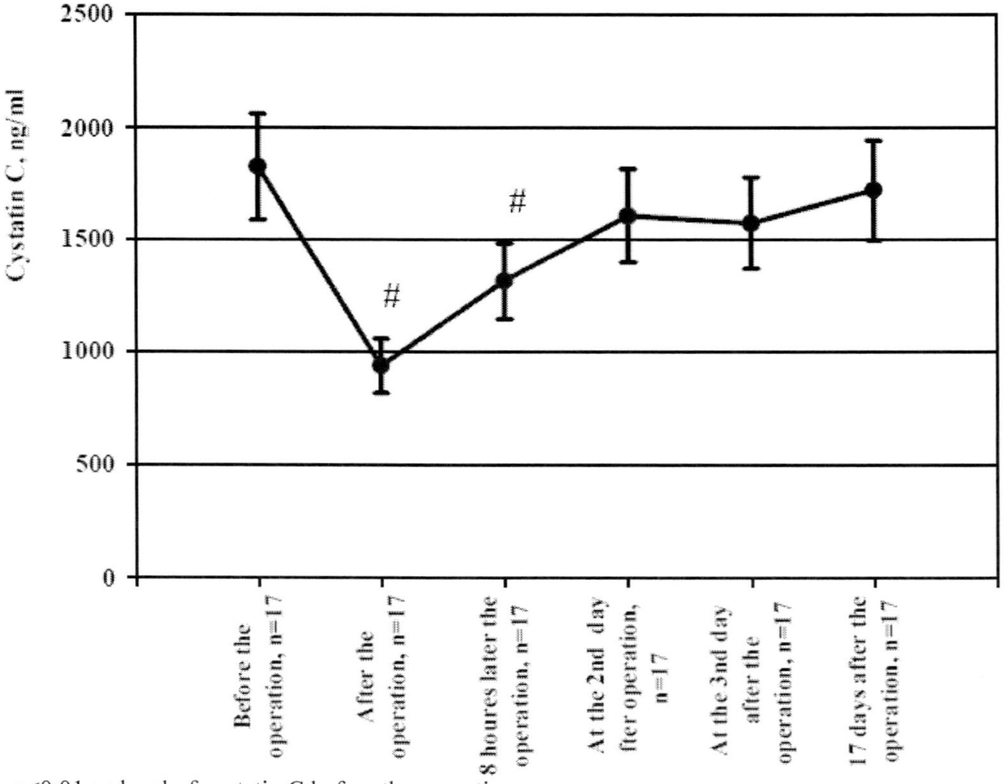

- p<0,01 vs level of cystatin C before the operation.

Figure 2. Serum cystatin C concentration in patients after cardio bypass surgery.

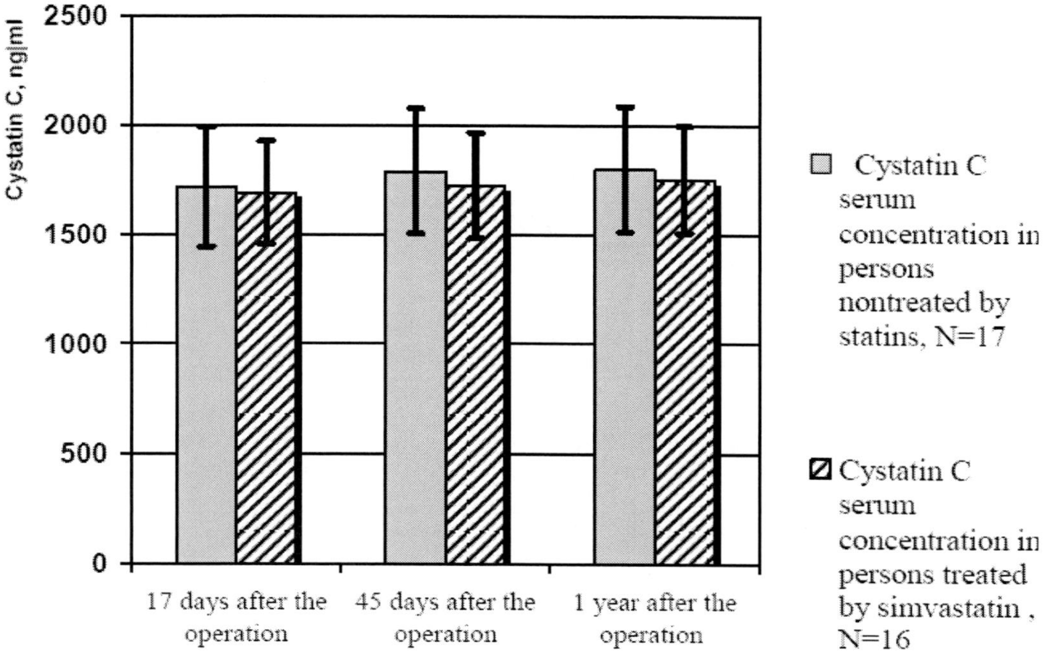

Figure 3. The level of cystatin C in persons with ischemic heart disease treated by simvastatin in postoperative state.

Cardio Bypass Surgery and Cystatin C

Cardio bypass surgery was followed by a drastic elevation in the hs-CRP level (about 70 times) (Korolenko et al., 2008) at the first day (immediately after the operation), then at the 2nd and 3rd days after the coronary bypass operation, the hs-CRP concentration significantly decreased, but not to its level before the operation. In the early postoperative state, the level of cystatin C decreased (Figure 2), followed by its restoration up to the 2nd and 3rd days after the operation up to a level of cystatin C for this group of patients before the operation (being increased as compared to the cystatin C level in the control group). At discharge (17th day after the cardio surgery), the cystatin C concentration was still elevated, just as it was before the operation in this group (Figure 2).

At the same time, the early postoperative period was characterized by a drastic elevation in MMP activity, especially 8 hr after (increase about 12 times), followed by a rapid decrease at day 2, reaching its level before the operation during the period of 7-17 days after the surgery. Chitotriosidase activity was elevated constantly during this period after the operation (Korolenko et al., 2008). The cystatin C level was moderately elevated during the postoperative period and statins had no significant effect on this index (Figure 3, Table 4).

One month after the operation, and even one year after the operation, there was an increased level of hs-CRP, MMP, and chitotriosidase activity. Moreover, MMP activity was more elevated one year after the operation (Korolenko et al., 2008). Simvastatin treatment had some positive effect on hs-CRP and MMP activity in this group one year after the operation (but not on the cystatin C concentration) (Figure 3, Table 4). One can conclude that according to the results of this longitudinal study, the recurrence of symptoms of atherosclerosis development was demonstrated.

Table 3. Effect of simvastatin treatment on serum cystatin C, hs-CRP concentrations, chitotriosidase and MMP activity in patients with high risk of CVD

Groups, the number of patients	hs-CRP, mg/L	Chitotriosidase activity, nmol MUF/mL per hour	MMP activity, μmol MCA/L per hour	Cystatin C, ng/mL
Patients of high risk CVD, nontreated by statins, n=25	6.09±0.07	261.8 ± 14.8	88.5±4.16	1704.5±235.0
Patients of high risk CVD, treated by simvastatin, n=25	4.71±0.07*	399.3±8.35*	65.7±3,64*	1687.9±292,\.3

*- p<0,001 as compare to untreated by statins group.

Table 4. Serum cystatin C concentrations in persons with ischemic heart disease treated by symvastatin in postoperative period

Groups, the number of persons	*Cystatin C,* ng|ml
1. Ischemic heart disease (IHD), before the operation, n=32	1824,7±256,9
2. IHD, immediately after the operation, n=32	933,5±100,8 P_{1-2} <0.05
3. IHD, 8 hrs after the operation, n=31	1312,2±181,3 P_{1-3} <0.05
4. IHD, 2nd day after operation, n=32	1572,2±293,5
5. IHD, 3rd day after operation, n=32	1605,0±244,3
6. IHD, 17 days after the operation, n=17	1719,8±201,7
7. IHD, 17 days after the operation, *treated by statins,* n=15	1692,6±209,9
8. IHD, 45 days after the operation, *n*=17	1792,5±268,8
9. IHD, 45 days after the operation, *treated by statins*, n=15	1726,9±241,6
10. IHD, 1 year after the operation, n=17	1801,2±306,1
11. IHD, 1 year after the operation, *treated by statins* n=15	1754,9±263,1

Changes of Cystatin C and Some Non-Lipid Markers and Predictors of Atherosclerosis

Atherosclerosis is closely related to inflammation (Anderson, 2005; Artieda et al., 2007). The initiation of atherogenesis involves activation of endothelial cells, monocyte infiltration into the vessels, followed by monocytes differentiation into macrophages, which accumulate lipids from the bloodstream and form lipid-laden macrophages. We have shown increased hs-CRP in elderly persons and, especially, in patients with atherosclerosis, as was documented in several previous investigations (Evrin et al., 2005). We have shown that an elevated serum level of cystatin C, MMPs and chitotriosidase had been observed in an elderly group, while lipid-lowering therapy by statins significantly decreased only the activity of MMPs and the hs-CRP level. So, elderly patients with arterial hypertension and atherosclerosis development were characterized by an elevated level of all indexes studied, which were related to inflammation. Atherosclerosis patients with IHD were also characterized by elevation of all these indexes. Moreover, MMP and chitotriosidase activity, as well as hs-CRP values, were higher than in group 2 (Tables 1, 2).

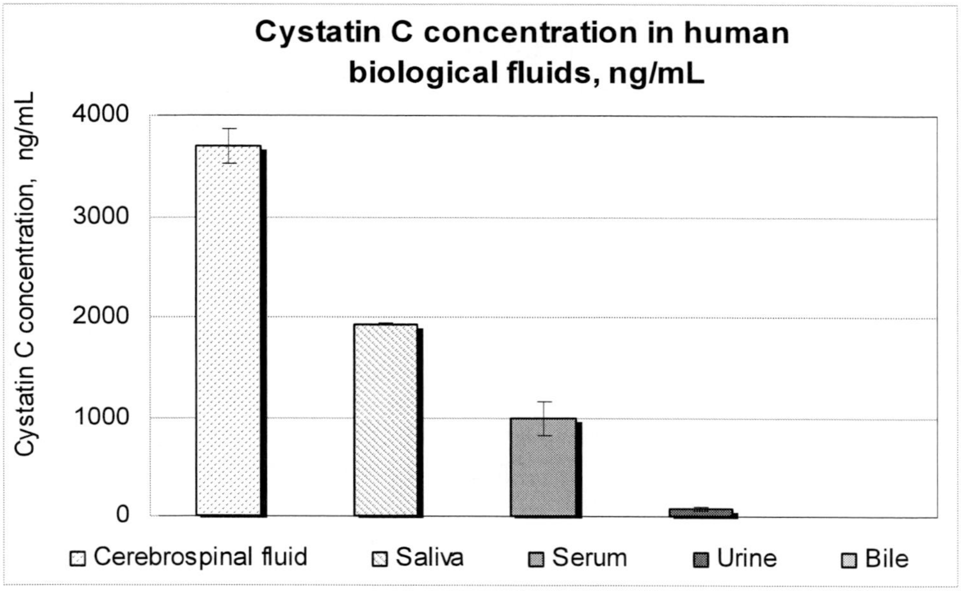

Figure 4.

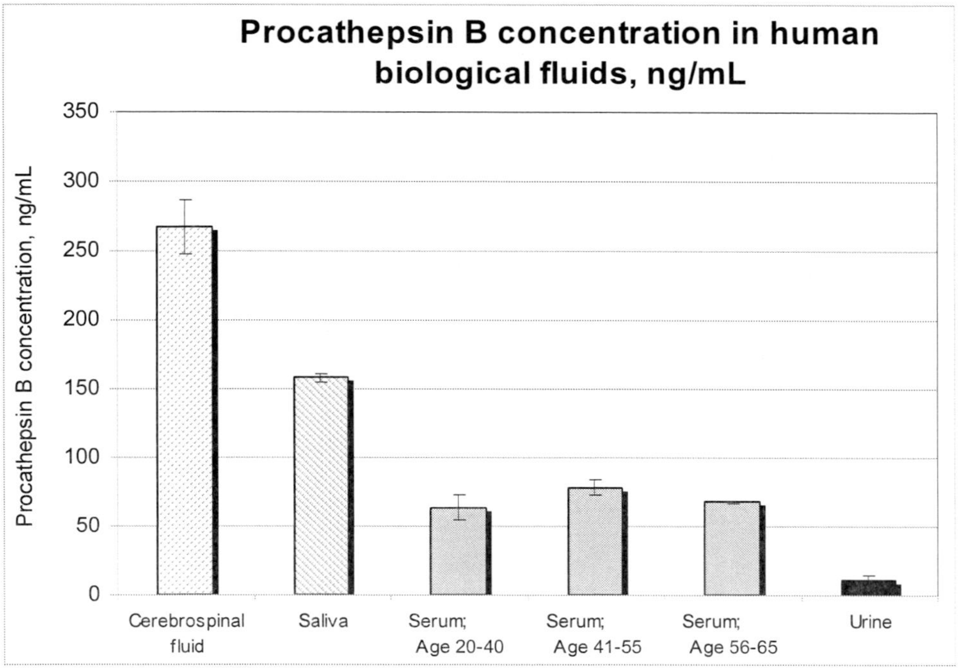

Figure 5.

All indexes studied in this work: namely, chitotriosidase, MMPs, as well as cystatin C are known to be secreted by activated macrophages (Nagase et al., 2006; Korolenko et al., 2008). Macrophages are key players in the pathogenesis of atherosclerosis, beginning from the onset of early symptoms of atherosclerosis, and also in the final stages, and are connected with plaque formation and disruption (Anderson, 2005; Brown and Bittner, 2008; Frendeus et al., 2009). Close interaction between macrophages and ECM was modified in atherosclerosis during formation of lipid-laden macrophages and their function was changed during lipid overloading (Tuomainen et al., 2007).

MMPs, catalytically active enzymes against ECM components, are essential for cellular migration and tissue remodeling (Nagase et al., 2006). *In vitro* expression of MMP-1, -3, and –9 in macrophages was enhanced by pro-inflammatory mediators and secreted from T-lymphocytes and monocytes (like TNF-α and IL-1β) (Turk et al., 2002). The activity of MMPs is essential for atherosclerotic plaque formation, including infiltration of inflammatory cells (macrophages). According to data obtained, serum MMP activity increased steadily during atherosclerosis development, reflecting macrophage stimulation and secretion *in vivo*. The positive correlation between MMP activity and hs-CRP confirms this suggestion.

Cystatins were shown to be very potent inhibitors of cysteine proteases of the papain superfamily. They form equimolar, tight, and reversible complexes with cysteine proteases (cathepsins L, S and in less degree B and H). Cystatin C, localized predominantly in the extracellular space, in the biological fluids of patients has been used as a diagnostic index in several inflammatory and tumor diseases (Korolenko et al., 2008; Frendeus et al., 2009). Cystatin C concentrations in serum were shown to increase both in elderly and atherosclerosis with and without IHD, and statin treatment had no effect on this index.

Cystatin C, also a serum measure of renal function, has been reported as a strong predictor of risk of death and cardiovascular events in elderly people. An elevated cystatin C level was independently associated with both ischemic and hemorrhagic stroke, and cystatin C was a strong predictor for the risk of cardiovascular events and death. At the same time, low levels of serum cystatin C precedes clinically manifested Alzheimer disease in elderly men free of dementia at baseline, and may be a marker of future risk of Alzheimer's disease (Lindholt et al., 2001; Larsson et al., 2005;Ni et al., 2007; Mares et al., 2009).

The cellular origin of cystatin C, MMP and chitotriosidase in serum are still not clear. Activated lipid-laden macrophages are thought to be the main source of increased chitotriosidase activity in serum, as was shown in Gaucher disease, an inborn lysosomal storage disease, followed by the formation of large lipid-loaded macrophages (Malaguarnera, 2006) which secrete chitotriosidase. Taken together, the results provide evidence that *in vivo* macrophage secretion of chitotriosidase, different proteases and their inhibitors, contribute to atherogenesis at an early stage of disease as a result of macrophage loading by lipids, followed by a change of balance between pro-atherogenic and anti-atherogenic plasma activity factors.

Cystatin C, Impairment in Atherosclerosis and Cardiovascular Events

Cystatins are competitive inhibitors of cysteine proteases, which bind to the active site of these enzymes, forming tight complexes. Cystatins refer to the so-called "emergency" inhibitors (Turk et al., 2002) whose main physiological function is believed to protect the host from the harmful effects of free cysteine proteases by rapidly trapping these enzymes in complexes. Cystatins are localized mainly in the extracellular space, separately from cysteine proteases (which are localized inside of lysosomes). At the same time "regulatory" inhibitor, co-localized in the same cellular compartment with active enzymes may include enzymatically inactive proenzymes (like procathepsin B etc.), which are able to regulate cysteine proteases more precisely (Turk et al., 2002).

Cystatin C in Human Biological Fluids

Cystatin C is preferentially localized outside of cells and has been found in all human biological body fluids, and this was confirmed in our study (Figure 4). The highest concentration of cystatin C is detected in cerebrospinal fluid, lower concentration is in blood serum and a significantly low concentration is detected in urine and bile (Figure 4). Cystatin C also resides in other biological body fluids like seminal fluid (semen), tears, oral liquid (saliva), breast milk (Seronie-Vivien et al., 2008). According to our results, a similar distribution of cystatin C was noted for the inactive proform of cathepsin B - procathepsin B was present in the same biological fluids studied (Figure 5). Procathepsin B, as is known to date, possesses inhibitory activity for cysteine protease (cathepsin B) (Turk et al., 2002). There were no changes of procathepsin B level connected with age (Figure 5).

During recent years the level of serum cystatin C is used for estimation of glomerular filtration (and this test has become more popular than traditional determination of serum creatinine). In normally functioning mammals (including humans), small molecular mass proteins are totally reabsorbed and subjected to lysosomal degradation in the proximal tubule cells of the kidneys.

Since cystatin C is constantly synthesized in cells, its determination in serum may be considered as the most optimal for evaluation of glomerular filtration rate in routine laboratory practice. Besides an ELISA method, cystatin C may be determined in human biological body fluids by other rapid and less expensive immune turbidimetric and immune nephelometric methods, which can determine cystatin C level in many samples. It should be noted that the use of ELISA methods provides high specificity, accuracy, and reliability of determinations.

The cellular origin of cystatin C in biological fluids studied is still not clear. It was shown that different types of cells (macrophages, monocytes, brain cells, and cell lines) continually secrete cystatin C (Warfel et al., 1987; Turk et al., 2002). The degradation of cystatin C by tubular cells of the kidney (inside of lysosomes) was suggested as one of the important routes of regulation of serum levels of cystatin C. Down-regulation was noted if cells were incubated with LPS, so immunomodulatory activity was suggested to relate to cystatin C (Turk et al., 2002; Keppler, 2006).

Cystatin C in Aging and Effect of Statins

We have found that the serum cystatin C concentration in healthy volunteers (20–45 year old) is 991.1±163.1 ng/ml; this is consistent with the literature (570–1790 ng/ml) (Shlipak et al., 2006; Ognibene et al., 2006). The cystatin C concentration in cerebrospinal fluid from patients with non-inflammatory CNS diseases is about four times higher than in serum (Figure 4). Cystatin C concentration in urine and bile was the lowest (compared to CSF, saliva and serum) (Figure 4). Recently, in some reports, changes in urine cystatin C was used as an early biomarker of acute kidney injury in humans (Uchida and Gotoh, 2002; Koyner et al., 2008).

We have found that the serum cystatin C concentration in patients 50–65 years old, without clinical manifestation of IHD, is higher than in basically healthy people of a younger age (20–40 year old). This is consistent with some other studies (Ericsson et al., 2004; Loew et al., 2005; Weinert et al., 2010) and may be considered as early manifestation of risk for the development of cardiovascular diseases (Table 2). On the contrary, some authors suggested that plasma cystatin C is not associated with a risk of systemic atherosclerosis development (Albert et al., 2001). During all time intervals after coronary bypass operation (except the immediate period after operation), cystatin C concentration was higher than in healthy individuals of younger age.

Human atherosclerotic lesions are accompanied by increased expression of elastolytic proteases, cathepsins S and K in the plaque, and in sites of arterial elastin damage, there is a reduced level of endogenous inhibitors (first of all cystatin C). This results in increased degradation of the arterial wall (Helske et al., 2006; Goncalves et al., 2008). Experimental data supporting the hypothesis on cystatin C deficit at sites of vascular damage formation were obtained in experiments employing cystatin C-deficient mice, which were crossbred with apolipoprotein E-deficient mice. Maintenance of these animals on an atherogenic diet caused fragmentation of elastic lamina of the vascular wall and dilatation of thoracic and abdominal aorta. This demonstrates the importance of an impaired ratio of cysteine proteases and their inhibitors for damage of arterial wall structure and extracellular matrix remodeling in experimental atherosclerosis.

Cystatin C is the most effective inhibitor of cathepsins S, K, and L, which also possess elastase and collagenase activities (Turk et al., 2002; Lutgens et al., 2007). At sites of the atherosclerotic plaque formation and aortic aneurysm, there was a sharp increase in expression of cathepsins S and K, and a marked decrease of the protease inhibitor expression in macrophages and smooth muscle cells (as well as modest change in tissue inhibitor of matrix metalloprotease type 1, TIMP-1). Mice deficient in cystatin C did not exhibit clear impairments in TIMP-1 expression, and the most important vascular changes occurred in the case of double knockout mice deficient in cystatin C and ApoE during their maintenance on an atherogenic diet (Bengtsson et al., 2005).

These results indicate that impairments in cystatin C expression may be important for atherosclerosis formation. Altered concentrations of cystatin C in various human biological body fluids (besides serum) are also associated with nervous system diseases (ischemic stroke, Creutzfeldt-Jakob disease), the vision organs, and other diseases, in which diagnostic and prognostic value of cystatin C is now being investigated (Mares et al., 2009).

Compared to human serum, a low cystatin C level was observed in urine in our study and also noted in the literature (Hellerstein et al., 2004), possibly as a result of degradation of this protein. The ratio of urinary cystatin C to urinary creatinine was suggested as a more efficient index for detecting decreased GFR in children with kidney pathology (Hellerstein et al., 2004).

In cerebrospinal fluid of healthy persons, the high cystatin C concentration was shown (3-12.5 mg/L vs serum cystatin C 0.63-1.44 mg/L, estimated by BioVendor ELISA kits, Czechia), however the diagnostic role in neurodegenerative diseases and Alzheimer disease is not evident (Mares et al., 2009).

In older patients (45-55 year old) and patients with atherosclerosis before cardio surgery, the serum cystatin C concentration is higher than in the control group, possibly because of the extracellular secretion of cystatin C (as result of macrophage stimulation). Moreover, the cystatin C level can predict progression of subclinical coronary atherosclerosis in individuals with type 1 diabetes (Maahs et al., 2007) and predict early renal function changes in patients after cardiac surgery (Artunc et al., 2005; Heise et al., 2009). It was suggested that the postsurgical inflammatory response was not directly related to increased serum cystatin C levels (Akerfeldt et al., 2010).

Kidney dysfunction increases the risk of death in cardiovascular disease. Several studies have found that increased levels of cystatin C are associated with the risk of death, cardiovascular diseases (including myocardial infarction, stroke, heart failure, peripheral arterial diseases and metabolic syndrome), and healthy aging (Lindholt et al., 2001). Therefore, further investigation is necessary in this field.

Cystatin C levels are decreased in atherosclerotic (so-called 'hardening' of the arteries) and aneurysmal lesions of the aorta (Mares et al., 2009). Genetic and prognostic studies also suggest a role for cystatin C in the breakdown of parts of the vessel wall. These changes are thought to result from an imbalance between different types of proteases, which tend to be increased (cysteine proteases and matrix metalloproteases) and their inhibitors, which tend to be decreased (cystatin C, TIMPs). Protease inhibitors are generally regarded as atheroprotective, because proteases participate in matrix degradation, a process regarded primarily as atherogenic. Cystatin C most likely has its origin in different cell types present in aortic lesions, such as SMCs, endothelial cells, and macrophages, which are known to produce cystatin C. Further studies are required in the investigation of the role of protease inhibitors in atherosclerosis.

References

Akerfeldt T., Helmersson J., Larsson A. Postsurgical inflammatory response is not associated with increased serum cystatin C values. *Clin. Biochem.* 2010; Jul 17 (Epub ahead of print).

Albert M.A., Rifai N., Ridker P.M. Plasma levels of cystatin-C and mannose binding protein are not associated with risk of developing systemic atherosclerosis. *Vascular Medicine,* 2001; 6:145-149

Anderson L. Candidate-based proteomics in the search for biomarkers of cardiovascular disease. *J. Physiol.* 2005; 563(Pt 1): 23 – 60.

Artieda M, Cenarro A, Ganan A.,Lukic A, Moreno E, Puzo J. et al. Serum chitotriosidase activity, a marker of activated macrophages, predicts new cardiovascular events independently of C-reactive protein. *Cardiology* 2007; 108(4): 297-306.

Artunc F.H., Fisher I.U., Risler T., Erley C.M. Improved estimation of CFR by serum cystatin C in patients with undergoing cardiac catheterization. *Int. J. Cardiol.* 2005: 102(2): 173-178.

Bengtsson E., To F., Hakansson K., Grubb A., Branen L., Nilsson J., Jovinge S. Lack of the cysteine protease inhibitor cystatin C promotes atherosclerosis in apolipoprotein E-deficient mice. *Arterioscler. Thromb.Vasc. Biol.* 2005; 25(10):2151-2156.

Bokenkamp A., Herget-Rosenthal S., Bokenkamp R. Cystatin C, kidney function and cardiovascular disease. *Pediatr Nephrol.* 2006; 21(9): 1223–1230.

Brown T.M., Bittner V. Biomarkers of atherosclerosis: clinical applications. *Curr. Cardiol. Rep.* 2008; 10(6): 497-504

Chew J.S.C, Saleem M., Florkowski Ch.M., George P.M. Cystatin C – a paradigm of evidence based laboratory medicine. *Clin. Biochem. Rev.* 2008; 29: 47-62

Ericsson P., Deguchi H., Samnegard A., Lundman P., Boquist S., Tornvall P., Ericsson C. -G., Bergstrand L., Hansson L.-O., Ye S., Hamsen A. Human evidence that the cystatin C gene is implicated in focal progression of coronary artery disease. *Arterioscler. Thromb. Vasc. Biol.* 2004; 24(3): 551-557.

Evrin P.E., Nilsson S.E., Oberg T,, Malmberg B. Serum C-reactive protein in elderly men and women: association with mortality, morbidity and various biochemical values. *Scand. J. Lab. Invest.* 2005; 65(1): 23 – 31.

Frendeus K.H., Wallin H., Janciauskiene S., Abrahamson M. Macrophage response to interferone-gamma are dependent on cystatin C levels. *Int. J. Biochem. Cell Biol.* 2009; 41(11): 2262-2269.

Goncalves I., Ares M., Moberg A., Moses J., To F., Montan J., Pedro L.M., Dias N., Fernandes J.F., Fredrikson G.N., Nilsson J., Jovinge S., Bengtsson E. Elastin- and collagen-rich human carotid plaques have increased levels of the cysteine protease inhibitor cystatin C. *J. Vasc. Res.* 2008; 45:395-401.

Heise D., Waeschle R.M., Schlobohm J., Wessels J., Quintel M. Utility of cystatin C for assessment of renal function after cardiac surgery. *Nephron Clin. Pract.* 2009;112:107–114

Hellerstein S., Berenbom M., Erwin P., Wilson N., DiMagio S. The ratio of urinary cystatin C to urinary creatinine for detecting decreased CFR. *Pediatr. Nephrol.* 2004; 19 (5): 521-525. PMID: 15015062

Helske S., Syvaranta S., Linstedt K.A., Lapalainen J., Oorni K., Mayranpaa M.I., Lommi J., Turto H., Werkkala K., Kupari M., Kovanen P.T. Increased expression of elastolytic cathepsins S, K, and V and their inhibitor cystatin C in stenotic aortic valves. *Arterioscler. Thromb. Vasc. Biol.* 2006; 26(8): 1791-1798.

Keppler D. Towards novel anti-cancer strategies based on cystatin functions. *Cancer Lett.* 2006; 235(2): 159-176

Knight C.G., Willenbrock F., Murphy G. A novel coumarin-labelled peptide for sensitive continuous assays of the matrix metalloproteinases. *FEBS Lett.* 1992; 296(3): 263-266. [PubMed: 1537400]

Korolenko T.A., Filatova T.G., Cherkanova M.S,, Khalikova T.A., Bravve I.Yu. Cystatins: cysteine proteases regulation and disturbances in tumors and inflammation. Biochemistry (Moscow) Suppl. Series B. *Biomedical Chemistry.* 2008; 2(3): 293-297.

Koyner J.L., Bennett M.R., Worchester E.M., Ma Q., Raman J., Jeevanandam V., Kasza K.E., O'Connor M.F., Konczal D.J., Trevino S., Devarajan P., Murray P.T. Urinary cystatin C as an early biomarker of acute kidney injury following adult cardiothoracic surgery. *Kidney International* 2008; 74:1059-1069.

Kurt I, Abasli D, Cihan M, Serdar MA, Olgun A, Saruhan et al. Chitotriosidase levels in healthy elderly subjects. *Ann. N. Y. Acad. of Sci.* 2007; 1100: 185-188

Larsson A., Helmersson J., Hansson L.-O., Basu S. Increased serum cystatin C is associated with increased mortality in elderly men. *Scand. J. Clin. Lab. Invest.* 2005; 65:301-306.

Lindholt J.S., Erlandsen E.J., Henneberg E.W. Cystatin C deficiency is associated with the progression of small abdominal aortic aneurisms. *British Journal of Surgery.* 2001; 88:1472-1475.

Loew M., Hoffman M.M., Koenig W., Brenner H., Rothenbacher D. Genotype and plasma concentration of cystatin C in patients with coronary heart disease and risk for secondary cardiovascular events. *Arterioscler. Thromb.Vasc.Biol.* 2005; 25:1270-1474.

Lutgens S.P.M., Cleutjens K.B.J.M., Daemen M.J.A.P., Heeneman S. Cathepsin cysteine proteases in cardiovascular disease. *FASEB J.* 2007; 21:3029-3041.

Maahs D.M., Ogden L.G., Kretowski A., Snell-Bergeon J.K., Kinney G.L., Berl T., Rewers M. Serum cystatin C predicts progression of subclinical coronary atherosclerosis in individuals with type 1 diabetes. *Diabetes* 2007; 56:2774-2779.

Malaguarnera L. Chitotriosidase: the yin and yang. *Cell Mol. Life Sci.* 2006; 63(24): 3018-29.

Mares J., Kanovsky P., Herzig R., Vavrouskova J., Hlustik P., Vranova S., Burval S., Zapletalova J., Pidrman V., Obereigneru R., Suchy A., Vesely J., Podivinsky J., Urbanek K. The assessment of beta amyloid, tau protein and cystatin C in cerebrospinal fluid: laboratory markers of neurodegenerative diseases. *Neurol. Sci.*, 2009:30 (1):301-307.

Mussap M., Plebani M. Biochemistry and clinical role of human cystatin C. Critical Reviews in Clinical Laboratory Sciences 2004; 41(5-6): 467-550.

Nagase H, Fields CG, Fields GB. Design and characterization of a fluorogenic substrate selectively hydrolyzed by stromelysin 1 (Matrix Metalloproteinase-3). *J. Biol. Chem.* 1994; 269(33): 20952-20957

Nagase H, Visse R, Murphy G. Structure and function of matrix metalloproteinases and TIMPs. *Cardiovasc. Res.* 2006; 69(3): 562-573.

Naruse H., Ishii J., Kawai T., Hattori K., Ishikawa M., Okumura M., Kan S., Nakano T., Matsui S., Nomura M., Hishida H., Ozaki Y. Cystatin C in acute heart failure without advanced renal impairment. *Am. J. Med.* 2009; 122(6): 566-573.

Ni L., Lu J., Hou L.B., Yan J.T., Fan Q., Hui R., Cianflone K., Wang D.W. Cystatin C, associated with hemorrhagic and ischemic stroke, is a strong predictor of the risk of cardiovascular events and death in Chinese. *Stroke*; 2007:38:5387-3288.

Ognibene A., Mennucci E., Caldini A., Terreni A., Brogi M., Bardini G., Sposato I., Mosconi V., Salvadori B., Rotella C.M., Messeri G. Cystatin C reference values and aging. *Clin. Biochem.* 2006; 39(6): 658-661.

Seronie-Vivien S., Delanaye P., Pieroni L., Mariat Ch., Froissart M., Cristol J.P. Cystatin C: current position and future prospects. *Clin. Chem. Lab. Med.* 2008; 46(12): 1664-1686.

Shlipak M.G., Katz R., Sarnak M.J., Fried L.F., Newman A.B., Stehman-Breen C., Seliger S.L. Kestenbaum B., Psaty B., Tracy R.P. Siskovick D.S. Cystatin C and prognosis for cardiovascular and kidney outcomes in elderly persons without chronic kidney disease. *Annals of Internal Medicine* 2006; 145:237-246.

Tuomainen AM, Nyyssonen K, Laukkanen JA, Tervahartiala T, Tuomainen TP, Salonen JT et al. Serum matrix metalloproteinase-8 concentrations are associated with cardiovascular outcome in men. *Arterioscler. Thromb Vasc. Biol.* 2007 Dec; 27(12): 2722-2728

Turk B., Turk D., Salvesen G.S. Regulating cysteine protease activity: essential role of protease inhibitors as guardians and regulators. *Current Pharmaceutical Design* 2002; 8:1623-1637.

Uchida K., Gotoh A. Measurement of cystatin-C and creatinine in urine. *Clin. Chim. Acta.* 2002; 323(1-2): 121-128.

Warfel A.H., Zucker-Franklin D., Frangione B., Ghiso J. Constitutive secretion of cystatin C (χ-trace) by monocytes and macrophages and its downregulation after stimulation. *J. Exp. Med.* 1987; 166(12): 1912-1917.

Weinert L.S., Prates A.B., do Amaral F.B., Vaccaro M.Z., Camargo J.L., Silveiro S.P. Gender does not influence cystatin C concentrations in healthy volunteers. *Clin. Chem. Lab. Med.* 2010; 48(3): 405-408.

Yang Z, Strickland DK, Bornstein P. Extracellular MMP–2 levels are regulated by the low–density lipoprotein–related scavenger receptor and thrombospondin 2. *J. Biol. Chem.* 2001; 276: 8403 –8408.

In: Cystatins: Protease Inhibitors …
Editors: John B. Cohen and Linda P. Ryseck
ISBN: 978-1-61209-343-7

Chapter 9

Cystatin C as a GFR Marker in Renal Transplantation: Promises and Challenges

Ingrid Masson, Pierre Delanaye and Christophe Mariat
Service de Néphrologie, CHU de St-Etienne, St-Etienne, France
Service de Néphrologie, CHU Sart Tilman, Liège, Belgique

1. Introduction

Glomerular filtration rate (GFR) is a key parameter to evaluate the function and thereby the quality of the transplanted kidney. Direct measures of the renal elimination of different exogenous GFR (e.g. inulin clearance) are the "gold standard" for assessing GFR. These techniques are however rarely implemented in routine clinical practice. As an alternative, a number of easy-to-use mathematical equations, incorporating different anthropometrical variables in addition to biological parameters, have been developed to predict ('estimated GFR'), rather than to directly measure GFR ('true GFR').

International guidelines recommended relying on serum creatinine for GFR estimation (1). It has become, however, increasingly evident that serum creatinine, alone or even incorporated into estimating equations, is not an ideal marker of the renal graft function. As a result, interest has arisen regarding alternative endogenous marker. Among them, cystatine C tends to be regarded as a better marker of GFR than serum creatinine in a variety of different patient populations, including renal transplant patients.

In this chapter, we will first review the main clinical studies that have questioned the relevance of serum-creatinine based GFR estimates in renal transplantation. We will then present the existing evidence favoring serum cystatine C over serum creatinine as a GFR marker in this context. Finally, we will discuss the different challenges that still have to be addressed in order to definitely legitimate a widespread use of cystatine C as a routine index of renal graft function.

2. Serum Creatinine-Based Equations in Renal Transplantation

Measurement of serum creatinine concentration is a simple, inexpensive and universally available method for estimating GFR. Serum creatinine alone has however limited utility. Studies utilizing surveillance biopsies suggest that changes in serum creatinine may be insensitive to detect acute or chronic allograft injury. For example, up to 30% of acute rejection episodes were found to occur with no apparent change in serum creatinine (2). In this regard, prediction of GFR from serum creatinine is preferable. Many equations have been developed for this purpose. Several of them have been directly tested in renal transplant patients, and three of these equations are currently used in this context, either because they show a better predictive performance when compared to the others or alternately just because they have been consecrated by a more or less long usage. These three equations are:

- The Cockcroft–Gault formula (3):

$[(140\text{-age}_{\text{(years)}} \text{ x weight}_{\text{(kg)}} / (0.814 \text{ serum creatinine}_{(\mu\text{mol/l})})](\text{x } 0.85, \text{ for women})$

Cockcroft and Gault first published their equation in 1976, which since, has been the most widely used GFR test in the clinical as well as in the research setting.

- The Nankivell equation (4):

$6.7/ (\text{serum creatinine}_{\text{(mmol/l)}})^{-1} + 0.25 \text{ X weight}_{\text{(kg)}} - 0.5 \text{ x urea}_{\text{(mmol/l)}} - 100/ \text{height}_{\text{(m)2}} + 35$ (25 for women).

The Nankivell formula is the only one that has been computed from a renal transplant population (against a direct measure of GFR using plasma ^{99m}Tc DTPA clearance). For this reason this equation has always been seen as more appropriate than any other for assessing renal graft function and thus has been deemed valid for clinical research. External validation of this equation was actually evaluated long after it has been used as the preferred method to estimate GFR in clinical trials. The superiority of this equation has since been questioned.

- The MDRD study equations (5):

Abbreviated Equation: $175 \times (\text{serum creatinine}_{\text{(mg/dl)}})^{-1.154} \times (\text{age}_{\text{(year)}})^{-0.203} \times (0.742$ if patients is female) $\times$ (1.21 if patients is black).

Levey and colleagues derived different predictive equations from the 1628 patients included in the modification of diet in renal disease (MDRD) study. Since their publication in 1999, the MDRD equations have been presented as a somewhat new standard in GFR prediction. Importantly, the MDRD equation relies on the use of a reference traceable serum creatinine and is thus less sensitive to analytical errors and more reproducible.

Over the last decade, data have accumulated pointing to some serious limitations of these GFR-estimating equations in giving an accurate evaluation of renal graft function.

Gera et al. compared "true" GFR with eGFR (MDRD and Cockcroft-Gault) to measure kidney graft function at different time points after transplantation and to assess changes in graft function over time (6). They analyzed the slope of eGFR as compared to that of measured GFR (cold iothalamate) in 360 kidney transplant recipients followed for at least 3 years. First, the relationship between GFR end eGFR was found to be weaker in transplant recipients than in patients with chronic kidney disease; secondly, there was a significant variability in bias, precision and accuracy of eGFR at different time points after transplantation. Finally, changes in eGFR over time (slope) were found to significantly underestimate the number of patients losing graft function. Among patients losing GFR at a rate faster than -1ml/min/1.73 m^2/year, only 50% were correctly identified by the MDRD slope as patients losing graft function.

We later confirmed that overall the predictive performances of creatinine-based equations were indeed particularly poor in transplant patients (7).

The performance of the MDRD study, Cockcroft-Gault, and Nankivell equations in kidney transplant recipients has also been evaluated in a systematic review performed by White et al. (8). Marked heterogeneity was found between studies in terms of equations bias and accuracy. All studies showed quite poor precision leading to quite modest accuracies. The pooled estimate of the 30% accuracy was only 73%, 76% and 68% for the Cockcroft-Gault, 4-variable MDRD study, and Nankivell equations, respectively. The mean bias of the Cockcroft-Gault equation ranged from –4 to 16 ml/min/1.73 m2 and that of the 4-variable MDRD equation from -11.4 to 9.2 ml/min/1.73m2. The Nankivell equation generally significantly overestimated GFR with a mean bias from -1.4 to 36 ml/min/1.73 m2.

In addition, the performance of these estimates is not constant throughout the whole spectrum of GFR, and instead, decreases with decreasing GFR (9,10).

A consistent finding in non transplant patients is the systematic underestimation of GFR by the MDRD equation in patients with relatively well-preserved kidney function. The new Chronic Kidney Disease Epidemiology Collaboration (CKD-EPI) equation proposed recently by Level and colleagues introduces a "correction" or spline term for patients with low creatinine values (11). White et al. examined the performance of this equation in kidney transplantation (12). The percentage of estimated GFR within 30% of the measured GFR was significantly higher with the CKD-EPI equation than with the 4-variable MDRD study equation (84% vs 77%), but not compared with the re-expressed MDRD Study equation (84% vs 79%). As expected, in the lower-GFR subgroup (GFR<60ml/min/1.73 m2 measured with 99mTc-DTPA clearance) the new CKD-EPI equation was not better than the 4-vatiable MDRD equation (79% vs 78%).

In brief, a multitude of data exists pointing to the lack of accuracy of creatinine-based estimates in renal transplantation. So, why GFR prediction from serum creatinine is so challenging in transplantation?

Many factors specific to transplant recipients may impact on the predictive performance of these equations. Variables such as the number of acute rejections, improve nutritional status after kidney transplantation, length of time spent on dialysis or cumulative steroid dose have been shown to be predictors of muscle mass index in transplant recipients, independent of body weight and other parameters usually included in the GFR estimates.

In the past, issues of creatinine assay calibration have also complicated the validity and reliability of creatinine-based GFR estimation equations (13,14).

In addition, the process of tubular secretion of creatinine might be different in transplant patients as compared to patients with native kidneys and might account at least partially for the difficulty to predicting GFR in this population.

3. Serum Cystatin C: A More Accurate GFR Marker in Renal Transplantation

In a recent study, Hall et al. showed that serum cystatine outperforms serum creatinine in differentiating between levels of early graft function after kidney transplantation (15). They have shown that absolute values of cystatine C during the first postoperative day correspond with the graft function (delayed or immediate) after transplantation, whereas serum creatinine values at these times poorly differentiate between groups. Furthermore, in their study serum cystatine C at the time of transplantation also demonstrates excellent potential for predicting graft function up to 3 months after transplant. Rule et al. showed that there was a better relationship between cystatine C and measured GFR than between serum creatinine and measured GFR (iothalamate clearance) for transplant recipients (16). One of the purported advantages of this marker is thought to be its relative independence from clinical parameters known to affect serum creatinine level, such as age, race, and body composition.

However, it is the cystatine C-based GFR, rather than the Cystatin C concentration by itself that is of greater interest to the clinician. In this regards, there have been several cystatine C prediction equations published (table 1). It should be noted that only two of these equations (Rule and Le Bricon) were derived exclusively from a population of kidney transplant recipients. Fillers cohort comprised children aged 1-18 years having a significant hyperfiltration with a mean GFR above 100ml/min/1.73m^{2}.

Several studies have compared the cystatine C-based GFR estimates to the standard creatinine-based GFR equations.

Pöge and al. tested the diagnostic performance of four methods of GFR estimation in kidney transplant recipients (MDRD, Larsson, Hoek and Filler formulas) against the gold standard-derived GFR (DTPA-clearance) (28). They demonstrate a superiority of those equations over the MDRD equation in terms of accuracy (the accuracy within 30% is 77% for Hoek and Larsson formulas and 67% for MDRD equation).

White et al. compared the accuracy, precision and bias of several cystatin C equations in 117 stable kidney transplant recipients (29). They found that the equations of Filler and Le Bricon had the best overall performance. In this study, the estimated GFR was compared with radio-isotopic GFR reference measurement (^{99m}Tc-DTPA clearance).

More recently, we compared cystatine-C based GFR equations to the MDRD equation in 120 kidney transplant patients (30). The 30% accuracy for the MDRD equation was only 58% compared to 82% and 81% for the Hoek and Rule equations, respectively.

In contrast to these 3 favorable studies on cystatine C, Zahran et al. did not show any benefit of cystatine C –based GFR equations in the kidney transplant population (31). They evaluated nine cystatin C equations, and the 30% accuracy ranged from only 43% to 59%. However, in this study, cystatine C was measured using an enzyme-linked immunosorbent assay methodology, which was not used in the development phase for any of the estimating equations. This is a major limitation to the results of this study.

Table 1. GFR predicting equations based on Cystatine C alone or in combination with creatinine

Reference	Sample, n	GFR measurement	Cystatin C	Population	Equations
Bokenkamp et al.(17)	83	Inulin	PETIA	Paediatrics	(162/CC)–30
Tan et al.(18)	40	Iohexol	PENIA	Diabetics and health	(87.1/CC)–6.87
Hoek et al(19)	47	Iothalamate	PENIA	Various	(80.35/CC)–4.32
Larsson et al. (20)	100	Iohexol	PENIA	Various	$77.24xCC^{-1.2623}$
			PETIA		$99.43xCC^{-1.583}$
Filler et al.(21)	536	^{99}Tc-DTPA	PENIA	Paediatrics	$91.62x(1/CC)^{1.123}$
Le Bricon et al.(22)	25	^{51}Cr-EDTA	PENIA	Tx	[(78x(1/CC)]+4
Sjostrom et al.(23)	381	Iohexol	PETIA	Various	(124/CC)–22.3
Grubb et al.(24)	536	Iohexol	PETIA	Various + paediatrics (n=85)	$84.69xCC^{-1.68}$x1.384 (if less than 14 years old)
Rule et al.(16)	204	Iothalamate	PENIA	Various excluding Tx	1) $66.8xCC^{-1.3}$ 2) $[(66.8xCC^{-1.3})x(273xSCr^{-1.22}xage^{-0.299}x0.738 \text{ if female})]^{0.5}$
Rule et al(16).	206			Tx	$76.6xCC^{-1.16}$
MacIsaac et al.(25)	125	^{99}Tc-DTPA	PENIA	Diabetics	(84.6/CC)–3.2
Bouvet et al(26).	67	^{51}Cr-EDTA	PENIA	Paediatrics	$63.2x(SCr/96)^{-0.35}x(CC/1.2)^{-0.56}x(weight/45)^{0.3}x(age/14)^{0.4}$
Stevens et al.(27)	3418	Iothalamate	PENIA	Chronic Kidney disease	1) $76.6xCC^{-1.19}$ 2) $127.7x(CC)^{-1.17}x(age)^{-0.13}$x(0.91 if female)x(1.06 if black) 3) $177.6x(SCr)^{-0.65}x(CC)^{-0.57}x(age)^{-0.20}$x(0.82 if female) x(1.11 if black)

(serum cystatine is expressed in mg/L, serum creatinine in mg/dL, age in years, weight in kg, CC, cystatine C; SCr, serum creatinine).

4. Cystatin C in Renal Transplantation: What's Next?

Several questions need to be further addressed before validating a wider utilization of cystatine C in renal transplantation:

What are the transplantation-related factors susceptible to significantly interfere with serum cystatine C values?

Rule reported a 19% higher GFR at the same cystatin C level among patients after renal transplantation in comparison to patients with native kidney disease, suggesting the intervention of factors specific to transplant patients (e.g. increase production of cystatine C secondary to inflammation, modification of cystatine C production by immunosuppressive drugs) (16,32). White et al. tried to determine the effect of several bio-clinical variables, including immunosuppressive medication, on serum Cystatin C concentration in a cohort of kidney transplant patients (33).They showed that cystatine C levels are essentially influenced by albumin level and sex. Other variables, such as prednisone dose (7.5mg/day), age, proteinuria, smoking, and specific immunosuppressive (mycophenolate mofetil, tacrolimus or cyclosporine) did not have a significant effect on cystatine C level. Stevens et al. found similarly that a lower serum albumin is associated with higher levels of Cystatin C (34). Additionally, in their study diabetes and C-reactive protein were associated with higher Cystatin C levels.

Which cystatin C equation is the best for renal transplant patients?

Among the different cystatin C equations, there are some discrepancies between studies regarding which one is likely to perform the best in renal transplantation. Similar to what was observed with creatinine, analytical problems including calibration in CysC measurement can have important consequences. Two admitted reference methods to measure serum cystatine C concentration exist, namely the particle-enhanced immunoturbidimetric assay (PETIA) and the immunonephelometric assay (PENIA). These 2 methods are however not strictly equivalent and it is unlikely that an equation constructed with serum cystatine measured by PENIA method can be used with cystatine C measured by PETIA. As the relationship between GFR and serum Cystatin C is exponential, the impact of the precision of the equations is, as for the MDRD equations, less with lower CysC values (20).

Are cystatine C-based equations good enough to replace direct measure of GFR?

Despite the improved GFR estimation by these Cyc-C based formulas, it should be noted that an ideal GFR equation as suggested by the National Kidney Foundation ought to cover 99% of all tests within 10% of true GFR. Such an ambitious objective is highly recommended in patients with renal grafts. However, from this point of view all tested equations are far from being ideal. Yet, by time and with increasing knowledge of Cys C-derived GFR equations in renal transplantation it may be possible to further improve the diagnostic performance of these equations. Nonetheless, we must bear in mind that currently, a gold standard clearance procedure cannot be replaced by these estimates.

Conclusion

Accurate knowledge of GFR is essential in the care of kidney transplant recipients. In particular, the ability to detect changes in GFR is critical to identify progressive graft dysfunction and then to turn to strategies susceptible to preventing further injury. Markers of kidney function, notably serum creatinine and creatinine-based estimates of GFR, are increasingly being used despite their inability to adequately detect change in GFR. Serum Cystatine C is now touted as a possible alternative to serum creatinine and numerous equations incorporating cystatine C has been developed from various populations to give a direct GFR estimate. The definitive implementation of these equations in renal transplant population is however currently premature as further research is needed to validate the superiority as well as the clinical utility of cystatin C used as a functional marker of renal graft.

REFERENCES

[1] KDIGO clinical practice guideline for the care of kidney transplant recipients. *Am. J. Transplant.* 2009 Nov;9 Suppl 3:S1-155.

[2] Rush DN, Henry SF, Jeffery JR, Schroeder TJ, Gough J. Histological findings in early routine biopsies of stable renal allograft recipients. *Transplantation.* 1994 Jan;57(2):208-211.

[3] Cockcroft DW, Gault MH. Prediction of creatinine clearance from serum creatinine. *Nephron.* 1976;16(1):31-41.

[4] Nankivell BJ, Gruenewald SM, Allen RD, Chapman JR. Predicting glomerular filtration rate after kidney transplantation. *Transplantation.* 1995 Juin 27;59(12):1683-1689.

[5] Levey AS, Bosch JP, Lewis JB, Greene T, Rogers N, Roth D. A more accurate method to estimate glomerular filtration rate from serum creatinine: a new prediction equation. Modification of Diet in Renal Disease Study Group. *Ann. Intern. Med.* 1999 Mar 16;130(6):461-470.

[6] Gera M, Slezak JM, Rule AD, Larson TS, Stegall MD, Cosio FG. Assessment of changes in kidney allograft function using creatinine-based estimates of glomerular filtration rate. *Am. J. Transplant.* 2007 Avr;7(4):880-887.

[7] Mariat C, Maillard N, Phayphet M, Thibaudin L, Laporte S, Alamartine E, et al. Estimated glomerular filtration rate as an end point in kidney transplant trial: where do we stand? *Nephrology Dialysis Transplantation.* 2008 Jan 1;23(1):33 -38.

[8] White CA, Huang D, Akbari A, Garland J, Knoll GA. Performance of creatinine-based estimates of GFR in kidney transplant recipients: a systematic review. *Am. J. Kidney Dis.* 2008 Juin;51(6):1005-1015.

[9] Mariat C, Alamartine E, Barthelemy J, De Filippis J, Thibaudin D, Berthoux P, et al. Assessing renal graft function in clinical trials: can tests predicting glomerular filtration rate substitute for a reference method? *Kidney Int.* 2004 Jan;65(1):289-297.

[10] Poggio ED, Wang X, Weinstein DM, Issa N, Dennis VW, Braun WE, et al. Assessing glomerular filtration rate by estimation equations in kidney transplant recipients. *Am. J. Transplant.* 2006 Jan;6(1):100-108.

[11] Levey AS, Stevens LA, Schmid CH, Zhang YL, Castro AF, Feldman HI, et al. A new equation to estimate glomerular filtration rate. *Ann. Intern. Med.* 2009 Mai 5;150(9):604-612.

[12] White CA, Akbari A, Doucette S, Fergusson D, Knoll GA. Estimating glomerular filtration rate in kidney transplantation: is the new chronic kidney disease epidemiology collaboration equation any better? *Clin. Chem.* 2010 Mar;56(3):474-477.

[13] Murthy K, Stevens LA, Stark PC, Levey AS. Variation in the serum creatinine assay calibration: a practical application to glomerular filtration rate estimation. *Kidney Int.* 2005 Oct;68(4):1884-1887.

[14] Hallan S, Asberg A, Lindberg M, Johnsen H. Validation of the Modification of Diet in Renal Disease formula for estimating GFR with special emphasis on calibration of the serum creatinine assay. *Am. J. Kidney Dis.* 2004 Juil;44(1):84-93.

[15] Hall IE, Doshi MD, Poggio ED, Parikh CR. A Comparison of Alternative Serum Biomarkers With Creatinine for Predicting Allograft Function After Kidney Transplantation. *Transplantation*. 2011 1;91(1):48-56.
[16] Rule AD, Bergstralh EJ, Slezak JM, Bergert J, Larson TS. Glomerular filtration rate estimated by cystatin C among different clinical presentations. *Kidney Int.* 2006 Jan;69(2):399-405.
[17] Bökenkamp A, Domanetzki M, Zinck R, Schumann G, Byrd D, Brodehl J. Cystatin C--a new marker of glomerular filtration rate in children independent of age and height. *Pediatrics*. 1998 Mai;101(5):875-881.
[18] Tan GD, Lewis AV, James TJ, Altmann P, Taylor RP, Levy JC. Clinical usefulness of cystatin C for the estimation of glomerular filtration rate in type 1 diabetes: reproducibility and accuracy compared with standard measures and iohexol clearance. *Diabetes Care*. 2002 Nov;25(11):2004-2009.
[19] Hoek FJ, Kemperman FAW, Krediet RT. A comparison between cystatin C, plasma creatinine and the Cockcroft and Gault formula for the estimation of glomerular filtration rate. Nephrol. *Dial. Transplant*. 2003 Oct;18(10):2024-2031.
[20] Larsson A, Malm J, Grubb A, Hansson LO. Calculation of glomerular filtration rate expressed in mL/min from plasma cystatin C values in mg/L. Scand. *J. Clin. Lab. Invest.* 2004;64(1):25-30.
[21] Filler G, Lepage N. Should the Schwartz formula for estimation of GFR be replaced by cystatin C formula? *Pediatr. Nephrol.* 2003 Oct;18(10):981-985.
[22] Le Bricon T, Thervet E, Froissart M, Benlakehal M, Bousquet B, Legendre C, et al. Plasma cystatin C is superior to 24-h creatinine clearance and plasma creatinine for estimation of glomerular filtration rate 3 months after kidney transplantation. *Clin. Chem.* 2000 Aoû;46(8 Pt 1):1206-1207.
[23] Sjöström P, Tidman M, Jones I. Determination of the production rate and non-renal clearance of cystatin C and estimation of the glomerular filtration rate from the serum concentration of cystatin C in humans. Scand. *J. Clin. Lab. Invest.* 2005;65(2):111-124.
[24] Grubb A, Nyman U, Björk J, Lindström V, Rippe B, Sterner G, et al. Simple cystatin C-based prediction equations for glomerular filtration rate compared with the modification of diet in renal disease prediction equation for adults and the Schwartz and the Counahan-Barratt prediction equations for children. *Clin. Chem.* 2005 Aoû;51(8):1420-1431.
[25] Macisaac RJ, Tsalamandris C, Thomas MC, Premaratne E, Panagiotopoulos S, Smith TJ, et al. Estimating glomerular filtration rate in diabetes: a comparison of cystatin-C- and creatinine-based methods. *Diabetologia*. 2006 Juil;49(7):1686-1689.
[26] Bouvet Y, Bouissou F, Coulais Y, Séronie-Vivien S, Tafani M, Decramer S, et al. GFR is better estimated by considering both serum cystatin C and creatinine levels. *Pediatr. Nephrol.* 2006 Sep;21(9):1299-1306.
[27] Stevens LA, Coresh J, Schmid CH, Feldman HI, Froissart M, Kusek J, et al. Estimating GFR using serum cystatin C alone and in combination with serum creatinine: a pooled analysis of 3,418 individuals with CKD. *Am. J. Kidney Dis.* 2008 Mar;51(3):395-406.
[28] Pöge U, Gerhardt T, Stoffel-Wagner B, Palmedo H, Klehr HU, Sauerbruch T, et al. Cystatin C-based calculation of glomerular filtration rate in kidney transplant recipients. *Kidney Int.* 2006 Juil;70(1):204-210.

[29] White C. Estimating Glomerular Filtration Rate in Kidney Transplantation: A Comparison between Serum Creatinine and Cystatin C-Based Methods. *Journal of the American Society of Nephrology*. 2005 12;16(12):3763-3770.

[30] Maillard N, Mariat C, Bonneau C, Mehdi M, Thibaudin L, Laporte S, et al. Cystatin C-based equations in renal transplantation: moving toward a better glomerular filtration rate prediction? *Transplantation*. 2008 Juin 27;85(12):1855-1858.

[31] Zahran A, Qureshi M, Shoker A. Comparison between creatinine and cystatin C-based GFR equations in renal transplantation. *Nephrology Dialysis Transplantation*. 2007 6;22(9):2659-2668.

[32] Risch L, Herklotz R, Blumberg A, Huber AR. Effects of glucocorticoid immunosuppression on serum cystatin C concentrations in renal transplant patients. *Clin. Chem*. 2001 Nov;47(11):2055-2059.

[33] White CA, Akbari A, Doucette S, Fergusson D, Ramsay T, Hussain N, et al. Effect of clinical variables and immunosuppression on serum cystatin C and beta-trace protein in kidney transplant recipients. *Am. J. Kidney Dis*. 2009 Nov;54(5):922-930.

[34] Stevens LA, Schmid CH, Greene T, Li L, Beck GJ, Joffe MM, et al. Factors other than glomerular filtration rate affect serum cystatin C levels. *Kidney Int*. 2009 Mar;75(6):652-660.

In: Cystatins: Protease Inhibitors …
Editors: John B. Cohen and Linda P. Ryseck
ISBN: 978-1-61209-343-7

Chapter 10

EFFECT OF TEMPERATURE, WAVELENGTH, PH, ION PAIR REAGENTS AND ORGANIC MODIFIERS' CONCENTRATION ON THE ELUTION OF CYSTATIN C. STABILITY OF MOBILE PHASE

Othman Ibrahem Yousef Al-Musaimi,[1*] Manar Khalid Fayyad[2] and Adel Khalil Mishal[3]
[1]Department of Chemistry, University of Jordan, Amman-Jordan
[2]Analytical Research Department, Hikma Pharmaceuticals- Amman
[3]Department of Chemistry, University of Ha'il, Ha'il –Saudia Arabia

ABSTRACT

Robustness of an analytical chromatographic method for separation of cystatin c has been verified. Changes in many parameters were carried out, such as, wavelength, column oven, mobile phase composition, chromatographic column.

Imperative changes have altered the efficiency of the chromatographic separation; such changes include pH alteration of the mobile phase as well as alkyl sulfonate molarity changing.

All robustness conditions showed no major effect on the chromatographic separation of the analyte except with the changes related to TFA and alkyl sulfonate ion pair reagents. Peak area RSD, asymmetry and No. of theoretical plates were < 0.7%, < 1.2 and > 10,000, respectively. Results obtained using mobile phase after 6 months of storage have proven its stability and possibility of use. Gradient elution mode was utilized to elute cystatin c with a UV detection of 224 nm. Ace and Waters C8 (150 x 4.6 mm i.d., 5 μm) as chromatographic columns were used.

Keywords: Cystatin C, Robustness, TFA, Stability of mobile phase

* Corresponding author: Othman Ibrahem Yousef Al-Musaimi. Telephone no.: +962 78 5782259, +966 54 1699129. E-mail Address: Musamiau@Gmail.com.

1. Introduction

Cystatin C (CC) is composed of 120 amino acid residues with a molecular weight of about 14 KDa and two intra-chain disulfide bonds. CC has an isoelectric point (PI) equal to 9.2; accordingly, and due to protonation of the acid group it has a positive charge at physiological pH. [1-5]

Ion pair liquid chromatography technique (Cation analysis) has been used for separation and quantification of CC, incorporating trifluoroacetic acid (TFA) and 1-hexane sulfonic acid sodium salt in the mobile phase.

Two ion pair reagents were included in the mobile phase. CC has a leader hydrophobic sequence which enhances the binding with the hydrophobic surface of the stationary phase. 1-hexane sulfonic acid sodium salt reagent was used to bind with the stationary phase instead of the analyte [5]. To accomplish an efficient separation, using of both ion pair reagents is deemed necessary. Typical working concentrations of both reagents were incorporated in the mobile phases [6].

Different factors render the use of TFA, it has low volatility at low pH values (in the present work pH 2.4), it sharpens the peak shape where the ionic interaction chance between the silanol groups of the stationary phase and CC is reduced and finally it has a low absorption at the working wavelength.

Organic modifiers can also compete with the ion pair for the stationary phase and then decreasing the effective capacity of the column. [7-9]

2. Experimental

2.1. Chemicals

CC protein was purchased from Scipac (UK), HPLC grade acetonitrile, methanol, 1-hexane sulfonic acid sodium salt, trifluoroacetic acid (TFA) and acetone were purchased from Merck (Germany).

2.2. Chromatographic Conditions

Shimadzu HPLC system (Shimadzu, Japan) with a degasser, low pressure gradient pump, column oven, an autosampler, a UV detector was used. Data acquisition was performed with LC-Solution 1.21 software.

A reversed phase Ace C8 (150 x 4.6 mm i.d., 5 μm) column was placed in the column oven at 25°C.

Mobile phase A of 0.01M 1-hexane sulfonic acid sodium salt plus 0.05% TFA, pH 2.4 (filtered through 0.45 μm Teflon filter) and mobile phase B (acetonitrile: methanol: mobile phase A) (300: 300: 225, v/v/v), pH 2.5) [10].

2.3. Stability of Mobile Phase Solutions

Mobile phase has been stored at 25°C for six months. RSD were determined from 5 consecutive injections of calibration urine containing CC at its pathological concentration.

2.4. Robustness

It is defined as per the ICH Guideline Q2 (R1) [11], a measure of the method to remain unaffected by deliberate changes in the parameters of the method.

Eight chromatography parameters were altered; these parameters were found to contribute significantly to the method: acetonitrile and methanol content in the mobile phases, column oven temperature, wavelength, TFA and 1-hexan sulfonic acid sodium salt content in the mobile phases which corresponds to the pH value and the buffer molarity, respectively. Chromatographic column from another supplier with the same specification, in addition, to another alkyl sulfonic acid sodium salt ion pair reagent were also used. Analysis was carried out using a calibration urine containing CC at its pathological concentration (Table 1).

3. Results and Discussion

3.1. Stability of Mobile Phase

Close retention time of CC peak, RSD < 0.7%, No. of theoretical plates was > 10,000 and the asymmetry factor was < 1.2, all prove the stability of mobile phase after six months of room temperature storage. (Table 2)

Table 1. Selected parameters and their variabilities*

Parameter	-	0	+
Methanol content in mobile phase, constant Acetonitrile (mL)	285	300	315
Acetonitrile content in mobile phase, constant Methanol (mL)	285	300	315
Mobile Phase A (mL)	210	225	240
Wavelength (nm)	229	224	221
Column Oven Temperature (°C)	20	25	30
TFA Content (mL)	0	0.5	1.0
Alkyl Sulfonic Acid Sodium Salt (M)	0	0.01	NP

* 0: Original condition; - and + are lower and upper limits, respectively.

Table 2. CC Using Stability Mobile Phase

Injection #	Peak Area Stability Mobile Phase (X 10^{-6})	Peak Area Fresh Mobile Phase (X 10^{-6})
1	6.24	6.65
2	6.24	6.64
3	6.26	6.63
4	6.31	6.60
5	6.27	6.54
Average	6.27	6.61
RSD %	0.47	0.67
No. of Theoretical Plates	10407	12726
Asymmetry	1.12	1.14
Retention time (t_r)	10.4	11.0

Table 3. CC at Different Wavelengths (± 5 nm)

Injection #	Peak Area (219 nm) (X 10^{-6})	Peak Area (229 nm) (X 10^{-6})
1	6.54	3.17
2	6.55	3.18
3	6.54	3.18
4	6.54	3.18
5	6.52	3.17
Average	6.54	3.18
RSD %	0.17	0.17

Table 4. CC at Different Column Oven Temperatures (± 5 °C)

Injection #	Peak Area (+ 5 °C) (X 10^{-6})	Peak Area (- 5 °C) (X 10^{-6})
1	4.97	4.80
2	5.00	4.84
3	5.01	4.84
4	5.01	4.85
5	5.01	4.85
Average	5.00	4.84
RSD %	0.35	0.43

3.2. Robustness

No. of theoretical plates, asymmetry factors, RSD of peak areas and retention times were used to evaluate the effects on the qualitative responses (Table 3 – 6).

Table 5. Using of 1-Pentane Sulfonic Acid Sodium Salt in the Mobile Phase

Injection #	Peak Area Using1-Pentane ion pair Reagent ($X 10^{-6}$)	Peak Area Using1-Hexan ion pair Reagent ($X 10^{-6}$)
1	6.23	6.65
2	6.26	6.64
3	6.29	6.63
4	6.27	6.60
5	6.25	6.54
Average	6.26	6.61
RSD %	0.36	0.67

Table 6. CC Using Waters Column

Injection #	Peak Area Using Ace Column ($X 10^{-6}$)	Peak Area Using Waters Column ($X 10^{-6}$)
1	5.76	6.65
2	5.82	6.64
3	5.86	6.63
4	5.87	6.60
5	5.87	6.54
Average	5.84	6.61
RSD %	0.81	0.67
No. of Theoretical Plates	13982	12726
Asymmetry	1.10	1.14
Retention time (t_R)	10.8	11.0

No major changes were observed as a result of the alterations made except when TFA and alkyl sulfonate ion pair reagents were eliminated from the mobile phases (Figure 1 - 3).

Alteration of organic modifiers content (acetonitrile and methanol) has affected only the retention time of CC peak. Nevertheless, the obtained RSD values for CC peak area did not exceed 0.6 % (Figure 4 - 5) (Table 7 – 8).

Interestingly, CC peak was distorted when no TFA is added to the mobile phase and the peak area of CC was reduced by about 67 %. This phenomenon can be attributed to the silanol groups in the stationary phase that have been activated as a result of high pH value (7.0). Moreover, CC protein might dissociate at high pH values due to denaturation process. [6, 12]

CC peak was distorted when no alkyl sulfonates reagent is added to the mobile phase in addition to retention time shifting was also observed. The interaction between the hydrophobic leader of CC sequence and the hydrophobic stationary phase is considered as a reason behind the observed distortion, whereas, in case of alkyl sulfonates-containing mobile such interaction occurs with the alkyl sulfonate reagent. The reduction in the retention time (6.9 minutes instead of 10 minutes) can be also attributed to the low concentration of the alkyl sulfonate in the mobile phase (Figure 6).

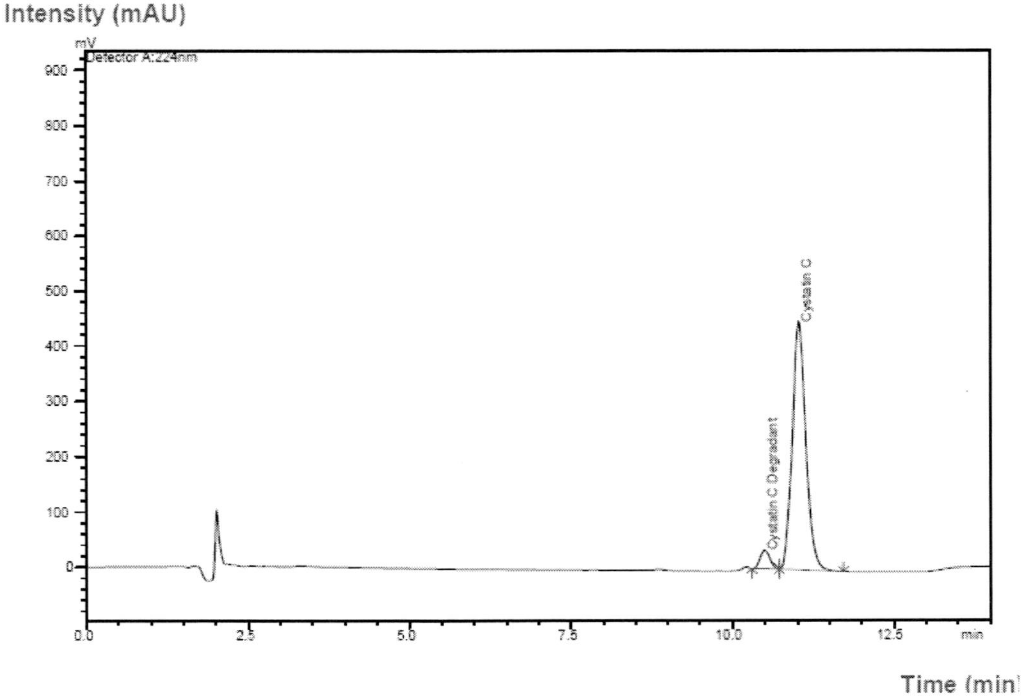

Figure 1. Chromatogram of CC at 224 nm and at 25°C Column Oven.

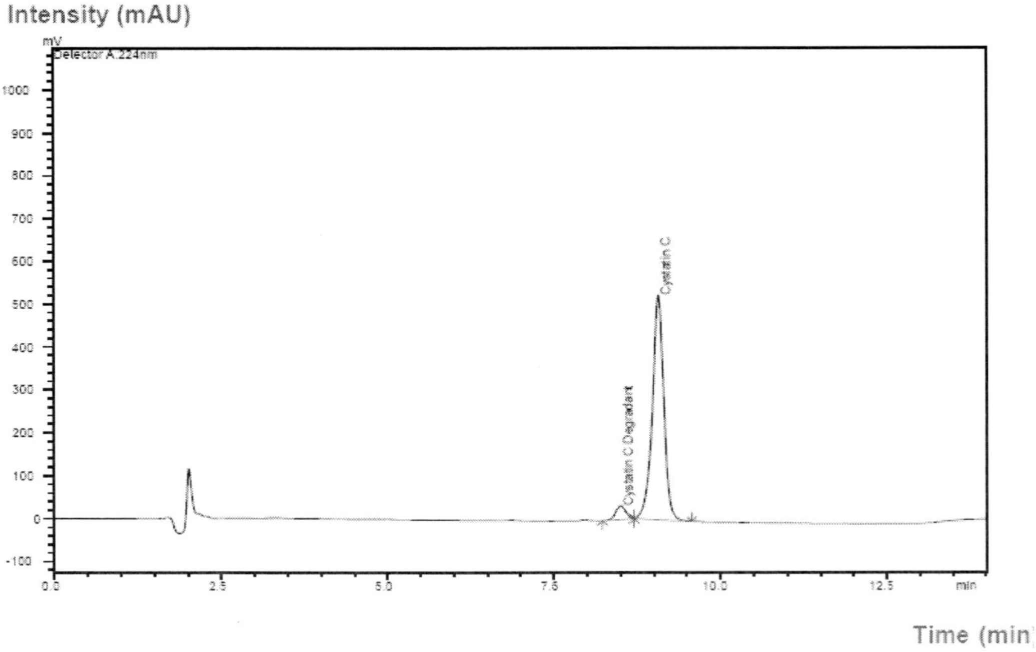

Figure 2. Chromatogram of CC, Using 1-Pentane Sulfonic Acid Sodium Salt.

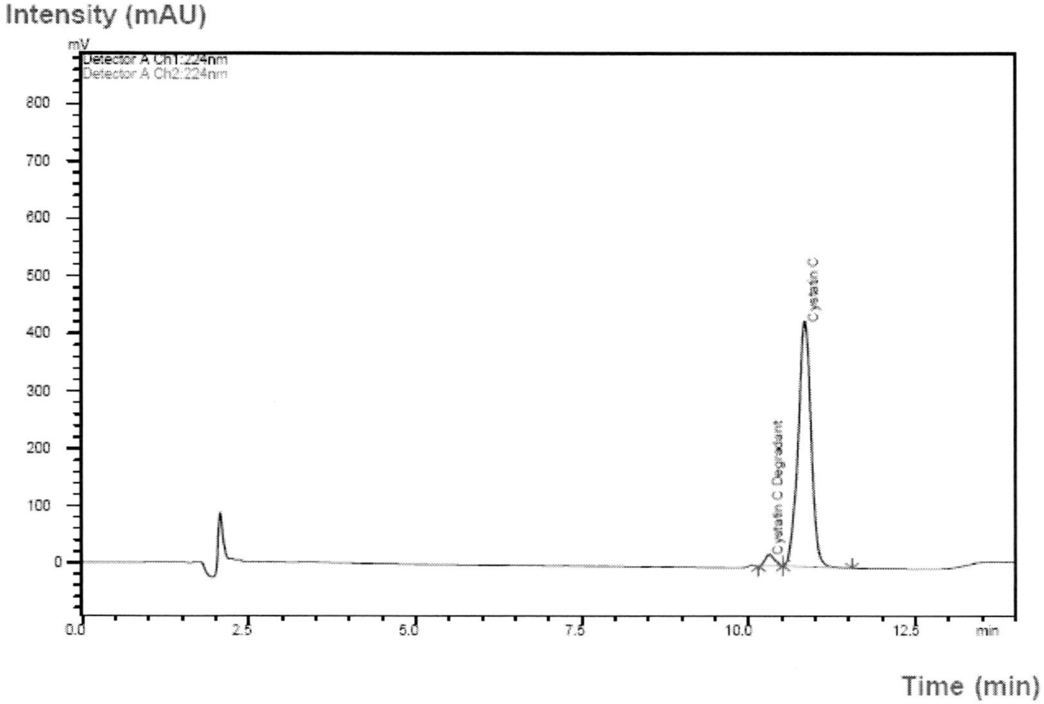

Figure 3. Chromatogram of CC, Using Waters C8 (15 cm*4.6 mm), 5μm Column.

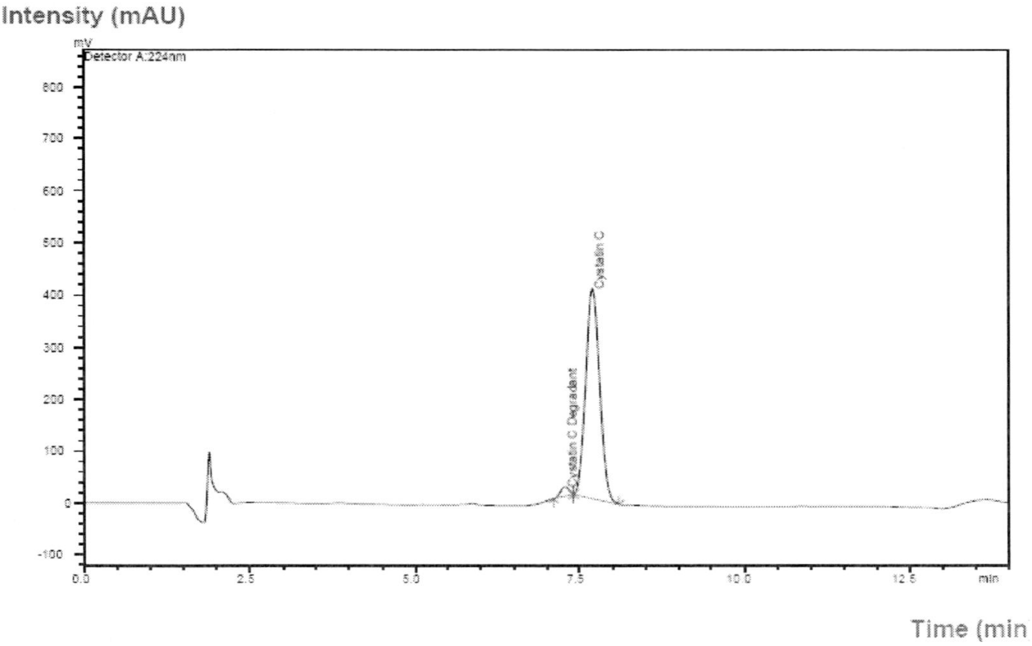

Figure 4. Chromatogram of CC, Mobile Phase + 5% Methanol.

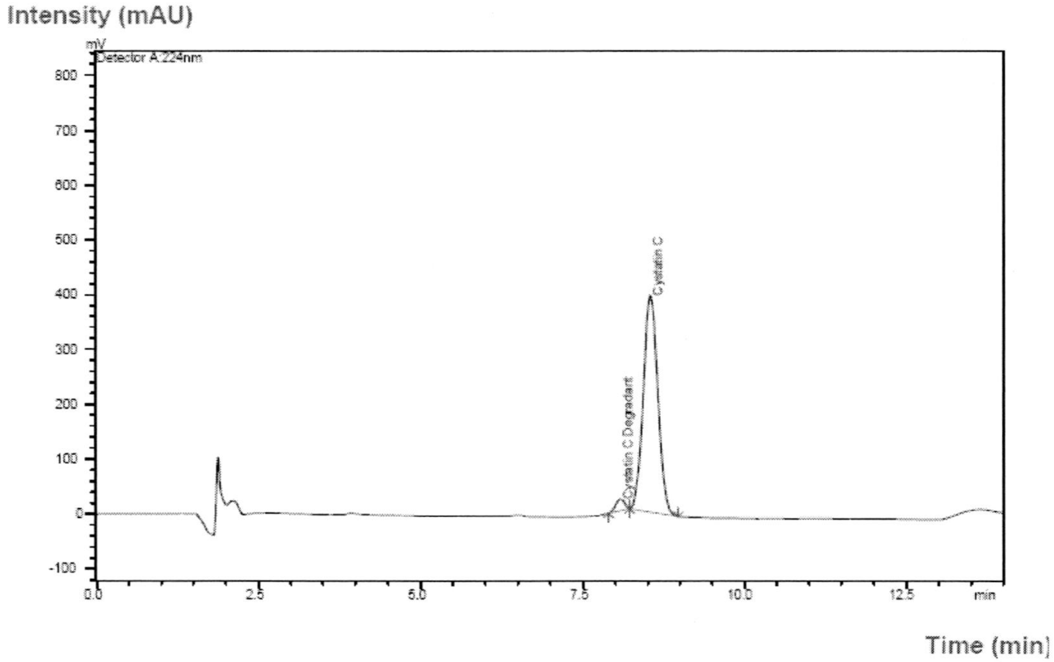

Figure 5. Chromatogram of CC 0.2 mg/mL, Mobile Phase - 5% Methanol.

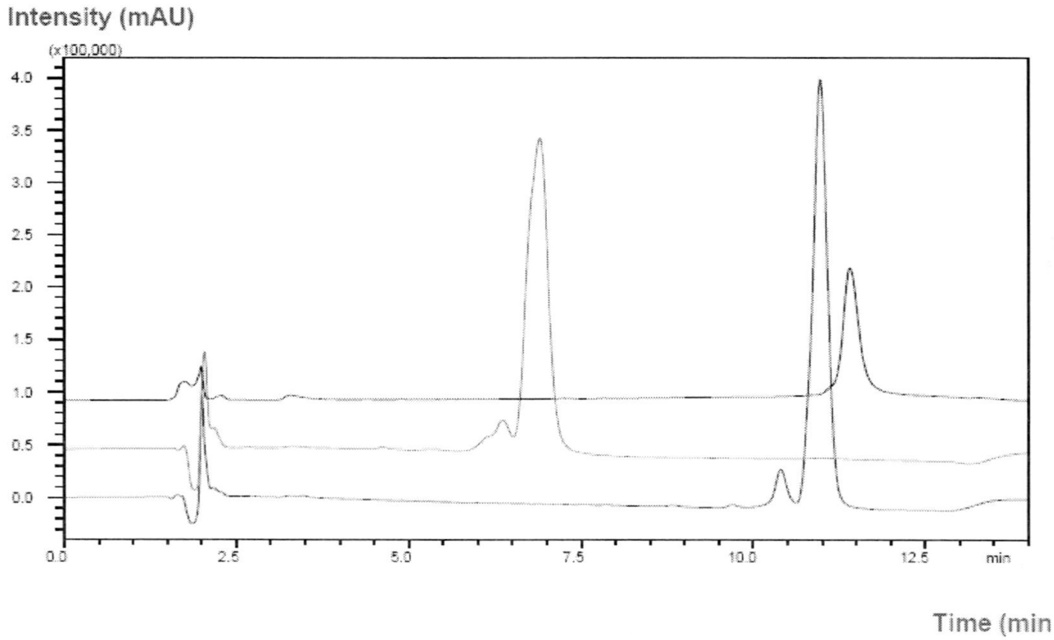

Figure 6. Overlay Chromatograms of CC, Comparison Among Three Mobile Phases: ▬ Original Mobile Phase, ▬ No TFA, No Alkylsulfonate.

Table 7. ± 5 % Acetonitrile Added to Mobile Phase, Constant Methanol

Injection #	Peak Area (+ 5 %) (X 10^{-6})	Peak Area (- 5 %) (X 10^{-6})
1	6.15	6.08
2	6.21	6.14
3	6.22	6.16
4	6.20	6.14
5	6.18	6.14
Average	6.19	6.13
RSD %	0.45	0.49

Table 8. ± 5 % Methanol Added to Mobile Phase, Constant Acetonitrile

Injection #	Peak Area (+ 5%) (X 10^{-6})	Peak Area (- 5 %) (X 10^{-6})
1	6.00	6.10
2	6.08	6.18
3	6.06	6.19
4	6.06	6.18
5	6.07	6.18
Average	6.05	6.17
RSD %	0.52	0.60

CONCLUSION

The present study has investigated the effect of varying some of the chromatography parameters of the previously developed and validated HPLC method. The obtained results proved the robustness of the method in addition to the stability of the mobile phase. According to the obtained results, the presence of the both ion pair reagents, TFA and the alkyl sulfonate, is extremely important to achieve the maximum separation and sensitivity.

ACKNOWLEDGMENT

Financial support has been furnished by Hikma Pharmaceuticals-Amman.

REFERENCES

[1] L. A. Bobek and M.J. Levine. *Crit. Rev. Oral Biol. M.* 3(4) (1992): 307-332

[2] P.J. Berti and A.C. Storer. *Biochem. J.* 302 (1994) 411-416

[3] M. Mussap , M. D. Vestra, P. Fioretto, A. Saller, M. Varagnolo, R. Nosadini and Plebani. *Kidney Int*, 61 (2002) 1453-146

[4] F. j. hoek, F. A. W. Kemperman and R. T. Krediet. *Nephrol. Dial. Transpl.*, 18 (2003) 2024-2031

[5] G. Filler, A. Bokenkamp, W. Hofmann, T. Le Bricon, C. Martinez-Bru and A. Grubb. *Clin. Biochem.*, 38 (2005). 1-8

[6] Y. Chen, A.R. Mehok, C.T. Mant and R.S. Hodges. *J. Chromatogr A*, 1043(1) (2004): 9-18

[7] L. R. Snyder, J. J. Kirklend and J. L. Glajch.. *Practical HPLC Method Development*, (1997) (2nd ed). USA: Library of Congress.

[8] M. Yasuda, T. Sonda, T. Hiraoka, A. Horita and M. tabata. *Anal. Sci.*, 19 (2003) 1637-1641

[9] P. S. T. Yuen, S. R. Dunn, T. Miyaji, H. Yasuda, K. Sharma and R. A. Star. *Am. J. Physiol. Renal. Physiol.*, 286 (2004): F1116-F1119

[10] O. I. Y. Al-Musaimi, M. K. Fayyad and A. K. Mishal, *J. Chromatogr. B* 877 (2009) 747-750

[11] Guidance for Industry: ICH Q2 (R1) Validation of Analytical Procedures: Text and Methodology, 1994, http://www.ich.org/LOB/media/MEDIA417.pdf

[12] J. Nawrocki. *J. Chromatogr. A.*, 779 (1997) 29-71

In: Cystatins: Protease Inhibitors …
Editors: John B. Cohen and Linda P. Ryseck

ISBN: 978-1-61209-343-7

Chapter 11

CYSTATINS IN HUMAN CANCER

Mysore S. Veena and Eri S. Srivatsan*

Department of Surgery, VA Greater Los Angeles Healthcare System, David Geffen School of Medicine at University of California Los Angeles, Los Angeles, CA

ABSTRACT

Cystatins are protease inhibitors that are specifically active against lysosomal cysteine proteases. Cystatins regulate proteases by the formation of reversible high affinity complexes. Members of this superfamily are classified into three subfamilies based on their amino acid homology 1) Type 1 cystatins (cystatin A and B) have a single cystatin domain, intracellular, and lack secretory signal, 2) type 2 cystatins are mainly secretory and comprised of cystatin C, D, E/M, F, S, SN, and SA, 3) type 3 cystatins are those with multi cystatin domains and represent kininogens, the plasma proteins. Cystatins are essential to maintain cell homeostasis. Impairment of cystatins and their substrate proteases leads to pathological conditions including cancer. While some of the cystatins are over-expressed in some cancers they are also down-regulated. Cystatin A is over-expressed in lung, breast, head and neck and prostate cancers and serves as an important prognostic biomarker. Cystatin B expression is elevated in lung, breast, prostate, and hepatocellular carcinoma while being down regulated in oesophageal cancer. In contrast to Type 1 cystatins, Type 2 cystatins are mostly down regulated in breast, lung, cervical, prostate, and brain cancers. Type 2 cystatins are inactivated by various mechanisms including deletion, promoter hypermethylaion, deacetylation, and mutations in the substrate binding regions. Type 3 cystatins, kininogens, have been shown to play a suppressive role in colon cancer. The functions of cystatins in different cancers are not limited to protease inhibition alone but also include cell cycle regulation, apoptosis, and cancer cell adhesion. In this chapter, details on the mechanisms of cystatin gene inactivation and their role in different cancers are summarized.

Keywords: Cystatins, Cathepsins, Cancer

---

* Correspondence: Eri S. Srivatsan, PhD., Department of Surgery, VAGLAHS/David Geffen School of Medicine at UCLA, 11301 Wilshire Blvd., Bldg. 304, Los Angeles, CA 90073, Tel: 310-268-3217, Fax: 310-268-3190, Email:esrivats@ucla.edu

INTRODUCTION

Cystatins are specific inhibitors of cysteine protease inhibitors, required for the maintenance of normal cell homoeostasis. The identification and characterization of the first cysteine protease inhibitor dates back to 1968 when chicken egg white cystatin was described [1-2]. Later, this protein was called "cystatin" and was shown to be a potent inhibitor of various cysteine proteases including cathepsins H and L [3-4]. Since then, various cystatins have been described and their functions are documented. Presently, cystatins are classified into a large superfamily subdivided into three families based on their location, size, and complexity of polypeptide chains [5-6]. Family 1 or type 1 cystatins (cystatin A and B), also called as stefins, have a single cystatin domain, are primariy intracellular in localization and lack secretory signal (Fig. 1). Family 2 or type 2 cystatins (Cystatin C, D, E/M, F, S, SA, and SN) are mainly secretory and most of them are found abundantly in body fluids and tissues. Among these, cystatin S, SA, and SN are abundant in saliva and are commonly grouped under salivary cystatins. They contain about 120 amino acid residues and two intra chain disulfide bonds. Family 3 cystatins or type 3 cystatins are characterized by the presence of multi cystatin domains and represent kininogens. They are mainly found in the plasma and body secretions. Three types of kininogens are known, they are: low-molecular weight, high molecular weight and T-kininogens. These proteins are single-chain glycoproteins with molecular weight ranging from 50 to 120 kDa. Some classification also includes family 4 or type 4 cystatin that is mainly comprised of fetuins. They have two cystatin domains but lack cysteine protease inhibitory activity. They are not reported to play any role in cancer.

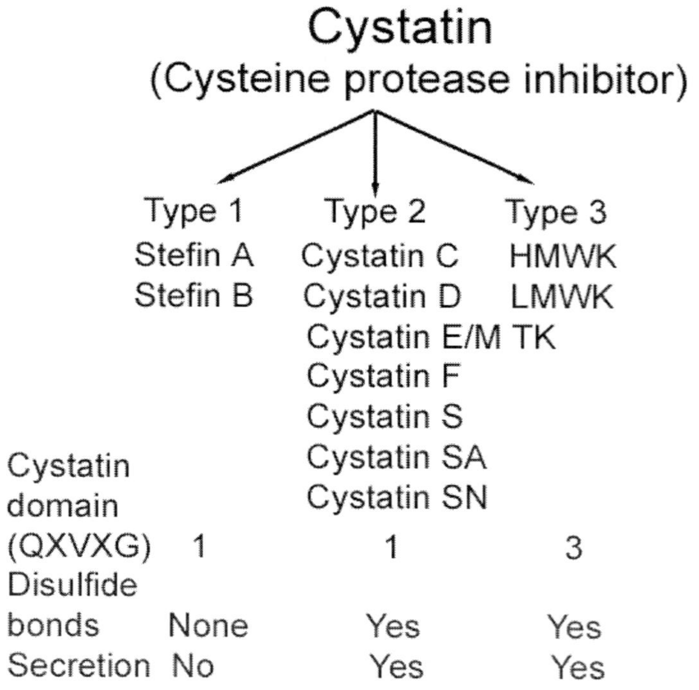

Figure 1. Description of cystatin family members. The amino acid sequence QXVXG represents conserved cystatin domain where X could be any amino acid.

All three types of cystatins exhibit protease inhibition property despite variation in their structure and distribution in the body indicating a common evolutionary origin. Protein sequence alignments of different cystatins from various organisms show conserved regions of substrate binding sites. In a normal cell, there is a balance between a protease and its inhibitor. A perturbation in the levels of either of these two components results in an imbalance leading to abnormal behavior of cells, the manifestation of which results in diseases including cancer.

TYPE 1 CYSTATINS IN CANCER

Stefin A: Stefin A is expressed abundantly in epithelial and lymphoid tissues. It is an acidic protein with a pI value ranging from 4.5 to 5.0. Stefin A is an inhibitor of cysteine proteases, particularly cathepsin L and cathepsin S with Ki values in the picomolar range whereas cathepsin B inhibition is weaker (Ki 10-8M). Stefin A seems to both promote and suppress cancer. Expression of stafin A is lost in skin cancer as well as in lung cancer and metastatic oral/pharyngeal squamous cell carcinoma [7, 8, 9]. Stefin A expression is also recognized as a biomarker for prognosis in patients with operable carcinoma of the head and neck, where the protein expression was observed in greater than 50% of the patients. However, the protein is down regulated in relapse patients. [10].

In some cancers, mechanism of stefin A mediated suppression of malignant progression was associated with the inhibition of cathepsins. As an example, in oesophageal squamous cell carcinoma, over-expression of stefin A inhibited cathepsin B activity thereby preventing invasiveness of cancer cells [11]. However, in a hepatoma cell line, ectopic expression of stefin A resulted in conferring resistance to cancer cells to cell death inducing agents [12]. Detailed studies on the expression of stefin A and its involvement in the cancer suppression or progression is required to gain knowledge on the biological role of stefin A in cancer development.

Stefin B: Stefin B is abundantly expressed in different tissues. Variations in expression of stefin B is found in tumor tissues compared to adjacent normal tissues [13,14]. Stefin B expression is suggested to be one of the potential diagnostic markers for invasive and aggressive meningiomas [15]. In atypical meningiomas, Stefin B was regulated at the transcriptional level as indicated by lower mRNA levels compared to benign meningiomas. In these cancer tissues, there was a corresponding increase in the expression level of its substrate, cathepsin B, in the cytosol. Recent studies show stefin B interaction with cathepsin L in the nucleus as well. A truncated form of cathepsin L is specifically localized to the nuclei of the cancer cells. The function of nuclear cathepsin L is to process transcription factors such as CUX1, resulting in enhanced transcription activity of genes involved in proliferation and metastasis. However, nuclear stefin B interaction inhibits the cathepsin L activity and subsequent processing of the transcription factors leading to delayed cell cycle progression [16].

Stefin B is also detected in human serum and other body fluids in trace amounts even though it is not known to be secreted [17-18]. In cancer patients, stefin B levels in biological fluids increased when compared to their normal counter parts. In bladder cancer, stefin B expression was elevated in the bladder tissue and in the urine of cancer patients. Increased expression of stefin B correlated to a shorter time to disease recurrence in comparison to patients with low level of urinary stefin B expression [19]. It could thus be a novel predictive biomarker for the development of transitional cell carcinoma of the bladder. In another study on colorectal cancer, serum levels of Stefin B correlated with cancer progression, with the maximum expression in patients in stage D disease.

Stefin B is also involved in biological functions other than protease inhibition. It is shown to be an inhibitor of TRAIL-induced apoptosis in melanoma cell lines [21]. Stefin B deficient melanoma cell lines displayed increased apoptosis associated with enhanced activation of caspase-8 induced by TRAIL. This was shown not to be related to the inhibitory effect of stefin B on cathepsin B or L. Sensitization of melanoma cells to TRAIL-induced apoptosis by inhibition of stefin B showed decreased stability of FLICE-inhibitory protein [FLIP (L)], a death receptor (TRAIL) antagonist. Over-expression of stefin B increased the levels of FLIP (L), decreased the amount of the E3 ligase Itch, resulting in reduced ubiquitination of FLIP (L). These results indicate that stefin B regulates Itch-mediated degradation of FLIP(L) thereby controlling TRAIL-induced apoptosis in melanoma cells.

Contradictory roles of stefins in cancer suppression and progression therefore indicate possible involvement of other factors influencing their expression and activity. Mutations and/or tissue micro-environment could greatly influence these properties. Studies in this field are therefore needed to understand the regulation of stefins and their role in cancer development.

Type 2 Cystatins in Cancer

Cystatin C: Cystatin C is one of the most abundantly expressed type 2 cystatins in most tissues including kidney, liver, pancreas, intestine, stomach, lung, placenta and seminal vesicles [22]. Cystatin C regulates bone resorption, neutrophil chemotaxis, and tissue inflammation as well as resistance to bacterial and viral infections. It is a potent inhibitor of cathepsin B and other human lysosomal cysteine proteases [23]. Cystatin C inhibition of cathepsin B has been described as a two-step process. Initially a weak interaction occurs between cystatin C and cathepsin B which is followed by a conformational change. This interaction most likely involves binding of the N-terminal region of the inhibitor (cystatin C) to the protease (cathepsin B). The subsequent conformational change is due to the inhibitor displacing the occluding loop of the cathepsin B partially obscuring the active site. The presence of this loop in cathepsin B, which allows the enzyme to function as an exopeptidase, thus complicates the inhibitory mechanism, rendering cathepsin B much less susceptible than other cysteine proteases to inhibition by cystatin C and other cystatins [24].

In many cancers, cystatin C is shown to have tumor suppressor function, and the mechanism of suppression in most cases, is mainly by the inhibition of cathepsin B. In melanoma and glioblastoma, overexpression of cystatin C suppresses the metastatic potential of cancer cells thereby inhibiting cancer progression [25-26]. In prostate cancer, loss of cystatin C and overexpression of androgen receptor (AR) have been suggested to be cooperative events involved in the progression of prostate cancer [27]. Tumor microarray studies on primary prostate cancer and benign prostatic tissue from 448 patients showed that tumors with low levels of cystatin C and high AR levels exhibited a worse clinical outcome compared to s tumors with high levels of cystatin C and low levels of AR. Secreted cystatin C has also been found to be a diagnostic marker in certain types of cancer. In prostate cancer patients with bone metastasis, serum cystatin C expression has been proposed as a useful marker of increased osteoblastic activity associated to bisphosphonate treatments [28]. In squamous carcinoma of the head and neck, higher cystatin C levels in the tumor tissues correlated with longer survival [29]. Abnormal serum levels of cystatin C or cathepsin B/cystatin C complex have also been suggested as diagnostic and prognostic indicators for skin, colon and lung cancers [30]. Additionally, serum cystatin C is shown to be a valid marker of accurate glomerular filtration rate (GFR) in several renal disorders. Monitoring of renal function in cancer patients is an ongoing challenge for clinical practice. In this context, serum cystatin C expression is a preferred marker than the expression level of serum creatinine. In studies carried out on a group of cancer patients, serum cystatin C was found to be superior to serum creatinine for the detection of decreased creatinine clearance and showed a greater potential for the estimation of GFR in cancer patients independent of the presence of metastases or chemotherapy [31]. In multiple myeloma, cystatin C expression is not only a sensitive marker of renal impairment but also reflects tumor burden and has shown a prognostic value. [32]. Cystatin C was essentially useful in patients under three years of age and had a better diagnostic value than routine serum creatinine monitoring [33]. Finally, in nonHodgkin B-cell lymphoma, serum levels of cystatin C serves as a predictive marker of disease relapse [34].

Cystatin D: Cystatin D differs from other members of the family in having a more restricted pattern of expression and a narrow protease inhibitory profile, being active against cathepsin S, L, and H and not against cathepsin B [35-36]. So far, secreted form of cystatin D has not been reported. Like most of the family members, cystatin D also exhibits tumor suppressive properties. In colon cancer specimens, cystatin D expression is lost in poorly differentiated tumors. Ectopic expression of cystatin D suppressed proliferation, migration and reduced tumorigenic potential of colon cancer cell lines [37]. Cystatin D also affected cellular phenotype and adhesive properties of colon cancer cell lines which are related, at least partially, to the repression of c-Myc oncogene and beta-cateinin and induction of E-cadherin. A link between vitamin D receptor (VDR) and cystatin D has also been proposed, as cystatin D expression was enhanced by treatment with vitamin D analogs. Cystatin D seems to have different mechanisms to regulate proliferation and migration of cancer cells. While the anti-proteolytic activity is mainly involved in inhibiting cell migration, the anti-proliferative activity is independent of the anti-proteolytic activity. Thus, it appears that cystatin D could be targeting different pathways to inhibit cell proliferation and migration.

Cystatin E/M: Cystatin E/M is expressed in different tissues including lung, heart, kidney, liver, ovary and brain [38]. Cystatin E/M was first identified as a down regulated transcript in a differential display between primary and metastatic breast cancers and was named cystatin M [38]. At the same time another group also cloned this gene from lung embryo and called it cystatin E [39]. It has since been reported in different cancers mainly as a tumor suppressor. It regulates different cysteine proteases and has high affinity for cathepsin L, and V [40]. A novelty of cystatin E/M lies in the fact that it can interact with both cathepsins and legumain simultaneously due to a secondary reactive site specific for papain inhibitory activity.

Cystatin E/M expression is mainly regulated by epigenetic mechanisms [41]. Promoter hypermethylation of cystatin E/M is seen in cancers of the breast, brain, gastric, cervix, and prostate [42, 43, 44, 45, 46]. In a co-culture study that involved breast epithelial cells and cancer associated fibroblasts (CAF), aberrant expression of AKT in CAF was shown to be the upstream initiator for hypermethylation of Cystatin E/M indicating an influence by the tumor microenvironment [47]. In addition, homozygous deletion and mutations in the cathepsin L binding sites have also been implicated in the inactivation of cystatin E/M in cervical cancer (Figure 2) [45]. In breast cancer, even though epigenetic mechanism seems to be the major mechanism of gene silencing, mutational inactivation can't be ruled out. We have recently observed stop codon in exon 1 of breast cancer specimens that could possibly be involved in silencing cystatin E/M [48]. Tumor suppressive function of cystatin E/M is confirmed in breast, cervical, and prostate cancers by the ectopic expression of cystatin E/M resulting in the suppression of proliferation, invasion and metastasis in vitro and in mouse tumor models [49, 50, 45, 46]. Contrary to tumor suppressive e function described above, cystatin E/M has also been reported to be over-expressed in a metastatic oral squamous cell carcinoma cell line [51]. In this situation, it is possible that dominant negative mutations may be playing a role, and detailed studies will be required to determine the precise nature of cystatin E/M overexpression in this tumor cell line.

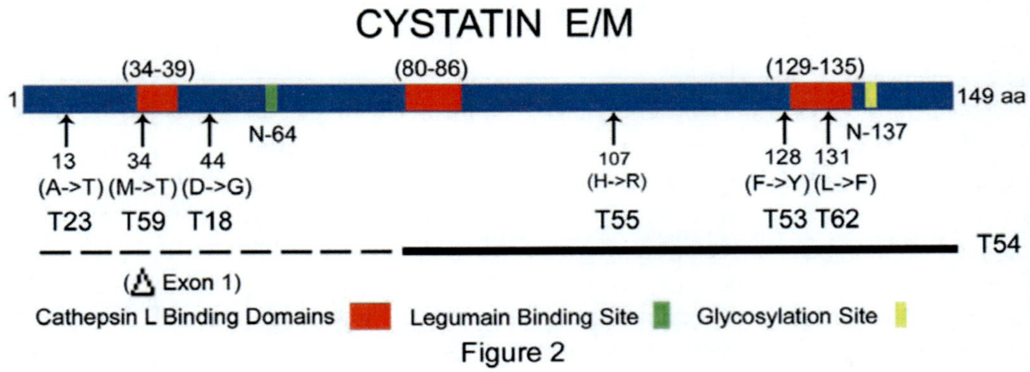

Figure 2. Schematic representation of cystatin E/M protein structure. The glycosylation site, and the cathepsin L and legumain binding sites are marked with yellow, red and green colors respectively. Exon 1 homozygous deletion (represented by Δ) and other somatic mutations of cystatin E/M identified in cervical cancer are also shown [45].

Cystatin F: Cystatin F is expressed selectively in thymus, spleen and immune cells. It is the only cystatin to be synthesised as an inactive disulphide-linked dimeric precursor. Cystatin F is also expressed in colorectal cancers, the highest expression correlating with higher metastatic potential [52].

Cystatin S: Possible involvement of cystatin S in the suppression of metastasis is suggested by a study in colon cancer. In a serial analysis of gene expression (SAGE) between a primary colon tumor cell line (SW480) and an isogenic lymph-node metastatic cell line (SW620), reduced expression of cystatin S was observed in the metastatic cell line. Similarly, Cystatin S expression was not detected in four other metastatic cell lines - LoVo, Colo201, T84 and 498LI while the expression was seen in the primary tumor cell lines, SW480 and SW1116 [53].

Cystatin SA: Cystatin SA expression has not been studied extensively in different cancers. It seems to be important in cancer as the expression was detected in pre therapy saliva of patients with stage I OSCC (oral squamous cell carcinoma) while the expression was lost in post therapy salivary samples [54]. A 14 kDa protein was detected in the pre-therapy saliva using the surface-enhanced laser desorption/ionization time-of-flight mass spectrometric (SELDI-TOF) analysis. The identification revealed the protein to be a truncated version of cystatin SA-I, with the deletion of three amino acids from the N-terminus. Thus, this truncated cystatin SA-I protein might serve as a useful tumor biomarker for OSCC. This investigation further underscores the importance of deletions, mutations and post-translational modifications leading to abnormal functioning of cystatins in cancer [54].

Cystatin SN: Cystatin SN, similar to cystatin SA, is also poorly studied in cancer. However, in a complex disease like cancer there is a constant need for novel biomarkers to specifically characterize the nature of the disease for effective treatment. In this context, cystatin SN could be an ideal candidate for further research in cancer. In a study of gastric cancer, cystatin SN showed high expression in the cancer cells. A significant correlation between high expression of cystatin SN and gastric cancer progression was observed [55]. Additionally, cystatin SN silencing reduced cell proliferation and increased proteolytic activity of cathepsins in gastric cancer. Finally, western blot analysis revealed upregulated expression of Cystatin SN in the urine of patients with colorectal cancer. These results suggest the possibility of utilizing this urinary excretory protein as a tumor marker for the mass screening of early stage colorectal cancers [56].

TYPE 3 CYSTATINS IN CANCER

Kininogens: Kininogens are plasma proteins categorized into three types, high molecular weight kininogen (HMWK), low molecular weight kininogens (LMWK) and T kininogens. Both types of kininogens are shown to be cysteine protease inhibitors [57]. HMWK consists of six domains. Each domain has been reported to possess a specific function: Domain 1 has a low affinity calcium-binding site [58]; domains 2 and 3 have specific sites functioning as cysteine protease inhibitors [59]; and domain 3 also has platelet and endothelial cell binding activity [60]; Domain 4 has the bradykinin (a vasodilator) sequence. Domain 5 (D5H) has cell-binding sites [61,62,63] and a negatively charged surface-binding site, and domain 6 (D6H) has prekallikrein- and factor XI-binding sites [64-65]. HMWK thus seems to be a multifunctional protein.

High-molecular-weight kininogen,(HMWK), is also known as the Williams-Fitzgerald-Flaujeac factor or the Fitzgerald factor or the HMWK-kallikrein factor. HMWK plays an important role in cancer by different mechanisms. They bind to a wide variety of cells in a specific, reversible, and saturable manner. The cell docking sites have been mapped to domains D3 and D5H of kininogens. In colon cancer, the D5 domain of high molecular weight kininogen specifically reduced cancer cell proliferation by interfering with G1 to S phase transition. Exogeneous addition of this kininogen as a recombinant fusion protein of D5 and glutathione S-transferase (GST-D5) reduced the proliferation of human colon cancer cells (HCT-116) in vitro. The G1/S phase transition of the cell cycle was down-regulated as evidenced by an increase in the percentage of cells in G0/G1 and a decrease in cells in the S phase by flow cytometry. There was also a correlation to a decrease in serine phosphorylation of the retinoblastoma protein Rb (p107) and reduced release of E2F-1 transcription factor.

Expression of high molecular weight kininogen down-regulated cyclin-dependent kinase protein activities while increasing the p27 suppressor protein activity [66]. The D5 domain was also found to be a critical factor for the suppression of metastasis in a breast cancer model [67]. However, in breast cancer patient samples, only a two-fold decrease in the expression of kininogen was observed in comparison to the normal tissues [68].

Low-molecular weight-kinonogens have not been studied extensively. It appears that LMWK have the opposing effect on HMWK in cancer. T-Kinonogen is shown to reduce cell proliferation by inhibiting cysteine proteases. Over-expression of T-kininogen in mouse fibroblasts results in diminished proliferative capacity. This inhibition was not overcome even after serum stimulation and cells failed to progress from G0 phase to the S phase of the cell cycle [69].

Conclusion

Today, cystatins have acquired a more important place in cell biology than their original house- keeping status. They are now known to participate in diverse biological functions. Variations in their expression levels, mutations, and post-translational modifications have resulted in various pathological conditions, including cancer. Today, even though we have gained substantial knowledge about cystatins and their role, lot more remains to be learnt, especially about their contradictory role in cancer suppression and promotion. Current knowledge of cystatins is mainly related to their anti-proteolytic activity as a major mechanism of tumor suppression. However, considering their capacity to participate in a multitude of biological functions, it is possible that cystatins may be regulating cancer cells by other mechanisms besides protease inhibition. We still do not have exact answers yet, also about their interaction with other members of the family and the sequence of events involved in of tumor suppression. Future studies addressing the biological activities of cystatins will therefore be valuable in cancer diagnosis and therapy.

Acknowledgments

The study was supported by funds from VAGLAHS, West Los Angeles Surgical Education Research Center, and Merit grant from the Veterans Administration, Washington, DC (E.S. Srivatsan).

References

[1] Fossum, K; Whitaker, JR. Ficin and Papain Inhibitor from Chicken Egg White. *Arch. Biochem. Biophys,* 1968 125, 367-375.

[2] Sen, LC; Whitaker, JR. Some Properties of a Ficin-Papain inhibitor from Avian Egg White. *Arch. Biochem. Biophys*, 1973 158, 623-632.

[3] Barrett, AJ. Cystatin, the Egg White Inhibitor of Cysteine Proteinases. *Meth. EnzymoL*, 1981 80,771-778.

[4] Anastasi, A; Brown, MS; Kembhavi, AA; Nicklin, MJH; Sayers, CA; Sunter, DC; Barrett, AJ. Cystatin, a Protein Inhibitor of Cysteine Proteinases. *Biochem. J,* 1983 211, 129-138.

[5] Barrett, AJ; Rawlings, N; Davies, M; Machleidt, W; Salvesen, G; Turk, V. Cysteine Proteinase Inhibitors of the Cystatin Superfamily. In: *Proteinase Inhibitors*. pp. 515-569. (A. Barrett and G. Salvesen, Eds.) Elsevier, Amsterdam.

[6] Barrett, AJ; Fritz, H; Grubb, A; Isemura, S; Jarvinen, M; Katunuma, N; Machleidt, W; Muller-Esterl, W; Sasaki, M; Turk, V. Nomenclature and Classification of the Proteins Homologous with the Cysteine-Proteinase Inhibitor Chicken Cystatin. *Biochem. J.* 236, 312.

[7] [Soderstrom, KO; Laato, M; Wu, P; Hopsu-Havu, VK; Nurmi, M; Rinnie, A. Expression of acid cysteine proteinase inhibitor (ACPI) in the normal human prostate, benign prostatic hyperplasia and adenocarcinoma. *Int. J. Cancer*, 1995 62, 1-4.

[8] Jarven, M; Rinnie, A; Hopsu-Havu, VK. Human cystatins in normal and diseased tissues – a review, *Acta Histochem*, 1987 82, 5-18.

[9] Blanchi, F; Hu, J; Pelosi, J; Cirincione, R; Feguson, M; Tatelififfe, C. et. al., Lung cancers detected by screening with spiral computed tomography have a malignant phenotype when analysed by cDNA microarray, *Clin. Cancer Res*, 2004 10, 6023-6028.

[10] Strojan, P; Anicin, A; Svetic, B; Pohar, M; Smid, L; Kos. Stefin A and Stefin B: markers for prognosis in operable squamous cell carcinoma of the heach and neck. *Int J Radiat Oncol Biol Phys,* 2007 68, 1335-1341.

[11] Li, W; Ding, F; Zhang, L; Liu, Z; Wu, Y; Luo, A; Wu, M; Wang, M; Zhan, Q; and Liu, Z. Overexpression of stefin A in human oesophageal squamous cell carcinoma cell inhibits tumor cell growth, angiogenesis, invasion and metastasis. *Clin Cancer Res*, 2005 11, 8753-8762.

[12] Jones, B; Roberts, PJ; Faubion, WA; Kominami, E; Gores, GJ. Cystatin A expression reduces bile salt-induced apoptosis in a rat hepatoma cell line. *Am. J. Physiol.* 1998 275, G 723-G730.

[13] Mirtti, T; Alanen, K; Kalljoki, M; Rinne, A; Soderstrom, K. Expression of cystatins, high molecular weight cytokeratin, and proliferation markers in prostatic adenocarcinoma and hyperplasia. *Prostate*, 2003 54, 290-298.

[14] Shiraishi, T; Mori, M; Tanaka, S; Sugimachi, K; Akiyoshi, T. Identification of cystatin B in human esophageal carcinoma, using differential displays in which the gene expression is related to lymph node metastasis. *Int. J. Cancer,* 1998 79, 175-178.

[15] Trinikaus, M; Vranic, A; Dolenc, VV; Lah, TT. 2005. Cathepsins B and L and their inhibitors stefin B and cystatin C as markers for malignant progression of benign meningiomas. *Int J Biol Markers*. 2005 20:50-59.

[16] Ceru, S; Konjar, S; Maher, K; Repnik, U; Krizaj, I; Bencina, M; Renko, M; Nepveu, A; Zerovnik, E; Turk, B; Kopitar-Jarala, N. Stefin B interacts with histones and cathepsin L in the nucleus. *J. Biol. Chem*, 2010 285, 10078-10086.

[17] Hopsu-Havu, AK; Joronen, I; Havu, S; Rinne, A; Jarvinen, M; Forstrom, J. Serum cystein proteinase inhibitors with specific reference to kidney failure. *Scand. J. Clin. La. Invest*. 1985 15, 11-16.

[18] Abrahamson, JJ; Barett, G; Salvesen, A; Grubb. Isolation of six cysteine proteinase inhibitors from human urine. Their physicochemical and enzyme properties and concentrations in biological fluids. *J. Biol. Chem*. 1986 261, 11282-11289.

[19] Feldman, AS; Banyard, J; Wu, CL; McDoughal, WS; Zetter, BR. *Clin Cancer Res.* 2009 5, 1024-1031.

[20] Kos, J; Krasovec, M; Cimerman, N; Nielson, HJ; Christenesen, IJ; Brunne, N. Cysteine proteinase inhibitors stefin A, stefin B, and cystatin C in sera from patients with colorectal cancer : relation to prognosis. *Clin Cancer Res*. 2000 6, 505-511.

[21] Yang, F; Tay, KH; Dong, L; Thorne, RF; Jiang, CC; Yang, E; Tseng, HY; Liu, H; Christopherson, R; Hersey, P; Zhang, XD. Cystatin B inhibition of TRAIL-induced apoptosis is associated with the protection of FLIP(L) from degradation by the E3 ligase itch in human melanoma cells. *Cell Death Differ*. 2010 17,1354-1367.

[22] Abrahamson, M; Olafsson, A; Palsodottir, M; Ulvsback, A; Lundwall, O; Jensson; Grubb, A. Structure and Expression of the human cystatin C gene. *Biochem J.* 1990 268, 287-294.

[23] Abrahamson, M; Alvarez-Fernandez, M; Nathanson, CM; Cystatins. *Biochem Soc Symp*, 2003, 179–199.

[24] Nycandera, M; Estradaa, S; Mortb, JS; Abrahamson, M; Björka, I. Two-step mechanism of inhibition of cathepsin B by cystatin C due to displacement of the proteinase occluding loop. *FEBS*, 1998 422, 61-64.

[25] Ervin, H; Cox, JL. Late stage inhibition of hematogenous melanoma metastasis by cystatin C over-expression. *Cancer Cell Int*, 2005 5, 14.

[26] Konduri, SD; Yanamandra, N; Siddique, K; Joseph, A; Dinh, DH; et al. Modulation of cystatin C expression impairs the invasive and tumorigenic potential of human glioblastoma cells. *Oncogene*, 2002 21, 8705–8712.

[27] Wegiel, B; Jiborn, T; Abrahamson, M; Helczynski, L; Otterbein, L; et al. Cystatin C Is Downregulated in Prostate Cancer and Modulates Invasion of Prostate Cancer Cells *via* MAPK/Erk and Androgen Receptor Pathways. *PLoS ONE*, 2009 4(11), e7953. doi:10.1371/journal.pone.0007953.

[29] Tumminello, FM; Badalamenti, G; Incorvaia, L; Fulfaro, F; D'Amico, C; et al. Serum interleukin-6 in patients with metastatic bone disease: correlation with cystatin C. *Med Oncol*, 2009 26, 10–15.

[30] Strojan, P; Oblak, I; Svetic, B; Mid, L; Kos, J. Cysteine proteinase inhibitor cystatin C in squamous cell carcinoma of the head and neck: relation to prognosis. *British J Cancer*, 2004 90, 1961–1968.

[31] Kos, J; Werle, B; Lah, T; Brunner, N. Cysteine proteinases and their inhibitors in extracellular fluids: markers for diagnosis and prognosis in cancer. *Int J Biol Markers*, 2000 15, 84–89.

[32] Tabuc, B; Vrhovec, L; Tabuc-ilih, M; Cizej, TE. Improved Prediction of Decreased Creatinine Clearance by Serum Cystatin C: Use in Cancer Patients before and during Chemotherapy. *Clinical Chemistry*, 2000 46,193-197.

[33] Terpos, E; Katodritou, E; Tsiftsaki, E; Kastritis, E; Christoulas, D; Puli, A; Michalis, E; Verrou, E; Anargyrou, K; Tsionos, K; Dimpoulos, MA; Zervas, K.. Cystatin-C is an inpendent prognostic factor for survival in multiple myloma an dis reduced by bortezomib administration. *Haematolgica*, 2009 94, 372-379.

[34] Lankisch, P; Weesalowski, R; Maisonneuve, P; Haghgu, D; Hermsen, D; Kramm, CM. Serum cystatin C is a suitable marker for routine monitoring of renal function in pediatric patients, especially of very young age. *Pediatr Blood Cancer*, 2006 46, 767-772.

[35] Mulamoerovic, A; Hallibasic, A; Cickusic, E; Zavasnik, Bergant, T; Begic, L; Kos, J. Cystatin C as a potential marker for relapse in patients with non Hodgkin B-cell lymphoma. *Cancer Let*, 2007 248, 192-197.

[36] Freije, JP; Abrahamson, M; Olafsson, I; Velasco, G; Grubb, A; López-Otín, C. Structure and expression of the gene encoding cystatin D, a novel human cysteine proteinase inhibitor. *J Biol Chem*, 1991 266, 20538-20543.

[37] Balbin, M; Hall, A; Grubb, A; Mason, RW; Lopez-Otín, C; Abrahamson, M. Structural and functional characterization of two allelic variants of human cystatin D sharing a characteristic inhibition spectrum against mammalian cysteine proteinases. *J Biol Chem*, 1994 269, 23156-23162.

[38] Alvarez-Díaz, S; Valle, N; García, JM; Peña, C; Freije, JM; Quesada, V; Astudillo, A; Bonilla, F; López-Otín, C; Muñoz, A. Cystatin D is a candidate tumor suppressor gene induced by vitamin D in human colon cancer cells. *J Clin Invest*, 2009 119, 2343-2358.

[39] Sotiropoulou, G; Anisowicz, A; Sager, R. Identification, cloning, and characterization of cystatin M, a novel cysteine proteinase inhibitor, down-regulated in breast cancer. *J Biol Chem*, 1997 272, 903-910.

[40] Ni, J; Abrahamson, M; Zhang, M; Fernandez, MA; Grubb, A; Su, J; Yu, GL; Li, Y; Parmelee, D; Xing, L; Coleman, TA; Gentz, S; Thotakura, R; Nguyen, N; Hesselberg, M; Gentz, R. Cystatin E is a novel human cysteine proteinase inhibitor with structural resemblance to family 2 cystatins. *J Biol Chem*, 1997 272,10853-10858.

[41] [40] Cheng, T; Hitomi, K; van Vlijmen-Willems, IM; de Jongh, GJ; Yamamoto, K; Nishi, K; Watts, C; Reinheckel, T; Schalkwijk, J; Zeeuwen, PL. Cystatin M/E is a high affinity inhibitor of cathepsin V and cathepsin L by a reactive site that is distinct from the legumain-binding site. A novel clue for the role of cystatin M/E in epidermal cornification. *J Biol Chem,* 2006 281, 15893-15899.

[42] Rivenbark, AG; Coleman, WB. Epigenetic regulation of cystatins in cancer. *Front Biosci*, 2009 14, 453-462.

[43] Qiu, J; Ai, L; Ramachandran, C; Yao, B; Gopalakrishnan, S; Fields, CR; Delmas, AL; Dyer, LM; Melnick, SJ; Yachnis, AT; Schwartz, PH; Fine, HA; Brown, KD; Robertson, KD. Invasion suppressor cystatin E/M (CST6): high-level cell type-specific expression in normal brain and epigenetic silencing in gliomas. *Lab Invest*. 2008 88, 910-925.

[44] Kioulafa, M; Balkouranidou, I; Sotiropoulou, G; Kaklamanis, L; Mavroudis, D; Georgoulias, V; Lianidou, ES. Methylation of cystatin M promoter is associated with unfavorable prognosis in operable breast cancer. *Int J Cancer*, 2009 125, 2887-92.

[45] Chen, X; Cao, X; Dong, W; Xia, M; Luo, S; Fan, Q; Xie, J. Cystatin M expression is reduced in gastric carcinoma and is associated with promoter hypermethylation. *Biochem Biophys Res Commun*, 2010 391, 1070-1074.

[46] Veena, MS; Lee, G; Keppler, D; Mendonca, M; Redpath, JL; Standbridge, EJ; Wilczynsky, SP; Srivatasan, ES. Inactivation of the cystatin E/M tumor suppressor gene in cervical cancer. *Genes Chromosomes Cancer*, 2008 47, 740-754.

[47] Pulukuri, SM; Gorantla, B; Knost, JA; Rao, JS. Frequent loss of cystatin E/M expression implicated in the progression of prostate cancer. *Oncogene*, 2009 28, 2829-2838.

[48] Lin, HJ; Zuo, T; Lin, CH; Kuo, CT; Liyanarachchi, S; Sun, S; Shen, R; Deatherage, DE; Potter, D; Asamoto, L; Lin, S; Yan, PS; Cheng, AL; Ostrowski, MC; Huang, TH.

Breast cancer-associated fibroblasts confer AKT1-mediated epigenetic silencing of Cystatin M in epithelial cells. *Cancer Res*, 2008 68, 10257-10266.

[49] Venkatesan, N; Shawnt, N; Ginther, C; Slamon, D; Srivatsan, ES; Veena, MS. Involvement of somatic mutations in silencing tumor suppressor gene, cystatin E/M in breast cancer. Abstract, American Association for Cancer Research (AACR), 2011, Accepted.

[50] Shridhar, R; Zhang, J; Song, J; Booth, BA; Kevil, CG; Sotiropoulou, G; Sloane, BF; Keppler, D. Cystatin M suppresses the malignant phenotype of human MDA-MB-435S cells. *Oncogene*, 2004 23, 2206-2215.

[51] Zhang J, Shridhar R, Dai Q, Song J, Barlow SC, Yin L, Sloane BF, Miller FR, Meschonat C, Li BD, Abreo F, Keppler D. Cystatin m: a novel candidate tumor suppressor gene for breast cancer. *Cancer Res*. 2004 64, 6957-6964.

[52] Vigneswaran, N; Wu, J; Zacharias, W. Upregulation of cystatin M during the progression of oropharyngeal squamous cell carcinoma from primary tumor to metastasis. *Oral Oncol*, 2003 9, 559-568.

[53] Utsunomiya, T; Hara, Y; Kataoka, A; Morita, M; Arakawa, H; Mori, M; Nishimura, S. N. Cystatin-like Metastasis-associated Protein mRNA Expression in Human Colorectal Cancer Is Associated with Both Liver Metastasis and Patient Survival. *Clin Cancer Res*, 2002 8, 2591-2594.

[54] Parle-McDermott , A; McWilliam, P; Tighe, O; Dunican, D; Croke, DT. Serial analysis of gene expression identifies putative metastasis-associated transcripts in colon tumour cell lines. *Br J Cancer*, 2000 83, 725-728.

[55] [54] Shintani, S; Hamakawa, H; Ueyama, Y; Hatori, M; Toyoshima, T. Identification of a truncated cystatin SA-I as a saliva biomarker for oral squamous cell carcinoma using the SELDI ProteinChip platform. *Int J Oral Maxillofac Surg*. 2010 39, 68-74.

[56] Choi, EH; Kim, JT; Kim, JH; Kim, SY; Song, EY; Kim, JW; Kim ,SY, Yeom, YI; Kim, IH; Lee, HG. Upregulation of the cysteine protease inhibitor, cystatin SN, contributes to cell proliferation and cathepsin inhibition in gastric cancer. *Clin Chim Acta*, 2009 406, 45-51.

[57] Yoneda, K; Iida, H; Endo, H; Hosono, K; Akiyama, T; Takahashi, H; Inamori, M; Abe, Y; Yoneda, M; Fujita, K; Kato, S; Nozaki, Y; Ichikawa, Y; Uozaki, H; Fukayama, M; Shimamura, T; Kodama, T, Aburatani, H; Miyazawa, C; Ishii, K; Hosomi, N; Sagara, M; Takahashi, M; Ike, H; Saito, H; Kusakabe, A; Nakajima, A. Identification of Cystatin SN as a novel tumor marker for colorectal cancer. *Int J Oncol*, 2009 35,33-40.

[58] Ohkubo, I; Kurachi, K; Takasawa, T; Shiokawa, H; Sasaki, M. Isolation of a human cDNA for .alpha.2-thiol proteinase inhibitor and its identity with low molecular weight kininogen. *Biochemistry*, 1984 23, 5691-5697.

[59] Higashiyama, S; Ohkubo, I; Matsuda, T; Nakamura, R. Heavy chain of human high molecular weight and low molecular weight kininogens binds calcium ion. *Biochemistry,* 1987 26, 7450-7458.

[60] Salvesen, G; Parkes, C; Abrahamson, M; Grubb, A; Barrett, A. J. Human low-Mr kininogen contains three copies of a cystatin sequence that are divergent in structure and in inhibitory activity for cysteine proteinases. *Biochem,* J. 1986 234, 429-434.

[61] Jiang, YP; Müller Esterl, W; Schmaier, AH. Domain 3 of kininogens contains a cell-binding site and a site that modifies thrombin activation of platelets. J. *Biol. Chem*, 1992 267, 3712-3717.

[62] Ahmad, A; Hasan, AK; Cines, DB; Herwald, H; Schmaier, AH; Müller-Esterl, W. Mapping the Cell Binding Site on High Molecular Weight Kininogen Domain 5. *J. Biol. Chem,* 1995 270, 19256-19261.

[63] Wachtfogel, YT; DeLa Cadena, RA; Kunapuli, SP; Rick, L; Miller, M; Schultze, RL; Altieri, DC; Edgington, TS; Colman, RW. High molecular weight kininogen binds to Mac-1 on neutrophils by its heavy chain (domain 3) and its light chain (domain 5). *J. Biol. Chem,* 1994 269, 19307-19312.

[64] Renné, T; Dedio, J; David, G; Müller-Esterl, W. High Molecular Weight Kininogen Utilizes Heparan Sulfate Proteoglycans for Accumulation on Endothelial Cells. *J. Biol. Chem*, 2000 275, 33688-33696.

[65] [64] Tait, JF; Fujikawa, K. Identification of the binding site for plasma prekallikrein in human high molecular weight kininogen. A region from residues 185 to 224 of the kininogen light chain retains full binding activity. *J. Biol. Chem,* 1986 261, 15396-15401.

[66] Tait, JF; Fujikawa, K. Primary structure requirements for the binding of human high molecular weight kininogen to plasma prekallikrein and factor XI. *J. Biol. Chem,* 1987 262, 11651-11656.

[67] Bior, AD; Pixley, RA; Colman, RW. Domain 5 of kininogen inhibits proliferation of human colon cancer cell line (HCT-116) by interfering with G1/S in the cell cycle. *J Thromb Haemost*, 2007 5, 403-411.

[68] Kawasaki, M; Maeda, T; Hanasawa,K; Ohkubo, I; Tani, T. Effect of His-Gly-Lys Motif Derived from Domain 5 of High Molecular Weight Kininogen on Suppression of Cancer Metastasis Both in Vitro and in Vivo. December 5, 2003 *J Biol Chemistry*, 278, 49301-49307.

[69] Gabrijelcic, D; Svetic, B; Spaić, D; Skrk, J; Budihna, J; Turk, V. Determination of cathepsins B, H, L and kininogen in breast cancer patients. *Agents Actions Suppl,* 1992 38 (Pt 2), 350-357.

[70] Torres, C; Li, M; Walter, R; Sierra, F. T-kininogen inhibits fibroblast proliferation in the G(1) phase of the cell cycle. *Exp Cell Res*, 2001 269:171-179.

In: Cystatins: Protease Inhibitors ... ISBN 978-1-61209-343-7
Editors: John B. Cohen and Linda P. Ryseck, pp.

Chapter 12

ROLE OF CYSTATIN C IN THE DEVELOPMENT, FUNCTION AND DEGENERATION OF THE NERVOUS SYSTEM: POSSIBLE USE AS A NERVOUS SYSTEM BIOMARKER.

Enrique Juárez Aguilar*[*1], Fabio García-García*[1]
***and María Teresa Croda Todd*[2,3]**
[1]Departamento de Biomedicina, Instituto de Ciencias de la Salud,
Universidad Veracruzana, Veracruz, Mexico
[2]Centro de Especialidades Médicas del Estado de Veracruz "Dr. Rafael Lucio"
Xalapa, Veracruz, Mexico
[3]Facultad de Bioanálisis, Universidad Veracruzana. Médicos y Odontólogos S/N.
Xalapa, Veracruz, México

ABSTRACT

Cystatin C (CysC) is an alkaline low molecular weight (13 kDa) cysteine protease inhibitor present in all tissues and biological fluids. This molecule has a broad spectrum of biological roles in numerous cellular systems, with growth promoting activity, down-regulating of inflammation, anti-viral and anti-bacterial properties. However, its function in the brain is unclear although it has been implicated in the processes of neuronal differentiation and neuroprotection as well as in the repair of nervous system after damage. Furthermore, deregulation of CysC activity has been implicated with the origin of several brain diseases. Particularly, CysC is genetically associated with late-onset Alzheimer`s disease (AD). Both CysC and amyloid beta (Aβ) protein co-localize in the parenchymal brain of patients with AD. On the other hand, CysC binds Aβ and inhibits formation of Aβ fibrils and oligomers both *in vitro* and in mouse models of amyloid deposition. Thus, deregulation of CysC could be considered a risk factor in the pathogenesis of AD. In the present chapter, we review the role of this inhibitor of proteases in the development, function and generation of neurodegenerative process and

---

[*] Corresponding author: Enrique Juárez Aguilar, enjuarez@uv.mx

weigh the evidence to consider the CysC determination as a useful diagnostic tool for brain disease.

INTRODUCTION

Proteases play significant roles in all aspects of cellular life, including cell division, metabolic and catabolic processes, protein translocation, immune defense, and apoptosis. To prevent proteolytic activity that might be harmful for the living organism, the activity of different proteases has to be tightly controlled, which is most commonly achieved by endogenous protease inhibitors (Rawlings et al., 2004). The cystatins are natural inhibitors of the papain-like cysteine proteases of family C1 (MEROPS; Rawlings et al., 2004), including the lysosomal enzymes cathepsin B, H, L, K, and S. Thus far, 12 human cystatins (A, B, C, D, E/M, F, G, S, SN, SA, and H- and L-kininogen) have been identified (Abrahamson et al., 2003), all of which are evolutionary-related and belong to the protease inhibitor family I25 (MEROPS; Rawlings et al., 2004). CysC is an endogenous inhibitor of cysteine protease belonging to Type 2 cystatin superfamily. Recently, it has been related to AD and other dysfunction of central nervous system.

CYSTATIN C

Human CysC is considered the physiologically most important inhibitor of endogenous papain-like cysteine proteases (Abrahamson et al., 1986). The mature, active form of human CysC is a single non-glycosylated polypeptide chain consisting of 120 amino acid residues, with a molecular mass of 13,343-13,359 Da, containing four characteristic disulfide-paired cysteine residues and is encoded by a 7.3-kb gene located on chromosome 20p11.2 (Mussap M and Plebani M, 2004). Cystatin C, also known as γ trace (Hochwald et al., 1967), is found in all mammalian body fluids and tissues (Bobek and Levine, 1992), being especially abundant in cerebrospinal fluid, seminal plasma and milk (Mussap M and Plebani M, 2004). The main catabolic site of CysC is the kidney: more than 99% of the protein is cleared from the circulation by glomerular ultrafiltration and tubular reabsorption. The diagnostic value of CysC as a marker of kidney dysfunction has been extensively investigated in multiple clinical studies (Mussap M and Plebani M, 2004).

In vitro experiments it has been shown that CysC can inhibit the cysteine proteases cathepsins B, H, K, L and S (Bernstein, 1996). In addition to being a protease inhibitor, CysC itself is a target of proteolysis (Rudensky et al., 1991, Rider et al., 1996) and is inactivated by proteolytic degradation by cathepsin D and elastase (Abrahamson et al., 1991, Lenarcic et al., 1991). Cathepsins are lysosomal proteases required for housekeeping function during protein turnover and they differ in structure, substrate-specificities and biochemical characteristics (Turk et al., 2000). Uncontrolled proteolysis as a result of imbalance between active proteases and their endogenous inhibitors has been associated with different diseases such as AD (Nakamura et al., 1991), ischemia (Nitatori et al., 1995, Palm et al., 1995, Ishimaru et al., 1996, Tsuchiya et al., 1999), rheumatoid arthritis (Trabandt et al., 1991), renal failure (Kabanda et al., 1995), multiple sclerosis (Bever and Garver, 1995), osteoporosis (Delaisse et al., 1991), muscular dystrophy (Sohar et al., 1988), inflammatory periodontal disease (Lah et al., 1993), inflammatory lung disease (Buttle et al., 1991), inflammation and trauma (Assfalg-Machleidt et al., 1990), and various types of cancer (Calkins and Sloane,1995; Duffy, 1996, Thomssen et al., 1995). CysC functions as an emergency inhibitor for cysteine protease activity during inflammation and tissue remodeling, e.g., during vascular injury and in bone resorption (Leung-Tack et al., 1990; Turk et al., 2002, Koenig et al., 2005). In addition, CysC is considered to take part in the innate immune response by acting as a scavenger of cysteine proteases from pathogenic microorganisms (Bjorck et al., 1990, Turk et al., 2002; Scharfstein, 2006) and has also been implicated in various diseases such as cerebral amyloid angiopathy (Levy et al., 2006) and cancer (Kos et al., 2000).

Cystatin C and Neurogenesis

As described above, CysC has a broad spectrum of activities related with its protease inhibition activity. However, CysC is also implicated in other cellular functions that are unrelated with this activity, such as the regulation of the phagocytotic activity and the chemotatic response of polymorphonuclear neutrophils (Leung-Tack et al., 1990), and the up-regulation of nitric oxide release from activated macrophages (Verdot et al., 1996). In the same way, CysC has been associated with the development and function of the central nervous system (CNS) through a different mechanism of action. The mammalian CNS is derived from a monolayer of germinal neuroepithelial cells, from which the common neural stem cell (NSC) arise, proliferate and differentiate to generate three major cell types in the brain: neuron, astrocyte and oligodendrocyte. The development of neurons and glial cells from NSC is regulated by cell-intrinsic factors as well as extrinsic factors (Temple, 2001). Recently, growing evidence suggest a relevant role of CysC on the glia cell development. In the developing CNS, CysC is weak expressed in the E14 cortex, but is elevated by E16 (Kumada et al., 2004). A role of CysC in the astrocytes development has been suggested by in vitro studies where this molecule interacts with the transforming growth factor-β (TGF-β) to promote astrocyte differentiation of the serum free mouse embryo (SFME) cell line, which has characteristics of CNS progenitor cells. The simultaneous treatment of TGF-β and CysC produces differentiation of astrocytes, accompanied by an increase of endogenous CysC and the downregulation of the neural precursor cell marker, nestin (Solem et al. 1990; Loo et al. 1994). The CysC was expressed 1 day before the expression of glial fibrillary acidic protein (GFAP) in SFME cells treated with TGF-β (Solem et al. 1990). It was also showed that TGF-β induces CysC and GFAP expression in AP16 cells (Kumada et al. 2004), an astrocyte progenitor-like cell line (Yoshida and Takeuchi 1993). Furthermore, CysC gene expression started earlier than that of GFAP in the mouse forebrain during development. CysC expression started in the ventricular zone at a similar time as (or slightly after) glutamate aspartate transporter (GLAST) expression, but before GFAP expression (Kumada et al. 2004). Hasegawa et al., (2007) showed that human CysC increases the number of neurospheres formed from embryonic brain, increasing the number of neural stem/precursor cells. These authors demonstrated that this is a specific effect of CysC since the addition of a neutralizing antibody greatly decreased the number GFAP and GLAST positive astrocytes. This decrease was reversed by the addition of CysC but not by another cysteine protease inhibitor. Thus, the promotion of astrocytes development by CysC appears to be independent of its protease inhibitor activity. Previous to this report, Taupin et al (2000) had identified CysC as a cofactor of fibroblast growth factor-2 (FGF-2) for proliferation of NSC in the adult brain. Neurogenesis, first thought to be limited to the prenatal period, occurs throughout adulthood in discrete regions of the brain, such as the subventricular zone (SVZ) (Lois and Alvarez Buylla, 1993; Luskin, 1993) and the subgranular zone (SGZ) of the dentate gyrus (DG) of the hippocampus of several species, including human (Altman and Das, 1965). Today is clear that maintenance of neural population is basically regulated by growth factors like the epidermal growth factor (EGF) or fibroblast growth factor (FGF). In addition of the characterization of CysC as a cofactor for NSC proliferation in vitro, these authors demonstrated that combined delivery of FGF-2 and CysC to the adult DG stimulate neurogenesis suggesting that this phenomenon is controlled by the cooperation between trophic factors (growth factors) and autocrine/paracrine cofactors like the CysC. Interestingly, Taupin's group reported that glycosylation of the rat CysC was essential to exert its proliferation activity, since the N-deglycosylated CysC form exhibited no mitogenic activity. Taking into account that N-linked carbohydrate moiety is localized in a different region than protease inhibitor domain; these authors concluded that this domain is not directly involved in the NSC regulation by CysC. Whether glycosylation of CysC is essential for its activity on neural stem cells seems to be controversial since the human CysC used in the Hasewage's work is not glycosylated. The effect of CysC on NSC proliferation is supported by the fact that total BrdU-positive cells in the dentate gyrus of CysC knockout mice (Huh et al.,1999) decreased to approximately 60% of those in the wild-type littermate controls (Taupin et al. 2000).

In addition of its effect on neurogenesis, the CysC is associated with the generation of new NSC. The role of CysC as a NSC generation factor comes from studies with embryonic stem cells (ES). The ES are totipotential stem cells which differentiate into all of the cell fates in a developing embryo. The generation of NSC from embryonic stem cells (ES) has been postulated as a source of cellular generation for a number of CNS disorders. Attempts to exclusively generate NSCs or neural progenitor cells from ES cells are restricted. The derivation of a specific phenotype from ES depends of several factors in the culture medium like the presence of fetal calf serum (FCS), co-culture with a "feeder" cell type that secretes unknown factors or the use of "differentiation factors" like retinoic acid (RA). Under these conditions both the quality and the quantity of ES-derived NSC are not sufficient. In 2006, Kato et al. described an efficient system for the generation of ES cell-derived NSC during coculture with dissociated neurosphere cells isolated from embryonic brains without a need for FCS or feeder cells. These authors postulated that commitment of ES to NSC was caused by an unknown factor in the conditioned medium (CM). Further characterization of the CM revealed that almost all the activity derives mainly from CysC. The substitution of CM by purified CysC generated NSC efficiently with self-renewal and multidifferentiation potentials. All together, these data suggest a relevant role of CysC on neuronal cell biology.

Cystatin C and Neuroprotection

In the adult CNS, CysC is synthesized by the choroid plexus and is found in the CSF (Thomas et al., 1989). This inhibitor of proteases has been detected in other parenchymal regions of the brain, such as cortical and hypothalamic neurons and astrocytes (Yasuhara et al., 1993). Although CysC function in the adult brain is unclear this protease inhibitor has been implicated in both the processes of neuronal degeneration and nervous system repair. In this sense, analysis of CysC activity has not been conclusive. While some authors attribute to this protease inhibitor a neuroprotection activity, others proclaim toxic effects. The neuronal protective action of specific and non specific protease inhibitors has been demonstrated following cerebral ischemia (Hoffmann et al., 1992; Lee et al., 1991). These observations have suggested that disregulation of proteolytic activity contributes to post-ischemic neuronal degeneration. In light of these results, it has been suggested a role of CysC in neuronal survival. *In vivo* studies using the two-vessel occlusion model with hypotension to induce a transient forebrain ischemia (TFI) in rat for 10 minutes have demonstrated a delayed in the neuronal death after TFI that was correlated with an up-regulation of CysC expression on pyramidal cells (Palm et al., 1995).

In the same way, enhanced CysC expression have been reported in human patients with epilepsy, in animal models of neurodegenerative conditions, and in response to injury, including facial nerve axatomy, noxious input to the sensory spinal cord, perforant panth transections, hypophysectomy, and induction of epilepsy (Levy et al., 2006). Thus, up-regulation of CysC expression in response to injury or degenerative disorders has been associated with a neuroprotective mechanism that may counteract progression of the disease. Recent reports suggest that CysC exerts its neuroprotective effect through induction of autophagy (Tizon et al., 2010).

These authors showed that CysC was able to protect murine primary cortical neurons and neuronal cell lines against several cytotoxic challenges, including nutritional-deprivation; colchicines (microtubule-depolymerizing agent), the apoptotic agent staurosporine and oxidative stress by inducing fully functional autophagy. Apparently, CysC is able to regulate the paghocytic pathway through enhancing the proteolytic clearance of substrates by lysosomes reducing the citotoxic damage. Specific blockers of phagocytosis inhibit the neuroprotection effect of CysC supporting a close interaction of this molecule with this intracellular process. Interestingly, cathepsin B was no required for the neuroprotective action of CysC through this mechanism since isoforms of this protease inhibitor with null or diminished activity exerts an equivalent neuroprotection effect.

Cystatin C and Neurodegeneration

Contrary to the neuroprotection theory of CysC, some authors have reported a neuronal cell death effect of this molecule *in vivo* and in cultured human CNS neurons through an enhanced apoptotic reaction (Nagai et al., 2002; Nagai et al., 2005; Nagai et al., 2008). These authors showed that this effect is inhibited by cathepsin B suggesting a misregulation in the CysC activity. Incubation of cell cultures or *in vivo* application into hippocampus increased the number of apoptotic cells and an up-regulation of active caspase-3 and DNA ladder. These results have suggested a possible role of CysC in the neurodegenerative process that could give rise brain diseases, like Alzheimer’s disease.

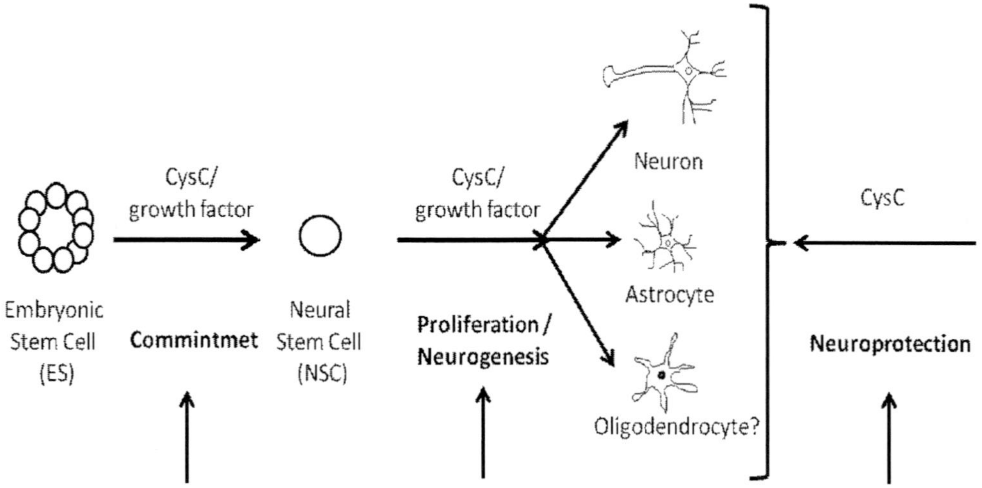

Figure 1. Resume of the possible effects of CysC in the biology of neuronal cells. CysC guide the embryionic stem cell commitment to neural stem cell and regulate the neurogenesis. Additionally, CysC acts as neuroprotective factor on differentiated cells.

CYSC AND ALZHEIMER'S DISEASE

AD is a progressive neurodegenerative disorder that represents the major cause of dementia in the world today (Jicha and Carr, 2010). Over 5 million persons in the US currently suffer from this fatal disease and that number is expected to quadruple by the year 2050, unless prevention efforts or disease modifying therapies are developed (Alzheimer's disease facts and figures, 2009; Jicha and Carr, 2010). AD is characterized clinically by the development of early amnestic and executive dysfunction that eventually spreads across cognitive domains, that leads to the complete incapacity and development of end-stage dementia (McKhann et al., 1984). The major pathological hallmarks of AD are extracellular amyloid-β (Aβ) plaque deposition and intraneuronal neurofibrillary tangle formation (Braak and Braak, 1991, Mirra et al., 1991). Both *in vivo* and *in vitro* reports describe a potent neurotoxic activity for soluble, nonfibrillar, oligomeric assemblies of Aβ. In this sense, synthetic Aβ peptides are toxic to hippocampal and cortical neurons (Lorenzo and Yankner, 1994, Geula et al., 1998). For example, $A\beta_{1\text{-}42}$ induces protein oxidation and lipid peroxidation both *in vitro* and *in vivo* and thus was suggested to play a central role as a mediator of free radical induced oxidative stress and neurotoxicity in AD brains (Butterfield and Boyd-Kimball, 2004). In addition, some reports have been suggested that progressive inflammation and increased oxidative stress by mitochondrial dysfunction play a key role in the early development of AD (Mirra et al., 1991, Swerdlow and Khan, 2004).

Familial forms of early-onset AD are caused by heterozygous mutations in the genes encoding the Aβ precursor protein (APP) and the presenilins PS1 and PS2, but the etiology of the much more prevalent forms of late onset AD is unknown (St George-Hyslop, 2000). Several susceptibility genes were reported to influence the risk of developing AD; of these, the effect of the apolipoprotein E gene (APOE) is now uniformly accepted, whereas additional putative genetic risk factors including α2-macroglobulin, interleukin 6, interleukin

1, and cathepsin D are matter of intense debate. As already mentioned, in the brain CysC is synthesized by neurons, astrocytes, and choroid plexus cells (Thomas et al., 1989; Yasuhara et a.l, 1993), and its levels increase in response to injury, including ischemia, axotomy, or surgery (Yasuhara et a.l, 1993; Palm et al., 1995; Ishimura et al., 1996; Miyake et al., 1996; Katakai et al., 1997). During the past decade, experimental (Maruyama et al., 1990; Wang et al., 1997; Sastre et al., 2004; Levy et al., 2001), genetic (Crawford et al., 2000; Kaeser et al., 2007; Mi et al., 2007) and clinical data (Levy et al., 2001; Deng et al., 2001, Chuo et al., 2007) have suggested that CysC activity in the brain may protect against the development of AD by inhibition of Aβ aggregation. For instance, it has been demonstrated that human CysC binds to Aβ and inhibits its oligomerization and amyloidogenesis, both *in vitro* and *in vivo* studies (Sastre et al., 2004; Kaeser et al., 2007; Mi et al., 2007). Although an *in vitro* study showed that CysC inhibit Aβ aggregation but not dissolve preformed Aβ fibrils or oligomers (Tizon et al., 2010). Recently, it has been showed that overexpression of human CysC in brains of APP-transgenic mice reduces cerebral Aβ deposition and that CysC binds to soluble Aβ, preventing its deposition in the brain (Kaeser et al., 2007; Mi et al., 2007). It has been suggested that CysC is a carrier of soluble Aβ in body fluids, such as CSF and blood, as well as in the neuropil. Sytemic or localized increase in CysC concentration, relative to that of Aβ, would serve to prevent Aβ aggregation and fibril formation (Levy, 2008). Therefore, CysC concentrations may modulate cerebral amyloidosis risk and provide an opportunity for genetic risk assessment and therapeutic interventions. It has been reported that human CysC binds to soluble Aβ in human brain and CSF. While monomeric form of Aβ is absent and stable CysC/Aβ complex is found in brains of non-demented individuals that do not have Aβ deposits, only the monomeric form of Aβ is found in AD brains (Mi et al., 2007). This suggests that the heightened tendency of Aβ to bind itself in AD brains competes with its binding to CysC. On the other hand, neuronal concentrations of CysC are increased in the brain the patient with AD, particularly in the cortical layers III and IV, entorhinal cortex and the hippocampus (Deng et al., 2001; Levy et al., 2001). This distribution of neuronal staining demarcated those neurons most susceptible to loss in AD (Deng et al., 2001). Therefore, it has been suggested that CysC may contribute to senile brain changes associated with AD. As already mentioned the physiological function of CysC as a protease inhibitor is regulated either by dimerization or by endoproteolysis mediated by cathepsin D, which is also increased in early endosomes in AD brains (Cataldo et al., 1997). Finally, there is a disorder called hereditary CysC amyloid angiopathy (HCCAA), this is a autosomal dominant disorder characterized by deposition of amyloid, primary in the CNS, although amyloid deposits have also been reported in other organs (Turk et al., 2008). CysC amyloid fibrils are accumulated in the walls of the brain arteries, leading to single or multiple strokes with fatal outcome (Jensson et al., 1987; Vinters et al., 1990; Levy et al., 1990).

Cystatin C as a Biomarker of the CNS Function

The CysC was initially used in clinic as a potential marker of kidney dysfunction. Nowadays, this protease inhibitor has been proposed as a potential tool in the diagnosis of a variety of diseases (Mares et al., 2003; Stejskal et al., 2005).

Table I. Assessment of CysC as a biomarker in different neurological diseases

Disease	Sample	n	CysC level	Biomarkers associated	Reference
Alzheimer's disease dementia	CSF/ serum	24	↔	-	Kálman et al., 2000
Ischemic type of vascular dementia		16			
Alzheimer 's disease dementia	CSF	38	↔		Brettschneider et al., 2004.
Idiopathic normal pressure hydrocephalus		19			
Pseudo-dementia by depression		13			
Vascular dementia		13			
Frontotemporal dementia		6			
Alzheimer's disease dementia	CSF	35	↓	β-trace, α(1)-antitrypsin (ATT), Transthyretin (TTR), Aβ–binding protein	Habsson et al., 2009.
Frontotemporal dementia		18	↓		
Alzheimer's disease dementia	CSF	69	↔	βA, tau-protein	Mares et al., 2009.
Alzheimer's disease dementia	CSF/ serum	79	↑	Tau protein	Mares et al 2009.
Dementia Lewy body disease	CSF	51	↓	Aβ-42	Maetzler et al., 2010.
Alzheimer's Disease dementia	CSF	9	↓	β-2microglobulin	Carrette et al., 2003
Alzheimer's disease	Serum	70 years= 1153	↓	-	Sundelöf et al., 2008.
Alzheimer's disease		77 years= 761			
Amyotrophic lateral sclerosis	CSF	23	↓	Transthyretin	Ranganathan et al., 2005.
Amyotrophic lateral sclerosis	CSF	36	↓	-	Pasinetti et al., 2006.
Amyotrophic lateral sclerosis	CSF	14	↓	-	Tsuji-Akimoto et al., 2009.
Amyotrophic lateral sclerosis	Spinal cord	9	↓	-	Mori et al., 2009.
Guillain-Narré syndrome	CSF	8	↓	-	Nagai et al., 2000.
Chronic Inflamatory Demyelinating Polyneuropathy (CIDP)		5	↓		
CNS inflammation	CSF/ serum	225	↓	Arginase-I, tau-protein, βA	Stejskal et al., 2005.
Ischemic white matter lesions	CSF/ tissue	23	↓	-	Umegae et al., 2008.

In accordance with evidence that suggest a relevant role of CysC in the function of CNS, this molecule has been proposed as a biomarker of brain dysfunction, as well.

Concentration of CysC in the cerebrospinal fluid (CSF) is 5.5-times higher than in the serum. Because serum CysC concentrations are less than 1 mg/l, CysC in CSF may preferably represent those produced in the CNS.

A constraint in the use of CysC as a brain biomarker is the lack of normal reference intervals in CSF. Moreover, current data are inconsistent and scarce. Using the same assay method as for serum CysC, Yamada et al., (2002) reported reference intervals ranged from 1.8 to 7.2 mg/l (2SD) with a mean of 3.6 mg/l in subjects without apparent disorders. In the other hand, Ndjole et al., (2010) reported even higher values of CSF-CysC in healthy population (reference value 2.42-14.33 mg/L). Further studies are necessary to determine the CysC reference intervals in CFS and to establish the factors that influence its secretion in normal or pathological conditions.

At the present time, several studies have demonstrated alteration of CysC levels in different neurological diseases (Table 1).

Taking into account its participation in the neurodegenerative process CysC has been proposed as an indicator of the progress of different types of dementia. Articles that report changes in CysC levels in this disorder are conflicting. While some authors report low levels of CysC in the AD dementia, frontotemporal dementia and Lewy body dementia (Habsson et al., 2009; Maetzler et al., 2010; Carrette et al., 2003), others reported no significant CysC alteration (see table 1) (Kálman et al., 2000; Brettschneider et al., 2004; Mares et al., 2009a).

In the other hand, Mares et al., (2009b) reported high concentration of CysC in CSF and serum of patient with AD dementia.

In contrast, low CSF levels of CysC have been consistently reported in early phases of AD as well as in amyotrophic lateral sclerosis, inflammation processes and in schemia (Mori et al., 2009; Nagai et al., 2000; Pasinetti et al., 2006; Ranganathan et al., 2005; Tsuji-Akimoto et al., 2009; Umegae et al., 2008) (Table 1). These facts suggest that CysC is not a reliable biomarker in the last phase of neurodegeneration but could be used in the early phases of this process and in traumatic events. Supporting this point of view, Sundelöf et al., (2008) reported low levels of CysC previous to clinical manifestation of AD in serum samples of elderly men free of dementia. In this work, lower serum levels of CysC were associated with higher incidence of AD. Sundelöf's work has the additional advantage of use a larger population (n=1,153, age70 years; n= 761, age 77 years,) to establish this association in comparison with others works. Furthermore, serum samples for CysC determination are more accessible than the CSF.

It is clear that while serum CysC determination could be useful in early diagnosis of AD, other CNS diseases could require direct analysis of CysC in the CSF. The decrease of CysC secretion in the early phases of AD is consistent with the proposed role of this molecule as a carrier of soluble Aβ in body fluids, such as CSF and blood. Less CysC in circulation increase insoluble Aβ that is finally deposited in the brain producing the histopathological manifestation of AD. The assessment of usefulness of CysC determination as CNS biomarker requires further studies with larger sample size and improved methods.

Conclusion

In conclusion, all these findings strengthen the evidence for a role for cystatin C in the development and maintenance of CNS and in the etiology of different neurological conditions. Further studies are necessary to understand the mechanisms involved in the action of CysC in the brain.

References

Alzheimer's disease facts and figures. *Alzheimers Dement* (2009) 5:234–270.

Abrahamson, M; Barrett, AJ; Salvesen, G; Grubb, A. Isolation of six cysteine proteinase inhibitors from human urine. Their physicochemical and enzyme kinetic properties and concentrations in biological fluids. *J. Biol. Chem.* (1986) 261:11282-11289.

Abrahamson, M; Buttle, DJ; Mason, RW; Hansson, H; Grubb, AO; Lilja, H; Ohlsson, K. Regulation of cystatin C activity by serine proteinases. *Biomed. Biochim. Acta* (1991) 50:587-593.

Abrahamson, M; Alvarez-Fernandez, M; Nathanson, CM. Cystatins. *Biochem. Soc. Symp.* (2003) 70:179-199.

Altman, J; Das, GD. Autoradiographic and histological evidence of postnatal hippocampal neurogenesis in rats. *J. Comp. Neurol.* (1965) 124:319-335.

Assfalg-Machleidt I; Jochum M; Nast-Kolb D; Siebeck M; Billing A; Joka T; Rothe G; Valet G; Zauner R; Scheuber HP; et al. Cathepsin B-indicator for the release of lysosomal cysteine proteinases in severe trauma and inflammation. *Biol. Chem. Hoppe Seyler* (1990) 371 Suppl: 211-222.

Bernstein HG; Kirschke H; Wiederanders B; Pollak KH; Zipress A; Rinne A. The possible place of cathepsins and cystatins in the puzzle of Alzheimer disease: a review. *Mol. Chem. Neuropathol.* (1996) 27:225-247.

Bever, CT Jr; Garver, DW. Increased cathepsin B activity in multiple sclerosis brain (1995). *J. Neurol. Sci.* 131:71-73.

Bjorck, L; Grubb, A; Kjellén, L. Cystatin C, a human protease inhibitor, blocks replication of Herpes simplex virus. *J. Virol.* (1990) 64: 941–943.

Bobek, LA; Levine, MJ. Cystatins-inhibitors of cysteine proteinases. *Crit. Rev. Oral Biol. Med.* (1992) 3:307-332.

Braak, H; Braak, E. Neuropathological stageing of Alzheimer-related changes. *Acta Neuropathol* (Berl) (1991) 82:239–259.

Brettschneider, J; Riepe, MW; Petereit, HF; Ludolph, AC; Tumani, H. Meningeal derived cerebrospinal fluid proteins in different forms of dementia: is a meningopathy involved in normal pressure hydrocephalus? *J. Neurol. Neurosurg Psychiatry* (2004) 75:1614-1616.

Butterfield, DA; Boyd-Kimball, D. Amyloid beta-peptide(1-42) contributes to the oxidative stress and neurodegeneration found in Alzheimer's disease brain. *Brain Pathol* (2004) 14:426-432.

Buttle, DJ; Abrahamson, M; Burnett, D; Mort ,JS; Barrett, AJ; Dando, PM; Hill, SL. Human sputum cathepsin B degrades proteoglycan, is inhibited by α 2-macroglobulin and is modulated by neutrophil elastase cleavage of cathepsin B precursor and cystatin C. *Biochem. J.* (1991) 276:325-331.

Calkins, CC; Sloane, B. Mammalian cysteine protease inhibitors: biochemical properties and possible roles in tumor progression. *Biol. Chem. Hoppe Seyler* (1995) 376:71-80.

Carrette, O; Demalte, I; Scherl, A; Yalkinoglu, O; Corthals, G; Burkhard, P; Hochstrasser, DF; Sanchez, JC. A panel of cerebrospinal fluid potential biomarkers for the diagnosis of Alzheimer's disease. *Proteomics* (2003) 3:1486-1494.

Cataldo, AM; Barnett, JL; Pieroni, C; Nixon, RA. Increased neuronal endocytosis and protease delivery to early endosomes in sporadic Alzheimer's disease: neuropathologic evidence for a mechanism of increased beta-amyloidogenesis. *J. Neurosci.* (1997) 17:6142-6151.

Chuo, LJ; Sheu, WH; Pai, MC; Kuo, YM. Genotype and plasma concentration of cystatin C in patients with late-onset Alzheimer disease. *Dement Geriatr Cogn. Disord.* (2007) 23:251–257.

Crawford, FC; Freeman, MJ; Schinka, JA; Abdullah, LI; Gold, M; Hartman, R; Krivian, K; Morris, MD; Richards, D; Duara, R; Anand, R; Mullan, MJ. A polymorphism in the cystatin C gene is a novel risk factor for late-onset Alzheimer's disease. *Neurology* (2000) 55:763–768.

Delaisse, JM; Ledent, P; Vaes, G. Collagenolytic cysteine proteinases of bone tissue. Cathepsin B, (pro)cathepsin L and a cathepsin L-like 70 kDa proteinase. *Biochem. J.* (1991) 279 (Pt 1):167-174.

Deng, A; Irizarry, MC; Nitsch, RM; Growdon, JH; Rebeck, GW. Elevation of cystatin C in susceptible neurons in Alzheimer's disease. *Am. J. Pathol.* (2001) 159:1061–1068.

Duffy, MJ. Proteases as prognostic markers in cancer. *Clin. Cancer Res.* (1996) 2:613-618.

Geula, C; Wu, CK; Saroff, D; Lorenzo, A; Yuan, M;Yankner, BA. Aging renders the brain vulnerable to amyloid beta-protein neurotoxicity. *Nat. Med.* (1998) 4:827-831.

Habsson, SF; Andrèasson, U; Wall, M, Skoog, I; Andreasen, N; Wallin, A; Zatterberg, H; Blennow, K. Reduced levels of amyloid-beta-binding proteins in cerebrospinal fluid from Alzheimer's disease patients. *J. Alzheimer Dis* (2009) 16:389-397.

Hasewaga, A; Naruse, M; Hitoshi, S; Iwasaki, Y; Takebayashi, H; Ikenaka, K. Regulation of glial development by cystatin C. *J. Neurochem.* (2007) 100:12-22.

Hochwald, GM; Pepe, AJ; Thorbecke, GJ. Trace proteins in biological fluids. IV. Physicochemical properties and sites of formation of γ trace and β trace proteins. *Proc. Soc. Exp. Med.* (1967) 124:961-966.

Hoffmann, MC; Nitsch, C; Scotti, AL; Reinhard, E; Monard, D. The prolonged presence of glia-derived nexin, an endogenous protease inhibitor, in the hippocampus after ischemia-induced delayed neuronal death. *Neuroscience* (1992) 49:397-408.

Huh, CG; Hakansson, K; Nathanson, CM; Thorgeirsson, UP; Jonsson, N; Grubb, A; Abrahamson, M; Karlsson, S. Decreased metastatic spread in mice homozygous for a null allele of the cystatin C protease inhibitor gene. *Mol. Pathol.* (1999) 52:332-340.

Ishimaru, H; Ishikawa, K; Ohe, Y; Takahashi, A; Maruyama, Y. Cystatin C and apolipoprotein E immunoreactivities in CA1 neurons in ischemic gerbil hippocampus. *Brain Res.* (1996) 709:155-162.

Jensson, O; Gudmundsson, G; Arnason, A; Blöndal, H; Petursdottir, I; Thorsteinsson, L; Grubb, A; Löfberg, H; Cohen, D; Frangione, B. Hereditary cystatin C (gamma-trace) amyloid angiopathy of the CNS causing cerebral hemorrhage. *Acta Neurol. Scand* (1987) 76:102-114.

Jicha, GA; Carr, SA. Conceptual evolution in Alzheimer's disease: Implications for understanding the clinical phenotype of progressive neurodegenerative disease. *J. Alzheimers Dis* (2010) 19:253–272.

Kabanda, A; Goffin, E; Bernard, A; Lauwerys, R; van Ypersele de Strihou, C. Factors influencing serum levels and peritoneal clearances of low molecular weight proteins in continuous ambulatory peritoneal dialysis. *Kidney Int* (1995) 48:1946-1952.

Kaeser, SA; Herzig, MC; Coomaraswamy, J, Kilger, E; Selenica, ML; Winkler, DT; Staufenbiel, M; Levy, E; Grubb, A; Jucker, M. Cystatin C modulates cerebral beta-amyloidosis. *Nat. Genet.* (2007) 39:1437–1439.

Kálman, J; Márki-Zay, J; Juhász, A; Sántha, A; Dux, L; Janka, Z. Serum and cerebrospinal fluid cystatin C levels in vascular and Alzheimer´s dementia. *Acta neurol.. Scand* (2000) 101:279-282.

Katakai, K; Shinoda, M; Kabeya, K; Watanabe, M; Ohe, Y; Mori, M; Ishikawa, K. Changes in distribution of cystatin C, apolipoprotein E and ferritin in rat hypothalamus after hypophysectomy. *J. Neuroendocrinol.* (1997) 9:247-253.

Kato, T; Heike, T; Okawa, K; Haruyama, M; Shiraishi, K; Yoshimoto, M; Nagato, M; Shibata, M; Kumada, T; Yamanaka, Y; Hattori, H. A neurosphere-dervided factor, cystatin C, supports differentiation of ES into neural stem cells. *Proc. Natl. Acad. Sci. USA* (2006) 103:6019-6024.

Koenig, W; Twardella, D; Brenner, H; Rothenbacher, D. Plasma concentrations of cystatin C in patients with coronary heart disease and risk for secondary cardiovascular events: more than simply a marker of glomerular filtration rate. *Clin. Chem* (2005) 51: 321–327.

Kos, J; Krasovec, M; Cimerman, N; Nielsen, H.J; Christensen, I.J; Brunner, N. Cysteine proteinase inhibitors stefin A, stefin B, and cystatin C in sera from patients with colorectal cancer: relation to prognosis. *Clin. Cancer Res* (2000) 6:505–511.

Kumada, T; Hasegawa, A; Iwasaki, Y; Baba, H; Ikenaka, K. Isolation of Cystatin C via functional cloning of astrocyte differentiation factors. *Dev. Neuosci.* (2004) 26:68-76.

Lah, TT; Babnik, J; Schiffmann, E; Turk, V; Skaleric, U. Cysteine proteinases and inhibitors in inflammation: their role in periodontal disease. *J. Periodon.* (1993) 64:485-491.

Lee, KS; Frank, S; Vanderklish, P; Arai, A; Lynch, G. Inhibition of proteolysis protects hippocampal neurons from ischemia. Proc Natl Acad Sci U S A (1991) 88:7233-7237.

Lenarcic, B; Krasovec, M; Ritonja, A; Olafsson, I; Turk, V. Inactivation of human cystatin C and kininogen by human cathepsin D. *FEBS Letters* (1991) 280:211-215.

Leung-Tack, J; Tavera, C; Gensac, MC; Martínez, J; Colle, A. Modulation of phagocytosis-associated respiratory burst by human cystatin C: role of the N-terminal tetrapetide Lys-Pro-Pro-Arg. *Exp. Cell Res.* (1990) 188:16-22.

Leung-Tack, J; Tavera, C; Martinez, J; Colle, A. Neutrophil chemotactic activity is modulated by human cystatin C, an inhibitor of cysteine proteases. *Inflammation* (1990) 14:247–257.

Levy, E; Carman, MD; Fernandez-Madrid, IJ; Power, MD; Lieberburg, I; van Duinen, SG; Bots, GT; Luyendijk, W; Frangione, B. Mutation of the Alzheimer's disease amyloid gene in hereditary cerebral hemorrhage, Dutch type. *Science* (1990) 248:1124–1126.

Levy, E, Sastre, M; Kumar, A, Gallo, G; Piccardo, P; Ghetti, B; Tagliavini, F. Codeposition of cystatin C with amyloid-beta protein in the brain of Alzheimer disease patients. *J. Neuropathol. Exp. Neurol.* (2001) 60:94–104.

Levy, E; Jaskolski, M; Grubb, A. The role of cystatin C in cerebral amyloid angiopathy and stroke: cell biology and animal models. *Brain Pathol* (2006) 16:60-70.

Levy, E. Cystatin C: a potential target for Alzheimer's treatment. *Expert Rev. Neurother* (2008) 8:687-689.

Lois, C; Alvarez-Buylla, A. Proliferating subventricular zone cells in the adult mammalian forebrain can differentiate into neurons and glia. *Proc. Natl. Acad. Sci. USA* (1993) 90: 2074-2077.

Loo, D.T; Althoen, MC; Cotamn, CW. Down regulation of nestin by TGF-beta or serum in SFME cells accompanies differentiation into astrocytes. *Neuroreport* (1994) 5:1585-1588.

Lorenzo, A; Yankner, BA. Beta-amyloid neurotoxicity requires fibril formation and is inhibited by congo red. *Proc. Natl. Acad. Sci. USA* (1994) 91:12243-12247.

Luskin, MB. Restricted proliferation and migration of postnatally generated neurons derived from the forebrain subventricular zone. *Neuron* (1993) 11:173-189.

Maetzler, W; Schmid, B; Synofzik, M; Schulte, C; Riester, K; Huber, H; Brockmann, K; Gasser, T; Berg, D; Melms, A. The CST3 BB genotype and low cystatin C cerebrospinal fluid levels are associated with dementia in Lewy body disease. *J. Alzheimers Dis.* (2010) 19:937-942.

Mares, J; Stejskal, D; Vavrouskova, J; Urbanek, K; Herzig, R; Hlustik, P. Use of Cystatin C determination in clinical diagnosis. *Biomed. Papers* (2003) 147:177-180.

Mares, J; Kanovsky, P; Herzig, R; Stejskal, D; Vavrouskova, J; Hlustik, P; Vranova, H; Burval, S; Zapletalova, J; Pidrman, V; Obereigneru, R; Suchy, A; Vesely, J; Podivinsky, J; Urbanek, K. The assessment of beta amyloid, tau and cystatin c in the cerebrospinalfluid: laboratory markers of neurodegenerative diseases. *Neurol. Sci.* (2009a) 30:1-7.

Mares, J; Kanovsky, P; Herzig, R; Stejskal, D; Vavrouskova, J; Hlustik, P; Vranova, H; Burval, S; Zapletalova, J; Pidrman, V; Obereigneru, R; Suchy, A; Vesely, J; Podivinsky, J; Urbanek, K. New laboratory markers in diagnosis of Alzheimer dementia. *Neurol. Res.* (2009b) 31:1056-1059.

Maruyama, K; Ikeda, S; Ishihara, T; Allsop, D; Yanagisawa, N. Immunohistochemical characterization of cerebrovascular amyloid in 46 autopsied cases using antibodies to beta protein and cystatin C. *Stroke* (1990) 21:397–403.

McKhann, G; Drachman, D; Folstein, M; Katzman, R; Price, D; Stadlan, EM. Clinical diagnosis of Alzheimer's disease: report of the NINCDS-ADRDA Work Group under the auspices of Department of Health and Human Services Task Force on Alzheimer's Disease. *Neurology* (1984) 34:939–944.

Mi, W; Pawlik, M; Sastre, M; Jung, SS; Radvinsky, DS; Klein ,AM; Sommer, J; Schmidt, SD; Nixon, RA; Mathews, PM; Levy, E. Cystatin C inhibits amyloid-beta deposition in Alzheimer's disease mouse models. *Nat. Genet* (2007) 39:1440–1442.

Mirra, SS; Heyman, A; McKeel, D; Sumi, SM; Crain, BJ; Brownlee, LM; Vogel, FS; Hughes, JP; van Belle, G; Berg, L. The consortium to establish a registry for Alzheimer's Disease (CERAD). Part II. Standardization of the neuropathologic assessment of Alzheimer's disease. *Neurology* (1991) 41479–486.

Miyake, T; Gahara, Y; Nakayama, M; Yamada, H; Uwabe, K; Kitamura, T. Upregulation of cystatin C by microglia in the rat facial nucleus following axotomy. *Brain Res. Mol.* (1996) 37:273-282.

Mori, F; Tanji, K; Miki, Y; Wakabayashi, K. Decreased cystatin c immunoreactivity in spinal motor neurons and astrocytes in amyotrophic lateral sclerosis. *J. Neuropathol. Exp. Neurol.* (2009) 68:1200-1206.

Mussap, M; Plebani, M. Biochemistry and clinical role of human cystatin C. *Crit. Rev. Clin. Lab. Sci.* (2004) 41: 467-550.

Nagai. A; Murakawa, Y; Terashima, M; Shimode, K; Umegae, N; Takeuchi, H; Kobayashi, S. Cystatin C and cathepsin B in CSF from patients with inflammatory neurologic diseases. *Neurology* (2000) 55:1828-32.

Nagai, A; Ryu, JK; Kobayash, S; Kim, SU. Cystatin C induces neuronal cell death in vivo. *Ann N Y Acad. Sci.* (2002) 977:315-321.

Nagai, A; Ryu, JK; Terashima, M; Tanigawa, Y; Wakabayashi, K; McLarnon, JG; Kobayashi, S; Masuda, J; Kim, SU. Neuronal cell death induced by cystatin C in vivo and in cultured human CNS neurons is inhibited with cathepsin B. *Brain Res.* (2005) 1066:120-128.

Nagai, A; Terashima, M; Sheikh, AM; Notsu, Y; Shimode, K; Yamaguchi, S; Kobayashi, S; Kim, SU; Masuda, J. Involvement of cystatin C in pathophysiology of CNS diseases. *Front Biosci.* (2008) 13:3470-3479.

Nakamura, Y; Takeda, M; Suzuki, H; Hattori, H; Tada, K; Hariguchi, S; Hashimoto, S; Nishimura, T. Abnormal distribution of cathepsins in the brain of patients with Alzheimer's disease. *Neurosci. Lett.* (1991) 130:195-198.

Ndjole, AM; Bodolea, C; Nilsen, T; Gordh, T; Flodin, M; Larsson, A. Determination of cerebrospinal fluid cystatin C on Architect ci8200. *J. Immunol. Methods* (2010) 360:84-88.

Nitatori, T; Sato, N; Waguri, S; Karasawa, Y; Araki, H; Shibanai, K; Kominami, E; Uchiyama, Y. Delayed neuronal death in the CA1 pyramidal cell layer of the gerbil hippocampus following transient ischemia is apoptosis. *J. Neurosci.* (1995) 15:1001-1011.

Palm, DE; Knuckey, NW; Primiano, MJ; Spangenberger, AG; Johanson, CE. Cystatin C, a protease inhibitor, in degenerating rat hippocampal neurons following transient forebrain ischemia. *Brain Res.* (1995) 691:1-8.

Pasinetti, GM; Ungar, LH; Lange, DJ; Yemul, S; Deng, H; Yuan, X; Brown, RH; Cudkowicz, ME; Newhall, K; Peskind, E; Marcus, S; Ho, L. Identification of potential CSF biomarkers in ALS. *Neurology* (2006) 66:1218-1222.

Ranganathan, S; Williams, E; Ganchev, P; Gopalakrishnan, V; Lacomis, D; Urbinelli, L; Newhall, K; Cudkowicz, ME; Brown, RH Jr; Bowser, R. Proteomic profiling of cerebrospinal fluid identifies biomarkers for amyotrophic lateral sclerosis. *J. Neurochem.* (2005) 95:1461-1471.

Rawlings, ND; Tolle, DP; Barrett, AJ. MEROPS: the peptidase database. *Nucleic Acids Res.* (2004) 32, D160–D164.

Rider, BJ; Fraga, E; Yu, Q; Singh, B. Immune responses to self peptides naturally presented by murine class II major histocompatibility complex molecules. *Mol. Immunol* (1996) 33:625-633.

Rudensky, A; Preston-Hurlburt, P; Hong, SC; Barlow, A; Janeway, CA Jr. Sequence analysis of peptides bound to MHC class II molecules. *Nature* (1991) 353:622-627.

Sastre, M; Calero, M; Pawlik, M; Mathews, PM; Kumar, A; Danilov, V; Schmidt, SD; Nixon, RA; Frangione, B; Levy, E. Binding of cystatin C to Alzheimer's amyloid beta inhibits in vitro amyloid fibril formation. *Neurobiol. Aging* (2004) 25:1033-43.

Scharfstein, J. Parasite cysteine protease interactions with alpha2-macroglobulin or kininogens: Differential pathways modulating inflammation and innate immunity in infection by pathogenic trypanosomatids. *Immunobiology* (2006) 211:117–125.

Sohar, I; Laszlo, A; Gaal, K; Mechler, F. Cysteine and metalloproteinase activities in serum of Duchenne muscular dystrophic genotypes. *Biol. Chem. Hoppe Seyler* (1988) 369 Suppl: 277-279.

Solem, M; Rawson, C; Lindburg, K; Barnes, D. Transforming growth factor beta regulates cystatin C in serum-free mouse embryo (SFME) cells. *Biochem. Biophys. Res. Commun* (1990) 172: 945-951.

St George-Hyslop, PH. Molecular genetics of Alzheimer’s disease. *Biol. Psychiatry* (2000) 47:183-199.

Stejskal, D; Vavrousková, J; Mares, J; Urbánek. K. Applications of new laboratory marker assay in neurological diagnoses- a pilot study. *Biomed. Pap Med Fac Univ Palacky* Olomouc Csech Repub (2005)149:265-266.

Sundelöf, J; Arnlöv, J; Ingelsson, E; Sundström, J; Basu, S; Zethelius, B; Larsson, A; Irizarry, MC; Giedraitis, V; Rönnemaa, E; Degerman-Gunnarsson, M; Hyman, BT; Basun, H; Kilander, L; Lannfelt, L. Serum Cystatin C and the risk of Alzheimer’s disease in elderly men. *Neurology* (2008) 371:1072-1079.

Swerdlow, RH; Khan, SM. A “mitochondrial cascade hypothesis” for sporadic Alzheimer's disease. *Med. Hypotheses* (2004) 63:8–20.

Taupin, P; Ray, J; Fischer, WH; Suhr, S; Hakansson, K; Grubb, A; Gage, FH. FGF-2-Responsive neural stem cell proliferation requires CCg, a novel autocrine/paracrine cofactor. *Neuron* (2000) 28:385-397.

Temple, S. Defining neural stem cells and their role in normal development of the nervous system. Rao M.S. ed. In *Stem cells and CNS Development*, New Jersey. Humana Press; 2001; 1-29.

Thomas,T; Schreiber, G; Jaworowski, A. Developmental patterns of gene expression of secreted proteins in brain and choroid plexus. *Dev. Biol. (*1989) 134:38-47.

Thomssen, C; Schmitt, M; Goretzki, L; Oppelt, P; Pache, L; Dettmar, P; Janicke, F; Graeff, H. Prognostic value of the cysteine proteases cathepsins B and cathepsin L in human breast cancer. *Clin. Cancer Res*. (1995) 1:741-746.

Tizon, B; Ribe, EM; Mi, W; Troy, CM; Levy, E. Cystatin C protects neuronal cells from amyloid-beta-induced toxicity. *J. Alzheimers Dis*. (2010) 19:885-894.

Trabandt, A; Gay, RE; Fassbender, HG; Gay, S. Cathepsin B in synovial cells at the site of joint destruction in rheumatoid arthritis. *Arthritis Rheum.* (1991) 34:1444-1451.

Tsuchiya, K; Kohda, Y; Yoshida, M; Zhao, L; Ueno, T; Yamashita, J; Yoshioka, T; Kominami, E; Yamashima, T. Postictal blockade of ischemic hippocampal neuronal death in primates using selective cathepsin inhibitors. *Exp. Neurol.* (1999) 155:187-194.

Tsuji-Akimoto, S; Yabe, I; Niino, M; Kikuchi, S; Sasaki, H. Cystatin C in cerebrospinal fluid as a biomarker of ALS. *Neurosci. Lett.* (2009) 452:52-55.

Turk, B; Turk, D; Turk, V. Lysosomal cysteine proteases: more than scavengers. Biochim *Biophys Acta* (2000) 1477:98-111.

Turk, B; Turk, D; Salvesen, GS. Regulating cysteine protease activity: essential role of protease inhibitors as guardians and regulators. *Curr. Pharm. Des*. (2002) 8:1623–1637.

Turk, V; Stoka, V; Turk, D. Cystatins: biochemical and structural properties, and medical relevance. *Front Biosci.* (2008) 13:5406-5420.

Umegae, N; Nagai, A; Terashima, M; Watanabe, T; Shimode, K; Kobayashi, S; Masuda, J; Kim, Su; Yamaguchi, S. Cystatin C expression in ischemic white matter lesions. *Acta Neurol. Scand* (2008) 118:60-67.

Verdot, L; Lalmanach, G; Vercruysse, V; Hartmann, S; Lucius, R; Hoebeke, J; Gauthier, F; Vray, B. Cystatins up-regulate nitric oxide release from interferon-gamma-activated mouse peritoneal macrophages. *J. Biol. Chem.* (1996) 271:28077-28081.

Vinters, HV; Nishimura, GS; Secor, DL; Pardridge, WM. Immunoreactive A4 and gamma-trace peptide colocalization in amyloidotic arteriolar lesions in brains of patients with Alzheimer's disease. *Am. J. Pathol.* (1990) 137:233-240.

Wang, ZZ; Jensson, O; Thorsteinsson, L; Vinters, HV. Microvascular degeneration in hereditary cystatin C amyloid angiopathy of the brain. *APMIS* (1997) 105:41–47.

Yamada, T; Mukaiyama, I; Miyake, N; Igari, J. Measurement of cystatin C in cerebrospinal fluid. *Rinsho Byori* (2002) 50:613-617.

Yasuhara, O; Hanai, K; Ohkubo, I; Sasaki, M; McGeer, PL; Kimura, H. Expression of cystatin C in rat, monkey and human brains. *Brain Res* (1993) 628:85-92.

Yoshida, T; Takeuchi, M. Establishment of an astrocyte progenitor cell line: induction of glial fibrillary acidic protein and fibronectin by transforming growth factor-beta 1. *J. Neurosci Res.* (1993) 35:129-137.

INDEX

A

B

C

D

E

F

G

H

I

J

K

L

M

N

O

P

Q

R

S

T